Lecture Notes in Mathematics

Edited by A. Dold and B. Eckmann

1140

Stephen Donkin

Rational Representations of Algebraic Groups:

Tensor Products and Filtrations

Springer-Verlag
Berlin Heidelberg New York Tokyo

Author

Stephen Donkin
School of Mathematical Sciences, Queen Mary College
Mile End Road
London E1 4NS, England

Mathematics Subject Classification (1980): 20G

ISBN 3-540-15668-2 Springer-Verlag Berlin Heidelberg New York Tokyo
ISBN 0-387-15668-2 Springer-Verlag New York Heidelberg Berlin Tokyo

Printing and binding: Beltz Offsetdruck, Hemsbach/Bergstr.
2146/3140-543210

To My Mother.

Table of Contents

Introduction

Let G be a connected, affine algebraic group over an algebraically closed field and B a Borel subgroup of G . For each one dimensional rational B-module L we have the induced G-module $Ind_B^G L$. These modules are of fundamental importance in the representation theory of G ; in characteristic 0 one obtains every simple rational G-module in this way and in arbitrary characteristic $Ind_B^G L$, when non-zero, has a simple socle and each simple G-module occurs as the socle of some such induced module. The formal character of $Ind_B^G L$ is independent of the characteristic, being given by Weyl's Character Formula, but the submodule structure depends very heavily upon characteristic and very little is known about this structure in characteristic p . For G semisimple, the induced modules also have an interpretation as the global sections of line bundles on the quotient variety G/B and so provide a bridge between the representation theory of G and the geometry of G/B .

We make the following key definition. A good filtration of a rational G-module V is an ascending chain of submodules $0 = V_0, V_1, V_2, \ldots$ of V such that V is the union of the V_i and, for each $i > 0$, V_i/V_{i-1} is either 0 or isomorphic to $Ind_B^G L$ for some rational one dimensional B-module L . It was shown in [25] that for G semisimple and simply connected each rationally injective indecomposable G-module has a good filtration. In this monograph we study, for a connected, affine algebraic group G over an algebraically closed field k , the following hypotheses.

Hypothesis 1 *For all rational* G*-modules* V,V' *which have a good filtration the tensor product* $V \otimes V'$ *has a good filtration.*

Hypothesis 2 *For every rational* G*-module* V *which has a good filtration and every parabolic subgroup* P *of* G *the restriction of* V *to* P *has a good filtration.*

The hypotheses are mainly of interest when G is semisimple (we reduce to the semisimple, simply connected case in Chapter 3) and hypothesis 1 has been considered in this case by Wang Jian-pan, [52]. In that paper hypothesis 1 is shown to be true when G is of type A , for the other classical groups when k has characteristic $p \geq h-1$ (h is the Coxeter number of G), for $p \geq 5$ in type G_2 , $p \geq 31$

in type F_4 , $p \geq 29$ in type E_6 , $p \geq 59$ in type E_7 and $p \geq 151$ in type E_8 . We show that both hypotheses are satisfied provided that either the characteristic of k is not 2 or G involves no component of type E_7 or E_8 . There are many technical advantages - which will emerge in due course - in allowing G to be an arbitrary (rather than semisimple) connected algebraic group and also in considering both hypotheses together.

Apart from the intrinsic appeal, there are several good reasons for trying to establish the hypotheses. It is enough to do this with $V = Ind_B^G L$, $V' = Ind_B^G L'$ for L and L' one dimensional B-modules. Thus we are attempting to say something about the tensor product of induced modules and the restriction of an induced module. Viewed from this standpoint what we are trying to do is to find substitutes, in the context of algebraic groups, for the Mackey Tensor Product Theorem and the Mackey Subgroup Theorem for finite groups ((44.2) and (44.3) of [21]). The search for Mackey type theorems is particularly tantalizing since one has, by the Bruhat decomposition (28.3 Theorem of [35]) an especially good indexing set for the double cosets of B in G , namely the elements of the Weyl group. As noted by Wang Jian-pan, [52], it follows from hypothesis 1 that the natural map $\Gamma(G/B,L) \otimes \Gamma(G/B,L') \to \Gamma(G/B,L \otimes L')$, between global sections of dominant induced line bundles on G/B , is surjective. When G is classical, that is G has type A, B, C or D, this is a result of Lakshmibai, Musili and Seshadri, [44] - we also obtain the result for F_4 , G_2 , E_6 in arbitrary characteristic and for E_7 and E_8 if the characteristic is not 2 . It seems likely that hypothesis 2 will be useful in the representation theory of reductive groups. We give a small application, in section 11.3 to homomorphisms between Weyl modules. Hypothesis 2 does seem to provide real insight into these modules and, together with hypothesis 1, makes possible many cohomological calculations. One can see this throughout the text but Chapter 7, treating F_4 , is particularly rich in examples of this kind. We also hope that, when k has prime characteristic, hypothesis 1 will be useful in connexion with the structure of the modules $St_n \otimes Y(\lambda)$ (St_n denotes the nth Steinberg module and $Y(\lambda)$ is the G-module induced from a one dimensional B-module λ). Components of these modules and their duals are frequently used ([10], [19], [24], [26], [36], [37], [41], [42]) to compare the representation theory of G with that of its infinitesimal subgroups.

When G is semisimple, the dual of a module induced from a one dimensional B-module is a Weyl module (see section 1 of [40]). The first time that I became aware of the hypotheses was in conversation with Humphreys and Jantzen at the Durham Symposium on Finite Simple Groups in 1978, where we discussed, for semisimple groups, hypothesis 1 in its dual formulation in terms of Weyl modules.

Before giving a synopsis of each chapter we briefly describe the overall strategy of the proof. We show that the hypotheses hold for a connected group G if and only if they hold for the quotient of G by its soluble radical. Moreover the hypotheses hold for a semisimple group G provided that they hold for the semisimple, simply connected group of the same type. Also, the hypotheses hold for a direct product of groups if and only if they hold for each factor. Thus we may assume G to be semisimple, simply connected with an indecomposable root system and by induction we may assume that the hypotheses hold for every proper parabolic subgroup. It suffices to prove that, for each fundamental dominant weight λ_i ($i = 1,2,\dots,\ell$ where ℓ is the rank of G) and each parabolic subgroup P, $Y(\lambda_i)|_P$ has a good filtration and, for each rational G-module V with a good filtration, $Y(\lambda_i) \otimes V$ has a good filtration. The first property is proved initially for P of largest possible dimension, so that the number of successive quotients in a good filtration of $Y(\lambda_i)|_P$ will be as small as possible (and assuming the result for $j < i$ in the classical case). Then it is proved for an arbitrary maximal parabolic subgroup Q, such that P and Q contain a common Borel subgroup, by examining the effect of restriction from G to $P \cap Q$ followed by induction from $P \cap Q$ to Q.

In Chapters 1, 2 and 5 we deal with general results on group cohomology and the derived functors of induction which are needed for the specific calculations in Chapters 4, 6, 7, 8, 9 and 10. The main purpose of Chapter 1 is to establish the notation and explain the relationship between various left exact functors. Chapter 2 contains results relating the cohomology of some modules for parabolic subgroups to the cohomology of other modules for other parabolic subgroups. This Chapter also contains a deduction from Kempf's Vanishing Theorem of Weyl's Character Formula for the character of $Ind_B^G L$ for a reduction group G and one dimensional B-module L. The formula is used extensively in the later calculations and we were unable to find a suitable reference (for reductive groups). The proof of Weyl's Character Formula given is quite short, it is not obtained by reduction from characteristic zero (c.f. section 1 of [40]) and, as in the paper [48] treating essentially the semisimple characteristic zero case, deals directly with the group rather than the Lie algebra (not even the Casimir operator is used). In deriving the results in Chapter 2 (and elsewhere) we use three different cases of the Grothendieck spectral sequence (section 2.4 of [30]) relating the derived functors of the composite of two left exact functors to the composites of the derived functors. The first application arises from the expression of Ind_H^G (the induction functor from a closed subgroup H to G) as the composite $Ind_K^G \circ Ind_H^K$ (transitivity of induction) for a closed subgroup K containing H. The second application is the Lyndon-Hochschild-Serre spectral sequence expressing G cohomology in terms of the cohomology of a closed normal subgroup N and the quotient G/N. The final application arises from the expression of the fixed point functor F_H, from rational H-modules to k-spaces, as $F_G \circ Ind_H^G$ (reciprocity of induction).

In Chapter 3 we make various reductions to the hypotheses so that they become susceptible to the case by case analysis which follows.

In Chapter 4 we prove the hypotheses for the classical groups. The argument here is independent of the characteristic and it has been possible to treat the groups of type B, C and D in a unified manner. Fortunately the restriction of $Y(\lambda_i)$ to a proper parabolic subgroup of maximal dimension has only 4 successive quotients in a good filtration (for i in "general position") and the module structure is much the same in all three types B, C and D.

Some additional homological algebra is needed to deal with the exceptional groups and this is given in Chapter 5. The hypotheses are proved for G_2 in Chapter 6 however it would not be difficult to treat this case without the benefit of Chapter 5. In Chapter 7, treating F_4, we found it necessary to consider separately the cases of odd and even characteristic. It takes just 11 pages to deal with F_4 for characteristic $p \neq 2$ but needs another 33 pages to give the additional arguments necessary to cover the case $p = 2$ (this may be omitted at a first reading). Briefly, the reason why $p = 2$ is a special case is that the exterior square of a module is not usually a summand of the tensor square. And the reason why we manage to prove the hypotheses in characteristic 2 is that we are able to use Andersen's strong linkage principle, [4], to analyse components of the tensor product of induced modules, the point being that, by Chapter 3, a summand of a module with a good filtration has a good filtration.

The subject, E_6, of Chapter 8 is altogether easier but again there is a division in the proof into odd and even characteristics.

Chapters 9 and 10 are devoted to the remaining exceptional groups E_7 and E_8. The procedure here is to analyse first the modules $Y(\lambda_i)$, corresponding to the terminal vertices α_i of the Dynkin-Diagram, and then exterior powers of these modules are used to deal with $Y(\lambda_r)$ for an arbitrary fundamental dominant weight λ_r. In fact one only ever needs to go as far as the fourth exterior power and the second stage of the procedure works very smoothly provided that the characteristic p is at least 5 (because the exterior power is a summand of the tensor power in this case). In characteristic 3 a consideration of exterior powers is insufficient and some block theory is needed. We have unfortunately not been able to prove the hypotheses in characteristic 2. In the case of E_7 and $p = 2$ we have at least, in Chapter 9, a satisfactory analysis of $Y(\lambda_i)$ for α_i a terminal vertex but for E_8 (in characteristic 2) we have not given an analysis of $Y(\lambda_8)$ (see section 8.1 for the labelling of the Dynkin-Diagrams of type E). Moreover we have no way of going from the terminal $Y(\lambda_i)$ to an arbitrary $Y(\lambda_r)$. At several points in the text (notably in sections 9.3 and 10.2) it is necessary to know that various dominant weights belong to different linkage classes (so the corresponding modules belong to different blocks) and this involves many routine but lengthy calculations which we have omitted. These calculations were done by hand and later checked, on a Sinclair ZX 81 Microcomputer, by J.F. Blackburn to whom I am extremely grateful.

Chapter 11 opens with an example of a reductive subgroup H of a reductive group G and a G-module V such that V has a good filtration but the restriction of V to H does not. The remainder of the chapter is devoted to applications of the hypotheses to rational cohomology, homomorphisms between Weyl modules, canonical products on induced modules and filtrations over $\mathbb{Z}$ of Weyl modules for Kostant's $\mathbb{Z}$-form $\mathcal{U}_{\mathbb{Z}}$ of the enveloping algebra $U(\underline{g})$ of a complex semisimple Lie algebra $\underline{g}$.

The final chapter is given over to a number of observations on issues not directly concerned with the hypotheses but which nevertheless have mainly arisen in the course of our work on the hypotheses. The issues discussed are the injective indecomposable modules for a parabolic subgroup of a reductive group, Kempf's Vanishing Theorem for rank 1 groups, Kempf's Vanishing Theorem in characteristic zero and the exactness of induction.

These notes are a substantially rewritten version of a winning Adams Prize Essay in Algebra for 1981/2. In content the notes differ from the essay only in that a proof of the hypotheses in the cases E_8 with $p = 3$ and 7 has been added and that the final chapter is new.

I wish to thank Tracy Kelly for the excellent job she has done in typing my manuscript.

1. Homological Algebra

1.1 Induction

We recall the induction functor for algebraic groups discussed more fully in [18].

Let G be an affine algebraic group over an algebraically closed field k. By a rational G-module we mean a left kG-module which is the union of its finite dimensional submodules and such that for each finite dimensional submodule W the induced map $G \to GL(W)$ is a morphism of algebraic groups. A morphism of rational G-modules is simply a kG-module homomorphism. We denote by M_G the category of rational G-modules.

Let H be a closed subgroup of G. For $V \in M_H$ we denote by $Map(G,V)$ the set of maps $f:G \to V$ such that the image $f(G)$ lies in a finite dimensional subspace W of V and the induced map $f:G \to W$ is a morphism of varieties. For $x \in G$, $f \in Map(G,V)$ the map $xf:G \to V$, defined by $(xf)(y) = f(yx)$ for $y \in G$, also belongs to $Map(G,V)$. Thus $Map(G,V)$ is naturally a G module which is, as one may easily verify, rational. We set

$$Map_H(G,V) = \{f \in Map(G,V) \mid f(hx) = hf(x) \quad \text{for all} \quad h \in H,\ x \in G\}.$$

The space $Map_H(G,V)$ is in fact a G submodule of $Map(G,V)$ called the induced module and more often written as $Ind_H^G V$. If $\phi:V \to V'$ is a morphism of rational H modules then the map $Ind_H^G \phi : Ind_H^G V \to Ind_H^G V'$ defined by $Ind_H^G \phi(f) = \phi \circ f$ is a G homomorphism and $Ind_H^G : M_H \to M_G$ is a left exact functor.

We now note for future reference the main properties of induction and of the derived functors of induction. Proofs of these properties of induction may be found in [18] (see also [23] for a more Hopf theoretic treatment). For a left exact functor $F:A \to B$ between abelian categories with A having enough injectives we denote by R^nF $(n \geq 0)$ the derived functors (defined by means of injective resolutions in A). For $V \in M_H$ let $e_V:Ind_H^G(V) \to V$ be the H-module map defined by $e_V(f) = f(1)$.

(1.1.1) *For each* $n \geq 0$, $R^n Ind_H^G$ *commutes with direct sums and direct limits.*

For $n = 0$ this follows fairly directly from the definition of induction. The argument involved in extending the result to arbitrary $n \geq 0$ is that used in Proposition 2.9, Ch.III of [32] to prove that sheaf cohomology on a Noetherian space commutes with direct limits.

(1.1.2) (Reciprocity of induction) *For any* $V' \in M_G$ *the map* $Hom_G(V', Ind_H^G(V)) \to Hom_H(V'|_H, V)$ *taking* θ *to* $e_V \circ \theta$ *is a k-isomorphism.*

(1.1.3) (Transitivity of induction) *For a closed subgroup* K *containing* H , Ind_H^G *is naturally isomorphic to* $Ind_H^G \circ Ind_H^K$.

(1.1.4) Ind_H^G *takes injective objects to injective objects;* $Ind_H^G(k[H])$ *is isomorphic to* $k[G]$.

The coordinate ring $k[G]$ is equal to $Map(G,k)$ and so has the structure of a rational left G-module as above. Taking $H = 1$, the identity subgroup, we see from (1.1.2) that $Ind_1^G(k) = k[G]$ is an injective object in M_G , moreover a rational G-module is injective if and only if it is a direct summand of a direct sum of copies of $k[G]$ (see section 1.5 of [29]) so that the first part of (1.1.4) is a consequence of the second part. The second part is a consequence of (1.1.3).

(1.1.5) *For a closed subgroup* K *containing* H *there is a Grothendieck spectral sequence* $E_r^{\ell,m}$ *converging to* $(R^n Ind_H^G)V$ *with* $E_2^{\ell,m} = (R^\ell Ind_K^G) \circ (R^m Ind_H^K)V$.

The existence of the Grothendieck spectral sequence (see section 2.4 of [30]) is guaranteed by the left exactness of induction, (1.1.3) and (1.1.4).

In particular we have the following.

(1.1.6) *Let* $V \in M_H$ *and* $n \in \mathbb{N}$ *be such that* $(R^m Ind_H^K)V = 0$ *for* $m \neq n$. *Then* $(R^m Ind_H^G)V$ *is zero for* $m < n$ *and is isomorphic to*

$(R^{m-n}Ind_K^G) \circ (R^n Ind_H^K)V$ *for* $m \geq n$.

(1.1.7) (<u>Tensor identity</u>) *For* $V' \in M_G$, $V \in M_H$ *and* $n \geq 0$, $(R^n Ind_H^G)(V' \otimes V) \cong V' \otimes (R^n Ind_H^G)V$.

For $n = 0$ see [18]. It may be proved for arbitrary $n \geq 0$ by dimension shifting or a Grothendieck spectral sequence argument.

(1.1.8) *There is a Grothendieck spectral sequence* $E_r^{\ell,m}$ *converging to* $H^n(H,V)$ *with* $E_2^{\ell,m} = H^\ell(G, R^m Ind_H^G)V)$, *where* H^n *denotes Hochschild cohomology* (see [33]).

Let $\mathcal{D} = Ind_H^G$, let $E: M_G \to \underline{k\text{-sp}}$ (the category of k-vector spaces) be the fixed point functor and let $F: M_H \to \underline{k\text{-sp}}$ be the H fixed point functor. Taking V to be the trivial one dimensional G-module in (1.1.2) gives a natural isomorphism $E \circ \mathcal{D} \to F$. Moreover E , $\mathcal{D}$ are left exact and $\mathcal{D}$ takes injective objects to acyclic objects (by (1.1.4)) so there is a Grothendieck spectral sequence (section 2.4 of [30]) as required.

In particular we have the following.

(1.1.9) *Let* $V \in M_H$ *and* $n \in \mathbb{N}$ *be such that* $(R^m Ind_H^G)V = 0$ *for* $m \neq n$. *Then* $H^m(H,V)$ *is* 0 *if* $m < n$ *and is isomorphic to* $H^{m-n}(G,(R^n Ind_H^G)V)$ *for* $m \geq n$.

Combining (1.1.9) and (1.1.7) one obtains:

(1.1.10) *Suppose that* $V_1 \in M_G$, $V_2 \in M_H$ *and* $n \in \mathbb{N}$ *such that* $(R^m Ind_H^G)V_2 = 0$ *for* $m \neq n$. *Then* $H^m(H, V_1 \otimes V_2)$ *is* 0 *if* $m < n$ *and is isomorphic to* $H^{m-n}(G, V_1 \otimes (R^n Ind_H^G)V_2)$ *if* $m \geq n$.

We call (G,H) a vanishing induction pair (VIP for the sake of brevity) if the trivial one dimensional module k satisfies $(R^n Ind_H^G)k = 0$ for $n > 0$ and is k for $n = 0$. From (1.1.7) we obtain:

(1.1.11) *If* (G,H) *is a VIP then for any* $V \in M_G$ *we have* $(R^n Ind_H^G)V = 0$ *for* $n > 0$ *and is* V *for* $n = 0$.

From (1.1.6) we have:

(1.1.12) *If* (K,H) *is a* VIP *then* $(R^m Ind_H^G)V = (R^m Ind_K^G)V$ *for any* $V \in M_K$ *and* $m \in \mathbb{N}$.

Putting $V = k$ we obtain:

(1.1.13) *If* (G,K) *and* (K,H) *are* VIP's *then* (G,H) *is a* VIP .

Let $\sigma: G \to G$ be an algebraic group automorphism of G . For a closed subgroup H of G we denote by H^σ the closed subgroup $\{\sigma(h): h \in H\}$. For $V \in M_H$ we denote by V^σ the k-space V viewed as an $H^{\sigma^{-1}}$-module with action given by $\sigma^{-1}(h)\, v = hv \, (h \in H \, , \, v \in V)$. It is easy to check:

(1.1.14) $(Ind_H^G V)^\sigma$ *is isomorphic to* $Ind^G_{H^{\sigma^{-1}}}(V^\sigma)$.

Suppose now that N is a closed normal subgroup of G contained in H . We have the inflation functor $Inf^H_{H/N}: M_{H/N} \to M_H$, for $V \in M_{H/N}$, $Inf^H_{H/N}(V)$ is the k-space V regarded as an H-module via the quotient map $H \to H/N$. For a morphism $\theta: V \to V'$ of H/N-modules, $Inf^H_{H/N}(\theta)$ denotes the same k-map θ but now regarded as a morphism of H-modules. Clearly $Inf^H_{H/N}$ is exact. It is easy to check that $Inf^G_{G/N} \circ Ind^{G/N}_{H/N}$ is isomorphic to $Ind_H^G \circ Inf^H_{H/N} = F$, say. Now $Ind^{G/N}_{H/N}$ takes injective objects to $Inf^G_{G/N}$ acyclic objects so that $(R^n F)V \cong Inf^G_{G/N}\,(R^n Ind^{G/N}_{H/N})V$ for any $V \in M_{H/N}$ and $n \geq 0$. We claim also that $Inf^H_{H/N}$ takes injective objects to Ind_H^G acyclic objects. To verify this, since every rationally injective H/N-module is isomorphic to a direct sum of copies of $k[H/N]$ (see section 1.5 of [29]), it is enough to verify that $(R^n Ind_H^G)\, Inf^H_{H/N}(k[H/N]) = 0$ for $n > 0$. It follows from the definitions that $Inf^H_{H/N}(k[H/N])$ is isomorphic to

$Ind_N^H(k)$. However N is normal in H , thus H/N is affine and Ind_N^H is exact by [18]. Thus

$$(R^n Ind_H^G) Inf_{H/N}^H (k[H/N]) \cong (R^n Ind_H^G) Ind_N^H(k) \cong (R^n Ind_N^G) k ,$$

which is 0 for $n > 0$ since G/N is affine. Thus $Inf_{H/N}^H$ takes injective objects to acyclic objects and so $(R^n F)V \cong (R^n Ind_H^G) Inf_{H/N}^H(V)$. Hence we have

(1.1.15) *For any* $V \in M_{H/N}$ *and* $n \geq 0$,

$$Inf_{G/N}^G (R^n Ind_{H/N}^{G/N}) V \cong (R^n Ind_H^G) Inf_{H/N}^H(V) .$$

We shall need another interpretation of the derived functors of induction. Let $\pi: G \to G/H$ be the quotient map, where G/H denotes the variety of right cosets Hg . For $V \in M_H$ and W open in G/H we denote by $L_V(W)$ the set of maps $f: \pi^{-1}W \to V$ such that the image of f lies in a finite dimensional subspace say V_0 of V , the induced map $f: \pi^{-1}W \to V_0$ is regular and such that $f(hu) = hf(u)$ for all $h \in H$, $u \in \pi^{-1}W$. Thus V determines a sheaf L_V of k-spaces - if $W \subseteq W'$ are open sets the map $L_V(W') \to L_V(W)$ is restriction. If $a \in O_{G/H}(W)$ (i.e. a is a regular function $\pi^{-1}W \to k$ such that $a(hu) = a(u)$ for all $h \in H$, $u \in \pi^{-1}W$) and $f \in L_V(W)$ then $af: \pi^{-1}W \to V$, defined by $(af)(u) = a(u)f(u)$ also belongs to $L_V(W)$. Thus L_V is in fact an $O_{G/H}$-module. Notice that $\Gamma(G/H, L_V)$ is exactly $Ind_H^G V$, which is the reason for our interest in this construction.

Now assume that π is locally trivial. It follows that L_V is a locally free sheaf of rank equal to $dim_k V$, in particular L_V is quasi-coherent and, if V is finite dimensional, L_V is coherent.

If $\phi: V \to V'$ is a morphism of H-modules we have the obvious morphism $L(\phi): L_V \to L_{V'}$ of $O_{G/H}$-modules and the functor L , sending V to $L_{V'}$ from M_H to $O_{G/H}$-modules, is exact. Moreover it is easy to check that $H^i(G/H, L_V) = 0$ for $i > 0$ and $V = k[H]$. It follows that L takes injective objects to $\Gamma(G/H,-)$ acyclic objects. Thus if $D = \Gamma \circ L$ we have $(R^n D)V \cong H^n(G/H, L_V)$ for $V \in M_H$ and $n \geq 0$. However $\mathcal{D} \cong F \circ Ind_H^G$ where F is the forgetful functor from G-modules to k-spaces. Of course F is exact so we have that $(R^n\mathcal{D})V \cong F(R^n Ind_H^G)V$. We have shown:

(1.1.16) *Assume that* $\pi : G \to G/H$ *is locally trivial. For any* $V \in M_H$ *and* $n \geq 0$, $F \circ (R^n Ind_H^G)V = H^n(G/H, L_V)$.

1.2 Injective modules for soluble groups

Let T be a torus and $X = X(T)$ be the set of algebraic group morphisms from T to k^* , the multiplicative group of the field. For a rational T-module V and $\lambda \in X$ we define the weight space

$$V^\lambda = \{v \in V : tv = \lambda(t)v \text{ for all } t \in T\} .$$

The module V is the direct sum of its weight spaces. An element of V^λ is called a weight vector of weight λ and λ is called a weight of V if $V^\lambda \neq 0$. Rational T-modules V, V' are isomorphic if and only if $\dim_k V^\lambda = \dim_k V'^\lambda$ for all $\lambda \in X$.

Let B be a connected soluble algebraic group over k (algebriacally closed), $U = R_u(B)$ the unipotent radical and T a maximal torus of B . The coordinate ring $k[U]$ may be viewed as a B-module via the adjoint action. For $f \in k[U]$, $b \in B$, $Ad(b)(f)$ is defined by $Ad(b)(f)(x) = f(b^{-1}xb)$. Let $M = \{f \in k[U] : f(1) = 0\}$, clearly M^r is a B-submodule of $k[U]$ for any $r \geq 0$ (we take $M^0 = k[U]$) .

(1.2.1) $k[U]$ *and* $\bigoplus_{r\geq 0} M^r/M^{r+1}$ *are isomorphic as* T*-modules.*

This is true since either there is an $r \geq 0$ such that $(M^r/M^n)^\lambda = 0$ for all $n > r$ (but then $(M^r)^\lambda = 0$ since $\bigcap_{n\geq 0} M^n = 0$ and $k[U]^\lambda = \bigoplus_{n\geq 0}(M^n/M^{n+1})^\lambda)$ or $(M^r)^\lambda \neq (M^{r+1})^\lambda$ for infinitely many r in which case $k[U]^\lambda$ and $\bigoplus_{r\geq 0}(M^r/M^{r+1})^\lambda$ both have countably infinite dimension.

Let $m = \dim_k M/M^2$ and let $f(m,r)$, for $r \geq 0$, denote the dimension of the space of homogeneous polynomials of degree r in m variables. Let $R = k[U]_M$, the localisation of $k[U]$ at M . Now R

is a regular local ring (Theorem A, section 5.3 of [35]) and $\dim_k MR/(MR)^2 = m$ so that $\dim_k (MR)^r/(MR)^{r+1} = f(m,r)$ for any $r \geq 0$. The natural map $k[U] \to R$ induces an isomorphism $M^r/M^{r+1} \to (MR)^r/(MR)^{r+1}$ so that $\dim_k M^r/M^{r+1} = f(m,r)$. It follows that, if V is a T-module complement to M^2 in M and $X_1, X_2, \ldots, X_m$ is a basis of V of weight vectors then the monomials in the X_i's of degree r form a k-basis of weight vectors of a complement to M^{r+1} in M^r. Thus M^r/M^{r+1} is isomorphic to the rth symmetric power of M/M^2. We obtain from (1.2.1):

(1.2.2) $k[U]$ *and* $\bigoplus_{r\geq 0} \mathrm{Symm}^r(M/M^2)$ *are isomorphic as* T-*modules.*

Moreover M/M^2 may be naturally identified with the dual $\mathrm{Lie}(U)^*$ of the Lie algebra of U (p.75 of [47]) and this identification respects the T-module structure. Thus we have:

(1.2.3) *as a* T-*module* $k[U]$ *is isomorphic to* $\bigoplus_{r\geq 0} \mathrm{Symm}^r(\mathrm{Lie}(U)^*)$.

For $\lambda \in X$ we also denote by λ the one dimensional B-module on which U acts trivially and T acts with weight λ. Thus $\{\lambda : \lambda \in X\}$ is a full set of simple rational B-modules. We denote by $E(\lambda)$ the injective envelope (see section 1.5 of [29]) of λ. For any $\mu \in X$ the B module $\mu \otimes E(\lambda)$ is injective (see Proposition 2.2 of [33]) and has B-socle $\lambda + \mu$. Hence $\mu \otimes E(\lambda)$ is isomorphic to $E(\lambda + \mu)$. Since all rational T-modules are completely reducible λ is injective as a T-module and so $\mathrm{Ind}_T^B(\lambda)$, by (1.1.4), is an injective B-module. However, for $\mu \in X$, we have, by (1.1.2), $\mathrm{Hom}_B(\mu, \mathrm{Ind}_T^B(\lambda))$ is isomorphic to $\mathrm{Hom}_T(\mu,\lambda)$ which is one dimensional if $\mu = \lambda$ (by Schur's Lemma) and zero otherwise. Thus $\mathrm{Ind}_T^B(\lambda)$ has a simple B socle λ and $E(\lambda) \cong \mathrm{Ind}_T^B(\lambda)$. We now seek a more concrete description of $\mathrm{Ind}_T^B(k)$.

Let $\phi : B \to T \times U$ be the inverse of the isomorphism $T \times U \to B$ given by multiplication. From the definition

$$\mathrm{Ind}_T^B(k) = \{f \in k[B] : f(tu) = f(u) \text{ for all } (t,u) \in T \times U\} = \phi^*(k \otimes k[U])$$

where $\phi^* : k[T] \otimes k[U] \to k[B]$ is the comorphism. Thus we have an isomorphism of k-spaces $\theta : k[U] \to \mathrm{Ind}_T^B(k)$, satisfying

$\theta(a) = \phi^*(1 \otimes a)$ $(a \in k[U])$. Regarding $k[U]$ as a left U-module in the natural manner and as a T-module via the adjoint action we see that θ is an isomorphism of B-modules. In particular since $E(0) = Ind_T^B(k)$ and $E(\lambda) \cong \lambda \otimes E(0)$ $(\lambda \in X)$, from (1.2.3) we obtain:

(1.2.4) *as a* T-*module* $E(\lambda)$ $(\lambda \in X)$ *is isomorphic to* $\lambda \otimes (\bigoplus_{r\geq 0} Symm^r(Lie(U)^*))$.

1.3 Reductive groups

The notation established in this section will be in force for the rest of this chapter. Let G be a reductive group over an algebraically closed field k , T a maximal torus of G and $W = N_G(T)/T$ the Weyl group of G with respect to T . Let Φ be the roots of G with respect to T (that is the non-zero weights of Lie(G) considered as a T module via the adjoint action). For $\alpha \in \Phi$ we set $Z_\alpha = C_G((\ker\alpha)^0)$ the centraliser of the connected component of the kernel of $\alpha: T \to k^*$. By Corollary 24.3 of [35], Z_α has semisimple rank one. By Theorem 25.3 of [ ], the Weyl group $N_{Z_\alpha}(T)/T$ of Z_α has order two ; we denote by s_α the non-identity element of $N_{Z_\alpha}(T)/T$.

We let $E = \mathbb{R} \otimes_{\mathbb{Z}} X$ and identify X with a subgroup of E via the map $\lambda \to 1 \otimes \lambda$. The Weyl group W acts on X with action satisfying $(nT)\lambda(t) = \lambda(n^{-1}tn)$ $(n \in N_G(T)$, $\lambda \in X$, $t \in T)$. This action gives rise to an $\mathbb{R}W$-module structure on E in the obvious manner. We denote by E_Δ the subspace of E spanned by Φ - an RW-submodule of E . Let $(,)$ be a real symmetric, bilinear, positive-definite, W invariant form on E . Then the triple $(E_\Delta, \Phi, (,)|_{E_\Delta})$ is a root system, moreover the map $s_\alpha \to$ reflection with respect to α $(\alpha \in \Phi)$ extends to an isomorphism from W to the Weyl group of the root system $(E_\Delta, \Phi, (,)|_{E_\Delta})$.

We now fix a Borel subgroup B of G containing T and let $\Phi^+ = \{\alpha \in \Phi: -\alpha \text{ is a weight of } Lie(B)\}$. Then Φ^+ is a system of positive roots ; let Δ be the base of simple roots determined by Φ^+ and let Φ^- be the set of negative roots. There is a partial order on

E (and hence on X): $\lambda \geq \mu$ if $\lambda - \mu = \sum_{\alpha \in \Phi^+} m_\alpha \alpha$ for non-negative integers m_α.

1.4 B cohomology

There is some overlap between the properties of B cohomology described here and in section 2 of [20].

For $\lambda \in X$ we denote by $P(\lambda)$ the cardinality of

$$\{(m_\alpha)_{\alpha \in \Phi^+} : \text{each } m_\alpha \text{ is a non-negative integer and } \lambda = \sum_{\alpha \in \Phi^+} m_\alpha \alpha\} .$$

The set of weights of the Lie algebra $\mathit{Lie}(U)$ of the unipotent radical $R_u(B)$ of B with respect to the adjoint action of T is precisely Φ^- (see 26.2 Corollary B (b) of [35]). Thus we obtain from (1.2.4):

(1.4.1) *for any* $\lambda, \mu \in X$, $\dim_k E(\lambda)^\mu = P(\mu - \lambda)$, in particular

(1.4.2) $\dim_k E(\lambda)^\lambda = 1$ *and* $\dim_k E(\lambda)^\mu = 0$ *if* $\mu \not\geq \lambda$.

Suppose that the B module V is generated by a non-zero vector of weight λ for some $\lambda \in X$. Then $V/J(V)$, where $J(V)$ denotes the radical of V, is also generated by a vector of weight λ. However, $V/J(V)$ is completely reducible and therefore trivial as a $U = R_u(B)$-module. It follows that $V/J(V)$ is isomorphic to λ as a B-module. Thus the dual module V^* has simple socle $-\lambda$ and thus embeds in $E(-\lambda)$. From (1.4.2) now follows:

(1.4.3) *if the* B*-module* V *is generated by a non-zero vector of weight* λ *then* $\dim_k V^\lambda = 1$ *and* $\mu \leq \lambda$ *for all weights* μ *of* V.

This simple property is important in dealing with the cohomology of rational B-modules.

(1.4.4) *If* $V \subset M_B$, V' *is an essential extension of* V *and* λ *is a weight of* V' *but not of* V *then* λ *is greater than some weight of* V.

To show this we pick an element $0 \neq v$ of V' of weight λ. Then kBv is an essential extension of $kBv \cap V$ so we may suppose that V' is generated by v. The result now follows from (1.4.3).

<u>Definition</u> For $M \in \mathcal{M}_B$ and $\lambda \in X$ we denote by M_λ the sum of the weight spaces M^μ with $\mu \leq \lambda$. By (1.4.3) this is a B submodule of M.

By dimension shifting and (1.4.4) (see also 2.4a and 2.5a of [20]) one obtains:

(1.4.5) *Suppose that* $M \in \mathcal{M}_B$ *and* $M_0 = 0$. *Then* $H^n(B,M) = 0$ *for all* $n \geq 0$.

(1.4.6) *Suppose that* $M \in \mathcal{M}_B$ *and* V *is a submodule containing* M_0. *Then* $H^n(B,M) = H^n(B,V)$ *for all* $n \geq 0$.

Finally we record:

(1.4.7) *For any* $\alpha \in \Delta$, $H^1(B,-\alpha) = k$.

Let $E = E(-\alpha)$. By (1.4.1) E_0 has weights $-\alpha$ and 0 with multiplicity one. Thus there is a short exact sequence $0 \to -\alpha \to E_0 \to k \to 0$ (k denotes the trivial B-module). However $H^i(B,E_0) = 0$ for all $i \geq 0$ by (1.4.6) so that the long exact sequence gives $H^1(B,-\alpha) = (E_0/-\alpha)^B = k$, as required.

<u>1.5 Induced modules</u>

For $\lambda \in X$ we define

$$Y(\lambda) = \{f \in k[G]: f(bx) = \lambda(b)f(x) \text{ for all } (b,x) \in B \times G\}.$$

This is the left G-module induced from the B-module λ. We have:

(1.5.1) $Y(\lambda) \neq 0$ *if and only if* $\lambda \in X^+$.

Here X^+ is the set of dominant weights, that is those $\lambda \in X$ such $(\lambda,\alpha) \geq 0$ for all $\alpha \in \Delta$. A proof of (1.5.1), for G semi-simple, is given in section 31.4 of [35]. The proof for G reductive is similar.

It is not difficult to check that the restriction map $\phi: Y(\lambda) \to k[B]$ is injective and that the image of ϕ lies in $\mathrm{Ind}_T^B(w_0\lambda) \cong E(w_0\lambda)$, where w_0 is the longest element of W. It follows:

(1.5.2) *For* $\lambda \in X^+$, *the* *B-socle of* $Y(\lambda)$ *is* $w_o\lambda$, $\dim_k Y(\lambda)^\lambda = 1$ *and if* τ *is any weight of* $Y(\lambda)$ *then* $w_o\lambda \le \tau \le \lambda$.

Incidentally since $Y(\lambda)$ is identified with a submodule of $E(w_o\lambda)$ we have, for a weight τ of $Y(\lambda)$, $\dim_k Y(\lambda)^\tau \le \dim_k E(w_o\lambda)^\tau = P(\tau - w_o\lambda)$ and since there are only finitely many weights of $Y(\lambda)$ (by (1.5.2)) $Y(\lambda)$ is finite dimensional. Thus it follows from left exactness of induction that, for any finite dimensional B-module V , $Ind_B^G V$ is finite dimensional. We do not know how to prove that $(R^n Ind_B^G)V$ is finite dimensional for $n > 0$ without recourse to sheaf cohomology on projective varieties (see (1.6.2)).

From (1.5.2) and the reciprocity of induction we obtain:

(1.5.3) *For* $\lambda , \mu \in X^+$, $End_G(Y(\lambda)) = k$ *and* $Hom_G(Y(\lambda),Y(\mu)) = 0$ *if* $\mu \not\geq \lambda$.

Now suppose that V is a rational G module with B-socle $w_o\lambda$ $(\lambda \in X)$. Then by (1.4.2), $w_o\lambda$ is the unique smallest weight of V . Hence (since W permutes the weight spaces) λ is the unique highest weight. It follows that $V' = \sum_{\mu<\lambda} V_\mu$ is a B submodule of codimension one. Hence $Hom_B(V,\lambda) = Hom_B(V/V',\lambda) = k$. Thus by (1.1.2) $Hom_G(V,Y(\lambda))$ is non-zero. Let θ be a non-zero element of $Hom_G(V,Y(\lambda))$. We claim that θ is a monomorphism. Since the B socle, say V_1 , of V is simple we need only prove that $\theta(V_1) \neq 0$. Suppose for a contradiction that $\theta(V_1) = 0$. Then we have an induced B module map $\bar\theta : V/V_1 \to Y(\lambda)$. However by hypothesis $V_1 \cong w_o\lambda$ and so, by (1.4.2), $(V/V_1)^{w_o\lambda} = 0$. Hence $(Im\bar\theta)^{w_o\lambda} = 0$. Thus by (1.5.2), $Im\bar\theta$ has zero intersection with the B socle of $Y(\lambda)$. Hence $Im\bar\theta$ is zero and therefore $\theta = 0$ which is a contradiction. Hence θ is a monomorphism.
We have shown:

(1.5.4) *If* V *is a rational* G *module with* B *socle* $w_o\lambda$ *then* V *embeds (as a* G *module) in* $Y(\lambda)$. *If in addition* $\dim_k V = \dim_k Y(\lambda)$ *then* V *is isomorphic to* $Y(\lambda)$.

1.6 Kempf's Vanishing Theorem

It is well known that the quotient map $G \to G/B$ is locally trivial and thus, using (1.1.16) we may read off the following properties of induction from properties of sheaves on projective varieties (Grothendick's Theorem, 2.7, Ch.III of [32] and Serre's Theorem, 5.2(a), Ch.III of [32]):

(1.6.1) $R^n Ind_B^G = 0$ *for all* $n > \dim G/B$;

(1.6.2) *for any finite dimensional rational* B-*module* V *and any* $n \geq 0$, $R^n Ind_B^G(V)$ *is finite dimensional.*

Recall (section 1.1) that each $\lambda \in X$ determines an invertible sheaf (that is a locally free rank one sheaf) L_λ on G/B . (If G is semisimple and simply connected then by section 2 of [43], each invertible sheaf on G/B is isomorphic to precisely one such L_λ) . Now suppose that λ is dominant thus $Y(\lambda)$ and so $\Gamma(G/B, L_\lambda)$ is non-zero ((1.5.1) and (1.1.16)). However by Kempf's Vanishing Theorem, [43] if L is any invertible sheaf on G/B with a non-zero global section then $H^i(G/B, L) = 0$ for all $i > 0$ (actually, in [43], G is semisimple but the quotient G/B is unaffected by replacing G by $G/R(G)$ where $R(G)$ is the soluble radical of G) . Thus $H^i(G/B, L_\lambda) = 0$ for all $i > 0$ when $\lambda \in X^+$ and from (1.1.16) we obtain the following which we shall refer to in the sequel as

(1.6.3) Kempf's Vanishing Theorem. *For* $\lambda \in X^+$ *we have* $R^i Ind_B^G(\lambda) = 0$ *for all* $i > 0$.

1.7 Parabolic subgroups

Let Ω be a subset of Δ . We define P_Ω to be the subgroup of G generated by B together with $\{n_\alpha : \alpha \in \Omega\}$, where, for $\alpha \in \Phi$, n_α is chosen to be an element of $N_G(T)$ such that $n_\alpha T = s_\alpha$. Then P_Ω is a parabolic subgroup of G containing B and any parabolic subgroup containing B is equal to P_Ω for precisely one subset Ω of Δ (see 29.3 Theorem (a) of [35]). Let $U^\Omega = R_u(P_\Omega)$, the unipotent radical of P_Ω . We define $G_\Omega = C_G(T^\Omega)$ where $T^\Omega = \bigcap_{\alpha\in\Omega} (\ker \alpha)^\circ$, the intersection of the connected components of the kernels of the maps $\alpha : T \to k^*$ $(\alpha \in \Omega)$. Then $P_\Omega = U^\Omega G_\Omega$ is the semidirect product of the algebraic groups U^Ω and G_Ω .

Let $\Phi_\Omega = \mathbb{Z}\Omega \cap \Phi$ and $\Phi^\Omega = \Phi^- \mathbb{Z}\Omega$, by the first paragraph of 30.2 of [35] the set of roots of U^Ω with respect to T is Φ^Ω . However, by 30.1 Theorem (b) of [35] the set of roots of P_Ω with respect to T is $\Phi_\Omega \cup \Phi^-$ thus, since P_Ω is the semidirect product of U^Ω and G_Ω , the set of roots of G_Ω with respect to T is Φ_Ω . Let W_Ω be the subgroup of W generated by $\{s_\alpha : \alpha \in \Omega\}$. It follows from the description $P_\Omega = BW_\Omega B$ (see section 29.2 of [35]) and the Bruhat decomposition (28.3 Theorem of [35]) that W_Ω is (naturally identified with) the Weyl group of G_Ω with respect to T . Let E_Ω be the $\mathbb{R}$-span of Φ_Ω in $E = \mathbb{R} \otimes_{\mathbb{Z}} X$. Then, as in section 1.6, $(E_\Omega, \Phi_\Omega, (\ ,\ )|_{E_\Omega})$ is a root system and the map $s_\alpha \to$ reflection with respect to α $(\alpha \in \Omega)$ extends to an isomorphism from W_Ω to the Weyl group of the root system $(E_\Omega, \Phi_\Omega, (\ ,\ )|_{E_\Omega})$

Let $B_\Omega = B \cap G_\Omega$, it is not difficult to show that B_Ω is a Borel subgroup of G_Ω and the roots of B_Ω with respect to T are the elements of $\Phi_\Omega^- = \Phi_\Omega \cap \Phi^-$. It follows that $\Phi_\Omega^+ = \Phi_\Omega \cap \Phi^+$ is a system of positive roots for Φ_Ω and Ω is the set of simple roots.

Let $X_\Omega = \{\lambda \in X : (\lambda, \alpha) \geq 0 \text{ for } \alpha \in \Omega\}$, called the set of Ω-dominant weights of T . We set $Y_\Omega(\lambda) = \mathit{Ind}_{B_\Omega}^{G_\Omega}(\lambda)$ for $\lambda \in X$. By (1.5.1) we have:

(1.7.1) $Y_\Omega(\lambda) \neq 0$ *if and only if* $\lambda \in X_\Omega$

Also, from Kempf's Vanishing Theorem (1.6.3) we have:

(1.7.2) $R^i \mathit{Ind}_{B_\Omega}^{G_\Omega}(\lambda) = 0$ *for all* $i > 0$, $\lambda \in X_\Omega$.

For $\lambda \in X_\Omega$ we also regard $Y_\Omega(\lambda)$ as a P_Ω module with trivial U^Ω action. Applying (1.1.15) now to (1.7.1) and (1.7.2) we now obtain:

(1.7.3) $Y_\Omega(\lambda)$ *is isomorphic to* $Ind_B^{P_\Omega}(\lambda)$;

and

(1.7.4) $R^i Ind_B^{P_\Omega}(\lambda) = 0$ *for all* $i > 0$, $\lambda \in X_\Omega$.

The quotient map $G \to G/P$ is locally trivial and we obtain (c.f. (1.6.1) and (1.6.2)) :

(1.7.5) $R^n Ind_{P_\Omega}^G = 0$ *for all* $n > dim G/P_\Omega$

and

(1.7.6) *for any finite dimensional rational* B*-module* V *and any* $n \geq 0$, $(R^n Ind_{P_\Omega}^G)V$ *is finite dimensional.*

We close this section with some notation. For $\lambda \in X_\Omega$, $Y_\Omega(\lambda)$ has a simple B_Ω socle and hence a simple P_Ω socle. We denote the P_Ω-socle by $L_\Omega(\lambda)$. Then $L_\Omega(\lambda)$ is a P_Ω (and G_Ω) module of highest weight λ and $\{L_\Omega(\lambda) : \lambda \in X_\Omega\}$ is a full set of simple P_Ω-modules (and a full set of G_Ω-modules.

2. More Homological Algebra

2.1 Applications of the Vanishing Theorem

We examine the impact of Kempf's Vanishing Theorem on the derived functors of induction and on the rational cohomology of parabolic subgroups. The results will be used in showing that certain modules have a good filtration. There is some overlap between the results here and those of Andersen, [1] - [9] and those of Cline, Parshall, Scott and van der Kallen, [20]. See also O'Halloran, [45].

We retain the notation of Chapter 1.

Let $\Omega \subseteq \Delta$. It follows from (1.7.3), (1.7.1) and (1.5.2) that $\mathrm{Ind}_B^{P_\Omega} k = k$ and, from (1.7.2), $(R^i \mathrm{Ind}_B^{P_\Omega})k = 0$ for $i > 0$. Thus (P_Ω, B) is a VIP. If $\Omega \subseteq \Gamma \subseteq \Delta$ then, by (1.1.12), $(R^i \mathrm{Ind}_{P_\Omega}^{P_\Gamma})k$ is k if $i = 0$ and 0 for $i > 0$ and so (P_Γ, P_Ω) is a VIP. Thus, if $\Omega \subseteq \Gamma \subseteq \Sigma \subseteq \Delta$ and $\lambda \in X_\Gamma$ we have, by (1.1.12),

(2.1.1) $(R^m \mathit{Ind}_{P_\Omega}^{P_\Sigma}) Y_\Gamma(\lambda) = (R^m \mathit{Ind}_{P_\Gamma}^{P_\Sigma}) Y_\Gamma(\lambda)$ *for all* m.

Hence, by Kempf's Vanishing Theorem and (1.1.6),

(2.1.2) $(R^m \mathit{Ind}_{P_\Omega}^{P_\Sigma}) Y_\Gamma(\lambda) = (R^m \mathit{Ind}_{B}^{P_\Sigma}) \lambda$ *for all* m.

We obtain, from (1.1.10) for $V \in \mathcal{M}_{P_\Omega}$, $\lambda \in X_\Omega$

(2.1.3) $H^m(B, V \otimes \lambda) = H^m(P_\Omega, V \otimes Y_\Omega(\lambda))$ *for all* m, *in particular* $H^m(B,V) = H^m(P_\Omega, V)$ *and, for* $\mu \in X_\Omega$,
$H^m(B, Y_\Omega(\mu) \otimes Y_\Omega(\lambda)) = H^m(B, Y_\Omega(\mu) \otimes \lambda) = H^m(B, \mu \otimes Y_\Omega(\lambda))$.

(See also section 3 of [20]). We record also the elegant corollary, (3.3) of [20], of Cline, Parshall, Scott and van der Kallen.

(2.1.4) *For* $\lambda, \mu \in X^+$, $H^m(G, Y(\lambda) \otimes Y(\mu)) = 0$ *for all* $m > 0$.

We may assume, by symmetry, that $\mu \not\leq -w_o\lambda$. By (2.1.3), $H^m(G,Y(\lambda) \otimes Y(\mu)) = H^m(B,Y(\lambda) \otimes \mu)$ which is 0 by (1.4.5).

Suppose that $\lambda \in X$, $(\lambda,\alpha^\vee) = -1$ for some $\alpha \in \Delta$. We have, by (1.7.3), (1.5.2), a short exact sequence of B-modules $0 \to \lambda \to Y_{\{\alpha\}}(\lambda + \alpha) \to \lambda + \alpha \to 0$. The resulting long exact sequence, obtained by applying the derived functors of $Ind_B^{P_{\{\alpha\}}}$ implies, by (1.7.1), (1.7.4) and (1.7.7), that $(R^m Ind_B^{P_{\{\alpha\}}})\lambda = 0$ for all m . (See also section 2 of [2]).

(2.1.5) *Suppose that* $\Omega \subseteq \Gamma \subseteq \Sigma \subseteq \Delta$, $\lambda \in X_\Gamma$ *and* $(\lambda,\alpha^\vee) = -1$ *for some* $\alpha \in \Sigma\backslash\Gamma$. *Then* $(R^m Ind_{P_\Omega}^{P_\Sigma})Y_\Gamma(\lambda) = 0$ *for all* m .

By (2.1.2), $(R^m Ind_{P_\Omega}^{P_\Sigma})Y_\Gamma(\lambda)$ is isomorphic to $(R^m Ind_B^{P_\Sigma})\lambda$. However $(R^m Ind_B^{P_{\{\alpha\}}})\lambda = 0$ for all m so (2.1.5) follows from (1.1.6).

(2.1.6) *Suppose that* $\Omega \subseteq \Sigma \subseteq \Delta$, $V \in M_{P_\Sigma}$, $\lambda \in X_\Omega$ *and* $(\lambda,\alpha^\vee) = -1$ *for some* $\alpha \in \Sigma\backslash\Omega$ *Then* $H^m(B,V \otimes Y_\Omega(\lambda)) = 0$*f for all* m , *in particular* $H^m(B,Y_\Omega(\lambda)) = 0$.

This is true by virtue of (1.1.9).

2.2 Euler Characteristics

We denote by $\mathbb{Z}X$ (resp. $\mathbb{Z}E$ where $E = \mathbb{R} \otimes_{\mathbb{Z}} X$) the integral group ring of X (resp. E) . We have a canonical $\mathbb{Z}$ - basis $\{e(\lambda):\lambda \in X\}$ (resp. $\{e(\lambda):\lambda \in E\}$) the elements of which multiply according to the rule $e(\lambda)e(\lambda') = e(\lambda + \lambda')$. The group ring $\mathbb{Z}X$ is identified with a subring of $\mathbb{Z}E$ in the obvious manner. For a finite dimensional rational T-module V we define the formal character $chV \in \mathbb{Z}X$ of V by

$$chV = \sum_{\lambda \in X} (dim_k V^\lambda)e(\lambda) .$$

The operation of taking the character is additive on short exact sequences and multiplicative on tensor products. The action of W on X (resp. E) gives rise to a $\mathbb{Z}$-linear action of W on $\mathbb{Z}X$ (resp. $\mathbb{Z}E$) in the obvious fashion. It is not difficult to show that, for $\Omega \subseteq \Delta$, ch defines an isomorphism between the Grothendieck ring of finite dimensional P_Ω-modules and the ring of W_Ω invariants $(\mathbb{Z}X)^{W_\Omega}$ and that $\{chY_\Omega(\lambda) : \lambda \in X_\Omega\}$ is a $\mathbb{Z}$-basis for $(\mathbb{Z}X)^{W_\Omega}$.

For $\Omega \subseteq \Gamma \subseteq \Delta$ and a finite dimensional P_Ω-module V we define the Euler characteristic

$$\chi_{\Gamma,\Omega}(V) = \sum_{m\geq 0} (-1)^m \, ch(R^m Ind_{P_\Omega}^{P_\Gamma})V .$$

Note that $\chi_{\Gamma,\Omega}$ is additive on short exact sequences and so may also be defined on the Grothendieck ring $(\mathbb{Z}X)^{W_\Omega}$ of finite dimensional P_Ω-modules.

Suppose $\Omega \subseteq \Gamma \subseteq \Sigma \subseteq \Gamma$ and V is a finite dimensional P_Ω-module. By (1.1.5) we have a Grothendieck spectral sequence $E_r^{m,n}$ converging to $(R^* Ind_{P_\Omega}^{P_\Sigma})V$ with $E_2^{m,n} = (R^m Ind_{P_\Omega}^{P_\Gamma})(R^n Ind_{P_\Omega}^{P_\Gamma})V$. Putting $\chi_r = \sum_{m,n} (-1)^{m+n} chE_r^{m,n}$ for $r \geq 2$, we have $\chi_2 = \chi_3 \cdots$. However χ_2 is $\chi_{\Sigma,\Gamma}(\chi_{\Gamma,\Omega}(V))$ whereas χ_r is $\chi_{\Sigma,\Omega}(V)$ for r large. We have demonstrated the well known property of Euler characteristics:

(2.2.1) $\chi_{\Sigma,\Omega} = \chi_{\Sigma,\Gamma} \circ \chi_{\Gamma,\Omega}$

(2.2.2) *For* $\phi \in (\mathbb{Z}X)^{W_\Omega}$, $\psi \in (ZX)^{W_\Gamma}$, $\chi_{\Gamma,\Omega}(\phi)\psi = \chi_{\Gamma,\Omega}(\phi\psi)$.

By additivity it is enough to prove (2.2.2) with $\phi = ch\, V$, $\psi = ch\, Z$ for $V \in M_{P_\Omega}$, $Z \in M_{P_\Gamma}$. However, for each m, we have $(R^m Ind_{P_\Omega}^{P_\Gamma})(V \otimes Z) = ((R^m Ind_{P_\Omega}^{P_\Gamma})V) \otimes Z$ by (1.1.7) so the Euler characteristics must agree. In particular writing χ for $\chi_{\Gamma,\Omega}$ with Ω empty, $\Gamma = \Delta$:

(2.2.3) *For* $\lambda \in X$, $\psi = \sum a_\mu e(\mu) \in (\mathbb{Z}X)^W$, $\chi(\lambda)\psi = \sum_\mu a_\mu \chi(\lambda+\mu)$

This formula was discovered by R. Brauer [14] in the context of semi-simple Lie groups.

(2.2.4) *Suppose, for* $\lambda \in X$, $\alpha \in \Delta$, $(\lambda,\alpha^\vee) = n \geq 0$. *Then*

$$\chi_{\{\alpha\}}(\lambda) = e(\lambda) + e(\lambda-\alpha) + \ldots\ldots\ldots + e(\lambda-n\alpha) .$$

For $\Gamma \subseteq \Delta$ we mean by χ_Γ the Euler characteristic $\chi_{\Gamma,\Omega}$, where Ω is the empty set. The result is well known (see section 2 of [2]). It may be extracted from the material gathered so far in the following manner. For $(\lambda,\alpha^\vee) = 0$ or 1 (2.2.4) follows from Kempf's Vanishing Theorem (implying that $\chi_{\{\alpha\}}(\lambda) = ch\, Y_{\{\alpha\}}(\lambda))$ and (1.5.2), (1.7.3). Also by (1.5.2), (1.7.3) $\chi_{\{\alpha\}}(\alpha) = e(\alpha) + r e(0) + e(-\alpha)$ for some $r \geq 0$ and, by the remark following (1.5.2), $r \leq 1$. We have, by (2.2.2), $\chi_{\{\alpha\}}(\alpha) = \chi_{\{\alpha\}}(\alpha) + r\, \chi_{\{\alpha\}}(0) + \chi_{\{\alpha\}}(-\alpha)$. Suppose, for a contradiction, that $r = 0$. Then $\chi_{\{\alpha\}}(-\alpha) = 0$. Since $Ind_B^{P_{\{\alpha\}}}(-\alpha) = 0$ and $dim\, P_{\{\alpha\}}/B = 1$, $(RInd_B^{P_{\{\alpha\}}})(-\alpha) = 0$. Thus, by (1.1.9), $H^1(B,-\alpha) = 0$ contrary to (1.4.7). One may now complete the proof by induction on $(\lambda,\alpha^\vee)$ using the formula

$$\chi_{\{\alpha\}}(\lambda-\alpha)\chi_{\{\alpha\}}(\alpha) = \chi_{\{\alpha\}}(\lambda) + \chi_{\{\alpha\}}(\lambda-\alpha) + \chi_{\{\alpha\}}(\lambda-2\alpha)$$

obtained from (2.2.2).

We denote by $\rho \in E$, half the sum of the positive roots. For $\lambda \in E$, $w \in W$, $w.\lambda$ is defined to be $w(\lambda+\rho) - \rho$. Note that, for $\lambda \in X$, $w.\lambda \in X$.

(2.2.5) *For any* $\lambda \in X$, $w \in W$, $\chi(w.\lambda) = \varepsilon(w)\chi(\lambda)$.

The symbol $\varepsilon(w)$ is $+1$ if w is a product of an even number of reflections s_α $(\alpha \in \Delta)$ and -1 if not. Since any $w \in W$ is a product of reflections s_α $(\alpha \in \Delta)$ in order to prove (2.2.5) it suffices to show that $\chi(s_\alpha.\lambda) = -\chi(\lambda)$. If $(\lambda,\alpha^\vee) = -1$ then $\chi(\lambda) = \chi(s_\alpha.\lambda) = 0$ by (2.1.5). Thus we may assume that $(\lambda,\alpha^\vee) \geq 0$ or $(s_\alpha.\lambda,\alpha^\vee) \geq 0$ and by symmetry we need only consider the case

$(\lambda,\alpha^\vee) = n \geq 0$. Now $\chi_{\{\alpha\}}(\lambda + \alpha) - \chi_{\{\alpha\}}(\lambda) = e(\lambda + \alpha) + e(\lambda - (n+1)\alpha)$ by (2.2.4), which is $\chi_{\{\alpha\}}(\lambda + \alpha) + \chi_{\{\alpha\}}(\lambda - (n+1)\alpha)$ by (2.2.3). Thus $-\chi_{\{\alpha\}}(\lambda) = \chi_{\{\alpha\}}(\lambda - (n+1)\alpha)$ i.e. $\chi_{\{\alpha\}}(s_\alpha . \lambda) = -\chi_{\{\alpha\}}(\lambda)$, as required.

For $\tau \in X$ we put $A(\tau) = \sum_{w \in W} \varepsilon(w) e(w\tau)$. It is now easy to prove the famous:

(2.2.6) Weyl's Character Formula *For any* $\lambda \in X$, *we have, in the quotient field of* $\mathbb{Z}E$, $\chi(\lambda) = A(\lambda + \rho)/A(\rho)$.

We have

$$A(\lambda + \rho)A(\rho) = \chi(0)\Big(\sum_{w_1,w_2} \varepsilon(w_1w_2)e(w_1(\lambda + \rho) + w_2\rho)\Big)$$

$$= \sum_{w_1,w_2} \varepsilon(w_1w_2)\chi(w_1(\lambda + \rho) + w_2\rho)$$

$$= \sum_{w_1,w_2} \varepsilon(w_1w_2)\chi(w_1.(\lambda + w_1^{-1}w_2\rho + w_1^{-1}w_2\rho))$$

$$= \sum_{w_1,w_2} \varepsilon(w_2)\chi(\lambda + w_1^{-1}w_2\rho + w_1^{-1}w_2\rho)$$

$$= \sum_{w_1',w_2'} \varepsilon(w_1'w_2')\chi(\lambda + w_1'\rho + w_2'\rho)$$

using (2.2.3), (2.2.5) and substituting $w_1' = w_1^{-1}w_2'$ $w_2' = w_1^{-1}$. A further application of (2.2.3) yields

$$\chi(\lambda)A(\rho)^2 = \sum_{w_1',w_2'} \varepsilon(w_1'w_2')\chi(\lambda + w_1'\rho + w_2'\rho) .$$

Hence $\chi(\lambda)A(\rho)^2 = A(\lambda + \rho)A(\rho)$. Since $A(\rho) \neq 0$ (the coefficient of $e(\rho)$ is 1) and $\mathbb{Z}E$ is an integral domain we have $\chi(\lambda) = A(\lambda + \rho)/A(\rho)$, as required.

Of course, by Kempf's Vanishing Theorem:

(2.2.7) *For* $\lambda \in X^+$, $ch Y(\lambda) = \chi(\lambda) = A(\lambda + \rho)/A(\rho)$.

We conclude this section with some remarks. Suppose that Γ is a subset of Δ which is the disjoint union of subsets Γ_1, Γ_2 with $(\alpha,\beta) = 0$ for all $\alpha \in \Gamma_1$, $\beta \in \Gamma_2$. Suppose that $\lambda \in X$ and

$\lambda = \lambda_1 + \lambda_2$ where $(\lambda_1,\alpha) = 0$ for all $\alpha \in \Gamma_2$ and $(\lambda_2,\alpha) = 0$ for all $\alpha \in \Gamma_1$. We let

$$\rho = \rho_\Gamma = \tfrac{1}{2} \sum_{\alpha\in\Phi_\Gamma^+} \alpha \, , \ \rho_1 = \rho_{\Gamma_1} \text{ and } \rho_2 = \rho_{\Gamma_2} \, .$$

Since any element of Φ_Γ is conjugate, via a sequence of simple reflections, to an element of Γ we have that Φ_Γ is the disjoint union of Φ_{Γ_1} and Φ_{Γ_2}. Hence ρ is equal to $\rho_1 + \rho_2$. Moreover W is the direct product of W_{Γ_1} and W_{Γ_2}. Thus

$$A_\Gamma(\lambda_1 + \rho) = \sum_{w\in W_\Gamma} \varepsilon(w)e(w\lambda_1 + w\rho_1 + \rho_2) = A_{\Gamma_1}(\lambda_1 + \rho_1)A_{\Gamma_2}(\rho_2) \, ,$$

$$A_\Gamma(\lambda_2 + \rho) = A_{\Gamma_2}(\lambda_2 + \rho_2)A_{\Gamma_1}(\rho_1)$$

and

$A_\Gamma(\rho) = A_{\Gamma_1}(\rho_1)A_{\Gamma_1}(\rho_2)$. Thus we obtain, from Weyl's character formula, that $\chi_\Gamma(\lambda) = ch\,Y_\Gamma(\lambda)$ is given by

(2.2.8) $\quad \chi_\Gamma(\lambda) = \chi_{\Gamma_1}(\lambda_1)\chi_{\Gamma_2}(\lambda_2)$.

Taking $\lambda = \lambda_1$, $\lambda_2 = 0$ we obtain

(2.2.9) $\quad \chi_\Gamma(\lambda_1) = \chi_{\Gamma_1}(\lambda_1)$

so that (2 2.8) gives

(2.2.10) $\quad \chi_\Gamma(\lambda) = \chi_{\Gamma_1}(\lambda_1)\chi_\Gamma(\lambda_2)$.

2.3 Some useful results

(2.3.1) *Suppose* $\alpha \in \Sigma \subseteq \Delta$, $\lambda \in X$ *and* $(\lambda,\alpha^\vee) = -2$. *Then* $Ind_B^{P_\Sigma}(\lambda) = 0$ *and, for* $n > 0$, $(R^n Ind_B^{P_\Sigma})(\lambda) = (R^{n-1} Ind_B^{P_\Sigma})(\lambda + \alpha)$.

Let $\Gamma = \{\alpha\}$. We have $\mathrm{Ind}_B^{P_\Gamma}(\lambda) = 0$ by (1.7.1) and $(R^i \mathrm{Ind}_B^{P_\Gamma})(\lambda) = 0$ for $i > 1$ since $dim\, P_\Gamma/B = 1$. By (2.2.5) applied to G_Γ , $\chi_\Gamma(\lambda) = -\chi_\Gamma(s_\alpha\cdot\lambda) = -\chi_\Gamma(\lambda+\alpha)$, which is $-e(\lambda+\alpha)$ by Weyl's Character Formula. Hence $(RInd_B^{P_\Gamma})(\lambda) = \lambda+\alpha$ and, by (1.1.6), $(R^n Ind_B^{P_\Sigma})(\lambda)$ is 0 for $n = 0$ and $(R^{n-1}Ind_{P_\Gamma}^{P_\Sigma})(\lambda+\alpha)$ for $n > 0$. The latter is $(R^{n-1}Ind_B^{P_\Sigma})(\lambda+\alpha)$ by (2.1.2).

(2.3.2) *Suppose* $\Gamma \subseteq \Sigma \subseteq \Delta$, $\lambda \in X_\Gamma$ *and* $(\lambda,\alpha^\vee) = -2$ *for some* $\alpha \in \Sigma\backslash\Gamma$. *Then* $\chi_\Sigma(\chi_\Gamma(\lambda)) = -\chi_\Sigma(\lambda+\alpha)$ *which, if* $\lambda+\alpha \in X_\Gamma$, *is* $-\chi_\Sigma(\chi_\Gamma(\lambda+\alpha))$.

By (2.1.2) and the vanishing theorem $\chi_\Sigma(\chi_\Gamma(\lambda)) = \chi_\Sigma(\lambda)$ which, by (2.3.1) is $-\chi_\Sigma(\lambda+\alpha)$. This is $-\chi_\Sigma(\chi_\Gamma(\lambda+\alpha))$ if $\lambda+\alpha \in X_\Gamma$ by the vanishing theorem and (2.1.2).

(2.3.3) *Suppose* $V \in M_{P_{\{\alpha\}}}$ *and* $\lambda \in X$ *with* $(\lambda,\alpha^\vee) = -2$. *Then* $(V\otimes\lambda)^B = 0$ *and* $H^n(B,V\otimes\lambda) = H^{n-1}(B,V\otimes(\lambda+\alpha))$ *for* $n > 0$.

By (2.3.1) and (1.9.1), $H^n(B,V\otimes\lambda) = 0$ for $n = 0$ and $H^{n-1}(P_{\{\alpha\}}, V\otimes Y_{\{\alpha\}}(\lambda))$ for $n > 0$. The latter is $H^{n-1}(B,V\otimes\lambda)$ by (2.1.3).

<u>Proposition 2.3.4</u> *Suppose that* $V \in M_B$ *and* $\Omega \subseteq \Delta$. *If* μ *is a weight of* $R^n Ind_B^{P_\Omega}(V)$ *for some* $n \geq 0$ *then for some weight* λ *of* V , $\lambda-\mu \in \mathbb{Z}\Omega$.

<u>Proof</u>. By (1.1.1) the induction functor commutes with direct limits so it is enough to prove the proposition with V finite dimensional. By the long exact sequence of derived functors it is enough to prove the proposition with V a simple B-module, that is, we must prove that if $\lambda \in X$ and μ is a weight of $R^n Ind_B^{P_\Omega}(\lambda)$ for some $n \geq 0$ then $\lambda-\mu \in \mathbb{Z}\Omega$. However, by (1.1.15), $R^n Ind_B^{P_\Omega}(\lambda)$ is isomorphic, as a T-module, to $R^n Ind_{B_\Omega}^{G_\Omega}(\lambda)$. Thus we may replace G by G_Ω . We now prove the proposition for $P_\Omega = G$ by induction on n . If $n = 0$ then for any $\lambda \in X$, and any weight μ of $Ind_B^G(\lambda)$ we have $\lambda-\mu \in \mathbb{Z}\Phi$

(by 1.5.2)). It follows from (1.1.1) and the left exactness of Ind_B^G that the proposition holds with $n = 0$ for any $V \in M_B$. Now suppose that $n > 0$ and the proposition holds for $n-1$. Let $V \in M_B$ and let E be the injective envelope of V. Then by the long exact sequence of derived functors we have an epimorphism

$$R^{n-1}Ind_B^G(E/V) \to R^n Ind_B^G(V) .$$

Let μ be a weight of $R^n Ind_B^G(V)$. Then by the inductive hypothesis there is a weight τ of E/V such that $\mu - \tau \in \mathbb{Z}\Phi$. By (1.4.4) there is a weight λ of V such that $\tau > \lambda$, in particular $\tau - \lambda \in \mathbb{Z}\Phi$. Hence we have

$$\lambda - \mu = -(\tau - \lambda) - (\mu - \tau) \in \mathbb{Z}\Phi$$

as required.

Corollary 2.3.5 *Suppose that* $H^n(B,V) \neq 0$ *for some* $V \in M_B$. *Then there exists a subset* Ω *of* Δ *and a weight* λ *of* V *such that* $|\Omega| \le n$ *and* $\lambda \in \mathbb{Z}\Omega$

Proof. We prove this by induction on n. It is clearly true if $n = 0$. Now suppose that $n > 0$ and the result holds for $n - 1$. Since B cohomology commutes with direct limits we need only consider V finite dimensional. Using the long exact sequence of cohomology and induction on $dim_k V$ we see that it is enough to prove the result for $V = \lambda$ where $\lambda \in X$. Since $H^n(B,\lambda) \neq 0$, λ is not dominant, by (2.1.3) and (2.1.4). Hence $(\lambda,\alpha^\vee) < 0$ for some $\alpha \in \Delta$. Let $\Sigma = \{\alpha\}$. We have, by (1.1.9), (1.6.1), (1.5.1) and (1.1.15) that $H^n(B,\lambda) \cong H^{n-1}(P_\Sigma,(RInd_B^{P_\Sigma})\lambda)$. The result now follows from Proposition 2.3.4 and the inductive hypothesis.

(2.3.6) *Suppose that* $\Omega \subseteq \Delta$, $\mu \in X$ *and* $(\mu,\alpha^\vee) = 0$ *for all* $\alpha \in \Omega$. *If* $w \in W_\Omega$ *then*

$$(R^n Ind_B^{P_\Omega})(w.\mu) = \begin{cases} \mu & \text{if } n = \ell(w) \\ 0 & \text{otherwise} . \end{cases}$$

Here $\ell(w)$ denotes the length of w. Replacing P_Ω by P_Ω/U^Ω and B by B/U^Ω we may assume, (1.1.15), that $\Omega = \Delta$ and $G = P_\Omega$. Since $\mu = \mu \otimes w.0$ we may assume, (1.1.7), that $\mu = 0$. This case follows from Proposition 2.5 and Theorem 2.6 of [5].

<u>Remark</u> Suppose that $(\alpha_1,\alpha_2,\ldots,\alpha_r)$ is a regular sequence of simple roots, i.e. $(\alpha_i,\alpha_{i+1}^\vee) = (\alpha_{i+1},\alpha_i^\vee) = -1$ for $1 \le i < r$ and $\Omega = \{\alpha_i : 1 \le i \le r\}$. Then, with μ as in (2.3.6), $\mu' = s_{\alpha_1} \ldots s_{\alpha_r}.\mu$ satisfies $\mu' = \mu - (r\alpha_1 + (r-1)\alpha_2 + \ldots + \alpha_r)$ and $(\mu',\alpha_i^\vee) = 0$ for $2 \le i \le r$ and $(\mu',\alpha_1^\vee) = -(r+1)$. Hence if $\lambda \in X$ is such that $(\lambda,\alpha_i^\vee) = \delta_{i,1}$ (the Kronecker delta) for $1 \le i \le r$ then $(R^n Ind_B^{P_\Omega})(-(r+1)\lambda)$ is $-(r+1)\lambda + r\alpha_1 + \ldots + \alpha_r$ if $n = r$ and 0 otherwise.

We conclude with a consequence of (1.7.5).

<u>Proposition 2.3.7</u> *Let* Σ, Γ *be subsets of* Δ *such that* $(\alpha,\beta) = 0$ *for all* $\alpha \in \Sigma$, $\beta \in \Gamma$. *Suppose that* $\Sigma = \{\alpha_1,\alpha_2,\ldots,\alpha_r\}$ *for a regular sequence* $(\alpha_1,\ldots,\alpha_r)$ *and* $\Omega = \Gamma \cup \Sigma$. *If* $\lambda \in X$ *is such that* $(\lambda,\alpha_i^\vee) \ge 0$ *for all* $1 < i \le r$ *and* $(\lambda,\beta^\vee) = 0$ *for all* $\beta \in \Gamma$ *then*

$$R^m Ind_B^{P_\Omega}(\lambda) = 0 \quad \textit{for all} \quad m > r .$$

<u>Proof</u>. By (1.1.15) we may assume that $\Omega = \Delta = \Sigma \cup \Gamma$. Let $\pi_0 = \{\alpha_2,\alpha_3,\ldots,\alpha_r\}$ and $\pi = \pi_0 \quad \Gamma$. Then $\dim G/B = |\Phi^+|$ and $\dim P_\pi/B = |\Phi_\pi^+|$ so that $\dim G/P_\pi = |\Phi^+| - |\Phi_\pi^+|$. Moreover $\Phi_\pi^+ = \Phi_{\pi_0}^+ \cup \Phi_\Gamma^+$ and $\Phi^+ = \Phi_\pi^+ \cup \Phi_\Gamma^+$ so that

$$\dim G/P_\pi = |\Phi_\Sigma^+| - |\Phi_{\pi_0}^+| = r(r+1)/2 - (r-1)r/2$$
$$= r$$

since Φ_Σ and Φ_{π_0} are root systems of type A. Thus $\dim G/P_\pi = r$ and, by (1.7.5), $R^m Ind_{P_\pi}^G(Y_\pi(\lambda)) = 0$ for $m > r$. However, by (2.1.2)

$$R^m Ind_B^G(\lambda) = R^m Ind_{P_\pi}^G (Y_\pi(\lambda))$$

and the proposition is proven.

3. Reductions

3.1 Good Filtrations

Let G be an affine algebraic group over an algebraically closed field k. By a filtration of a rational G module V we mean an ascending filtration, that is a sequence of submodules $0 = V_0, V_1, V_2, \ldots$ such that $V_i \subseteq V_{i+1}$ for each $i \geq 0$ and such that $V = \bigcup_{i \geq 0} V_i$.

Definition Let P be a set of finite dimensional rational G-modules containing the zero module. A P filtration of a rational G-module V is a filtration $0 = V_0, V_1, \ldots$ of V such that, for each $i > 0$, the quotient V_i/V_{i-1} is isomorphic to an element of P. By specializing P we obtain the key definition of this monograph. Let P_0 be the set of G modules which have the form $\mathrm{Ind}_B^G(L)$, for some Borel subgroup B of G and some one dimensional rational B module L, together with the zero module. By a good filtration we mean a P_0 filtration.

Let B' be a Borel subgroup of G. Then $B' = g^{-1}Bg$ for some $g \in G$ so that $B' = \sigma(B)$ where $\sigma : G \to G$ is conjugation by g. For any rational G-module M, M^σ is isomorphic to M so we have, from (1.1.14), that $Ind_{B'}^G(L') \cong Ind_B^G(L'^\sigma)$ for a one dimensional B'-module L'. Thus $0 = V_0, V_1, \ldots$ is a good filtration if and only if, for each $i > 0$, V_i/V_{i-1} is either 0 or induced from a one dimensional rational module for a fixed Borel subgroup.

Recall that P is a set of finite dimensional rational G-modules which contains the zero module. We say that a rational G-module V is poly P if there is a P filtration $0 = V_0, V_1, \ldots$ such that $V_n = V_{n+1}$ for all n sufficiently large.

Though we are mainly interested in finite dimensional modules it is reassuring to have the following result. I am grateful to D. Segal for the short proof (which he describes as folklore) given here.

Proposition 3.1.1 *Suppose that* V *is a rational* G-*module which has a filtration* $0 = V_0, V_1, \ldots$ *such that, for each* $i > 0$, V_i/V_{i-1} *has a* P *filtration. Then* V *has a* P *filtration. In particular if*

V_i/V_{i-1} *has a good filtration (for each* $i > 0$*) then* V *has a good filtration.*

Proof. Let D be a rational G-module and C a submodule such that C and D/C each has a $\mathcal{P}$ filtration. We shall show that D has a $\mathcal{P}$ filtration. This special case will be used in proving the general result.

Let $C_0, C_1, \ldots$ be a $\mathcal{P}$ filtration of C and $K_0, K_1, \ldots$ a filtration of D/C with $K_i = E_i/C$ for $i \geq 0$. Let X_i $(i \geq 0)$ be finite dimensional submodules of D such that $E_i = C + X_i$ and $X_i \subseteq X_{i+1}$ for each i. Let δ be a monotonic increasing function such that $X_i \cap C = X_i \cap C_{\delta(i)}$. Let $L_0 = (0)$, $L_1 = C_{\delta(1)}$ and $L_{2i} = C_{\delta(i)} + X_i$, $L_{2i+1} = C_{\delta(i+1)} + X_i$ for $i \geq 1$. It is easy to check that, for $i > 0$, $L_{2i}/L_{2i-1} \cong K_i/K_{i-1}$ and $L_{2i+1}/L_{2i} = C_{\delta(i+1)}/C_{\delta(i)}$. Thus $\{L_i : i = 0,1,\ldots\}$ is a filtration in which each successive quotient is poly $\mathcal{P}$ and so D has a $\mathcal{P}$ filtration.

We now turn to the general case. Let $0 = W_0, W_1, \ldots$ be a filtration of V by finite dimensional subspaces. (Since V has countable dimension such a filtration exists.) We choose inductively submodules F_i of V satisfying:

1) $0 = F_0 < F_1 < \ldots < F_i$ with each factor poly $\mathcal{P}$;

2) $W_i \cap V_i \subset F_i \subset V_i$;

3) V_i/F_i is a poly $\mathcal{P}$ rational G-module.

Suppose $F_0, F_1, \ldots, F_i$ have been chosen. By 3) and the special case, V_{i+1}/F_i has a $\mathcal{P}$ filtration. Thus there is a submodule F_{i+1}/F_i containing $((W_{i+1} + F_i) \cap V_{i+1})/F_i$ such that V_{i+1}/F_{i+1} is poly $\mathcal{P}$.

Hence V has a filtration with poly $\mathcal{P}$ quotients and so V has a $\mathcal{P}$ filtration.

Remark The argument of proof may be applied in other situations. Let R be a Noetherian ring and $\mathcal{P}$ a set of finitely generated R-modules which contains the zero module. The argument shows that if an R-module X has a filtration in which each successive quotient has a $\mathcal{P}$ filtration then X has a $\mathcal{P}$ filtration. The dual formulation in terms of descending filtrations may also be of some interest.

Except where contrary indication is given we shall hence forth use the term G-module to mean a finite dimensional rational G-module.

By the root system of a connected algebraic group G we mean the root system of G/S(G) where S(G) denotes the soluble radical of G (thus we regard the empty set as a root system, that of a connected soluble group).

The purpose of this section is to obtain various reductions regarding the following hypotheses:

$\underline{H}_1(G)$ - *For all rational G-modules V , V' with good filtrations, V ⊗ V' has a good filtration;*

$\underline{H}_2(G;P)$ - *For the parabolic subgroup P of G the restriction* $V|_P$ *has a good filtration for every G-module V which has a good filtration;*

$\underline{H}_2(G)$ - *The hypothesis* $\underline{H}_2(G;P)$ *is true for all parabolic subgroups P of G ;*

$\underline{H}(G)$ - *The hypotheses* $\underline{H}_1(G)$ *and* $\underline{H}_2(G)$ *are true;*

$\underline{H}_1(\Phi)$ - *For the root system* Φ *, the hypothesis* $\underline{H}_1(G)$ *is true for every connected group of type* Φ *.*

$\underline{H}_2(\Phi)$ - *For the root system* Φ *, the hypothesis* $\underline{H}_2(G)$ *is true for every connected group of type* Φ *.*

$\underline{H}(\Phi)$ - *For the root system* Φ *, the hypothesis* $\underline{H}_1(\Phi)$ *and* $\underline{H}_2(\Phi)$ *are true.*

We believe these hypotheses to be true for every connected group G and root system Φ .

<u>Proposition 3.1.2</u> *If for all one dimensional B-modules L and L' the tensor product* $Ind_B^G(L) \otimes Ind_B^G(L')$ *has a good filtration then* $\underline{H}_1(G)$ *is true.*

If for each one dimensional B-module L the restriction $Ind_B^G(L)|_P$ *has a good filtration (for a parabolic subgroup P of G containing*

B) *then* $\underline{H}_2(G;P)$ *is true.*

This is true since tensoring and restricting to a subgroup are exact functors.

For a morphism $\phi:G_1 \to G_2$ of algebraic groups and a G_2-module V we denote by V^ϕ the k-space V regarded as a G_1-module with action given by $x\,v = \phi(x)v$ for $x \in G_1, v \in V$. In particular we use this notation where P is a parabolic subgroup of G, N is a closed normal subgroup of G contained in P and $\phi:P \to P/N$ is the natural map.

Lemma 3.1.3 *Suppose that* N *is a closed normal subgroup of* G *contained in a Borel subgroup and* $\phi:G \to G/N$ *is the natural map. A* G/N-*module* V *has a good filtration if and only if* V^ϕ *has a good filtration.*

This is obvious because $\{Ind_{B/N}^{G/N}(M):M \in M_{B/N}^1\} = \{Ind_B^G(M):M \in M_B^1$, N acts trivially on $Ind_B^G(M)\}$ where $M_B^1 = \{M \in M_B:dim_k M = 1\}$.

Proposition 3.1.4 *Let* N *be a closed normal soluble subgroup of* G. *Then* $\underline{H}_1(G)$ *implies* $\underline{H}_1(G/N)$ *and, for any parabolic subgroup* P *of* G, $\underline{H}_2(G;P)$ *implies* $\underline{H}_2(G/N;P/N)$ *and hence* $\underline{H}_2(G)$ *implies* $\underline{H}_2(G/N)$. *Suppose* N *is such that for each character* λ *of* B *there is a character* μ *of* G *such that* $\lambda|_N = \mu|_N$. *Then* $\underline{H}_1(G/N)$ *implies* $\underline{H}_1(G)$ *and, for any parabolic subgroup* P *of* G, $\underline{H}_2(G/N;P/N)$ *implies* $\underline{H}_2(G;P)$ *and hence* $\underline{H}_2(G/N)$ *implies* $\underline{H}_2(G)$. *In particular this holds if* N *is unipotent.*

Proof. Suppose that $\underline{H}_1(G)$ is true and that M, M' are one dimensional B/N-modules. Then we have

$$(Ind_{B/N}^{G/N}(M) \otimes Ind_{B/N}^{G/N}(M'))^\phi = Ind_B^G(M^\phi) \otimes Ind_B^G(M'^\phi)$$

(where $\phi:G \to G/N$ is the natural map). By hypothesis $Ind_B^G(M^\phi) \otimes Ind_B^G(M'^\phi)$ has a good filtration and so, by Lemma 3.1.3, $Ind_{B/N}^{G/N}(M) \otimes Ind_{B/N}^{G/N}(M')$ has a good filtration. Hence, by Proposition

3.1.2, $\underline{H}_1(G/N)$ is true. If N satisfies the additional hypothesis then, for a character λ of B, $Ind_B^G(\lambda) = Ind_B^G(\lambda - \mu) \otimes \mu$, by the tensor identity, where μ is a character of G such that $\mu|_N = \lambda|_N$. Thus that $\underline{H}_1(G/N)$ implies $\underline{H}_1(G)$ may be obtained by reversing the steps of the above argument. The proofs of the remaining assertions are similar.

By letting N be the unipotent radical of G we obtain:

Corollary 3.1.5 *For a root system* Φ, *the hypothesis* $\underline{H}_1(\Phi)$ *(respectively* $\underline{H}_2(\Phi)$*) is true if and only if* $\underline{H}_1(G)$ *(respectively* $\underline{H}_2(G)$*) is true for all reductive groups* G *of type* Φ.

3.2 Good filtrations for reductive groups

Let G be a reductive group. We adopt the notation of chapter 1. Thus T is a maximal torus, B a Borel subgroup containing T, $X = X(T)$, $E = \mathbb{R} \otimes_{\mathbb{Z}} X$ and Δ a base for the root system Φ. For a subset Ω of Δ, P_Ω denotes the corresponding parabolic subgroup and $Y_\Omega(\lambda) = Ind_B^{P_\Omega}(\lambda)$ for $\lambda \in X$.

The following result is an application of Corollary (3.2) of [20] (the assumption in [20] that G is semisimple is not used here). The most important case for us is $n = 1$.

Lemma 3.2.1 *If* $Ext^n_{P_\Omega}(Y_\Omega(\lambda), Y_\Omega(\tau)) \neq 0$ *for* $\lambda, \tau \in X_\Omega$ *and* $n > 0$ *then* $\lambda > \tau$

An easy application of the lemma is the following.

Lemma 3.2.2 *Suppose that* V *is a* P_Ω*-module with a good filtration* $0 = V_0, V_1, \ldots$ *and* τ *is a minimal element of* $\{\lambda \in X_\Omega : Y_\Omega(\lambda) \cong V_i/V_{i-1}$ *for some* $i > 0\}$. *Then* V *has a submodule, say* Z, *isomorphic to* $Y_\Omega(\tau)$ *and, for any such* Z, V/Z *has a good filtration.*

The arguments of Jantzen, 5.2 of [41], go over to the slightly more general situation considered here giving the following three results.

Lemma 3.2.3 *Suppose that* V *is a* P_Ω*-module which has a good filtration. Let* λ *be a maximal weight of* V *and* V' *the unique maximal submodule of which* λ *is not a weight. Then* V' *has a good filtration and* V/V' *is isomorphic to a direct sum of copies of* $Y_\Omega(\lambda)$.

Proposition 3.2.4 *Suppose that* V *is a* P_Ω*-module with a good filtration and* V' *is a submodule which also has a good filtration. Then* V/V' *has a good filtration.*

Corollary 3.2.5 *Let* V' , V'' *be* P_Ω*-modules. The direct sum* $V' \oplus V''$ *has a good filtration if and only if* V' *and* V'' *each has a good filtration.*

The following result may be proved by induction and Lemma 3.2.2.

Proposition 3.2.6 *Let* V *be a* P_Ω*-module with a good filtration* $0 = V_1, V_2, \ldots, V_n = V$ *with* $V_i/V_{i-1} = Y_\Omega(\lambda_i)$ *for* $\lambda_i \in X_\Omega$ *and* $i = 1,2,\ldots,n$. *If* π *is any permutation of* $1,2,\ldots,n$ *such that, whenever* $\lambda_{\pi(i)} > \lambda_{\pi(j)}$, $\pi(i) > \pi(j)$, *then there is a good filtration* $0 = V_0', V_1', \ldots, V_n' = V$ *with* $V_i'/V_{i-1}' \cong Y_\Omega(\lambda_{\pi(i)})$ *for* $1 \le i \le n$.

We already know, from Proposition 3.1.4, that to prove our hypotheses for all connected algebraic groups it is enough to do so for reductive groups. The following result will help us to reduce to the semisimple case.

Proposition 3.2.7 *Let* G_1 *be the derived subgroup of* G , $B_1 = B \cap G_1$ *and* $X_1 = X(T_1)$ *where* $T_1 = T \cap G_1$. *Then we have:*

(i) *for any* $\lambda \in X$, $Ind_B^G(\lambda)|_{G_1} = Ind_{B_1}^{G_1}(\lambda|_{B_1})$;

(ii) *if* V *is a rational* G*-module such that* $V|_{G_1} \cong Ind_{B_1}^{G_1}(\tau)$ *for some* $\tau \in X_1$ *then* $V \cong Ind_B^G(\lambda)$ *for some* $\lambda \in X$;

(iii) *a finite dimensional rational* G*-module* V *has a good filtration if and only if* $V|_{G_1}$ *has a good filtration.*

Proof (i) We have natural maps $X \to \mathrm{Pic}(G/B)$ (where $\mathcal{Pic}(G/B)$ is the Picard group of G/B) and $X_1 \to \mathcal{Pic}(G_1/B_1)$ determined as in section 1.6. Moreover it is easy to check that the diagram

$$\begin{array}{ccc} X & \longrightarrow & X_1 \\ \downarrow & & \downarrow \\ \mathcal{Pic}(G/B) & \longrightarrow & \mathcal{Pic}(G_1/B_1) \end{array}$$

commutes, where the top map is restriction and the bottom map is the inverse image functor ϕ^* defined by the obvious isomorphism $\phi : G_1/B_1 \to G/B$. We thus have, for $\lambda \in X$,

$$\dim_k \mathit{Ind}_B^G(\lambda) = \dim_k H^0(G/B, L_\lambda) = \dim_k H^0(G_1/B_1, L_{\lambda_1}) = \dim_k \mathit{Ind}_{B_1}^{G_1}(\lambda_1) ,$$

where λ_1 is the restriction of λ to B_1.

It is clear from the definition that if $f \in \mathit{Ind}_B^G(\lambda)$ then $f|_{G_1} \in \mathit{Ind}_{B_1}^{G_1}(\lambda_1)$ and that restriction is a morphism of G_1-modules. If $f \in \mathit{Ind}_B^G(\lambda)$, $f|_{G_1} = 0$ and $g = zg_1$ with $z \in Z$, the centre of G_1, and $g_1 \in G_1$ then $f(zg_1) = \lambda(z) f(g_1) = 0$. By 25.3 Corollary (b) of [35], $G = ZG_1$ and so we have $f(g) = 0$ for all $g \in G$. Thus restriction is a monomorphism of G_1-modules and therefore, by dimensions, an isomorphism.

(ii) Since $\mathit{Ind}_{B_1}^{G_1}(\tau)$ has a simple B_1 socle (by (1.5.1)) and B_1 is a normal subgroup of B we have, by Clifford's Theorem, that V has a simple B socle, say $w_0\lambda$. Since the B_1 socle of $\mathrm{Ind}_{B_1}^{G_1}(\tau)$ is $w_0\tau$ it follows that $\lambda|_{B_1} = \tau$ and hence, by (i), $\mathit{Ind}_B^G(\lambda)|_{G_1} \cong \mathit{Ind}_{B_1}^{G_1}(\tau)$. Let $0 \neq v_0$ be an element of the B socle of V and let $0 \neq v_0'$ be an element of the B socle of $\mathit{Ind}_B^G(\lambda)$. Thus v_0 has weight $w_0\lambda$ and, v_0' also has weight $w_0\lambda$. Now, as a Z module, V is completely reducible. However, the Z isotypic components are G summands and V is indecompsoable so that Z acts scalarly on

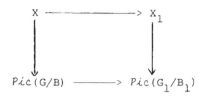

V, i.e. there is an algebraic group homomorphism $\xi : Z \to k^*$ such that $zv = \xi(z)v$ for all $z \in Z$, $v \in V$. However, $zv_o = (w_o\lambda)(z)v_o$ so that $\xi = w_o\lambda|_Z$. Also, for the same reason (using v'_o instead of v_o), $zv' = (w_o\lambda)(z)v'$ for all $z \in Z$, $v' \in Ind_B^G(\lambda)$. It is now easy to see that any G_1-map $V \to Ind_B^G(\lambda)$ is automatically a G-map (since $G = ZG_1$) and so $V \cong Ind_B^G(\lambda)$, as required.

(iii) If V has a good filtration then it is clear from (i) that $V|_{G_1}$ has a good filtration.

Now suppose that V is a finite dimensional rational G module and that $V|_{G_1}$ has a good filtration: we must show that V has a good filtration. By Corollary 3.2.5 we may assume that V is indecomposable. It follows, as in the proof of (ii), that Z acts scalarly on V and so any G_1-submodule of V is also a G-submodule of V. Thus, by (ii), any good G_1 filtration of V is also a good G filtration.

3.3 Canonical Filtration

We define a linear map $f_\Omega : E = \mathbb{R} \otimes_{\mathbb{Z}} X \to \mathbb{R}$ as follows: on $F = \mathbb{R}$-span of Φ, f_Ω is defined on the basis Δ by $f_\Omega(\alpha) = 0$ if $\alpha \in \Omega$ and $f_\Omega(\alpha) = 1$ if $\alpha \in \Delta \backslash \Omega$, we define $f_\Omega|_{F^\perp} = 0$, that is f_Ω is zero on the orthogonal complement of F. Notice that the restriction to X gives a group homomorphism $f_\Omega : X \to \quad$.

<u>Definition</u> For a rational T module V and a rational number r we denote by $F_\Omega^{(r)}(V)$ the sum of the weight spaces V^λ $(\lambda \in X)$ with $f_\Omega(\lambda) = r$. We denote by $F_\Omega^r(V)$ the sum of all $F_\Omega^{(s)}(V)$ with $s \leq r$.

Suppose that V is a rational G_Ω-module, r a rational number and $v \in F_\Omega^{(r)}(V)$ is a vector of weight λ. Then by (1.4.3) any weight of $kB_\Omega v$ has the form $\lambda - \sum_{\alpha \in \Omega} m_\alpha \alpha$ for some $m_\alpha \geq 0$. However $f_\Omega(\lambda - \sum_{\alpha \in \Omega} m_\alpha \alpha) = f_\Omega(\lambda)$ so that $F_\Omega^{(r)}(V)$ is a B submodule of V. Similarly, since $f_\Omega(s_\alpha \lambda) = f_\Omega(\lambda)$, for $\alpha \in \Omega$, $F_\Omega^{(r)}(V)$ is a $N_{G_\Omega}(T)$

submodule of V. But G_Ω is generated by B_Ω and $N_{G_\Omega}(T)$ (for example by the Bruhat decomposition, 28.3 of [35]) so that $F_\Omega^{(r)}(V)$ is a G_Ω-submodule of V.

Now suppose that V is a P_Ω module, r a rational number, λ a weight of $F_\Omega^{(r)}(V)$ and v an element of $F_\Omega^{(r)}(V)$ of weight λ. By (1.4.3) any weight of kBv has the form $\lambda - \sum_{\alpha\in\Delta} m_\alpha\alpha$ for some $m_\alpha \geq 0$ and $\delta_\Omega(\lambda - \sum_{\alpha\in\Delta} m_\alpha\alpha) \leq \delta_\Omega(\lambda)$. Hence kBv belongs to $F_\Omega^r(V)$. Thus $F_\Omega^r(V)$ is invariant under B and G_Ω and hence under the group P_Ω generated by B and G_Ω. We have now shown:

<u>Proposition 3.3.1</u> *For any rational* G_Ω*-module* V, $V = \sum_{r\in}^{\oplus} F_\Omega^{(r)}(V)$ *is a direct sum of* G_Ω*-modules. For any rational* P_Ω*-module* V, $F_\Omega^r(V)$ *is a* P_Ω*-submodule of* V.

The main use to which this will be put will be in finding good filtrations of $Y(\lambda)|_{P_\Omega}$ for $\lambda \in X^+$ (in the later chapters) but we give one application to be going on with.

<u>Proposition 3.3.2</u> *Suppose that* V *is a finite dimensional rational* P_Ω*-module. Then* V *has a good filtration if and only if* $V|_{G_\Omega}$ *has a good filtration.*

<u>Proof</u> We see from (1.7.3) that any good filtration of V as a P_Ω-module is also a good filtration of V as a G_Ω-module. Now suppose that V has a good G_Ω filtration, we shall prove that V has a good P_Ω filtration by induction on $\dim V$. If $V = 0$ the result is certainly true. Suppose that $\dim V > 0$ and the proposition is true for P_Ω-modules V' with $\dim V' < \dim V$. Let r be the smallest rational number such that $Z = F_\Omega^{(r)}(V) \neq 0$. Then $Z = F_\Omega^{(r)}(V) = F_\Omega^r(V)$ which is a P_Ω-submodule of V by Proposition 3.3.1. Suppose first that $Z \neq V$. By Proposition 3.3.1, V is isomorphic to $Z \oplus V/Z$ as a G_Ω-module so that, by Corollary 3.2.5, Z and V/Z each have a good filtration as a G_Ω module. Hence, by the inductive hypothesis, Z and

V/Z each have a good filtration as a P_Ω-module and so V has a good filtration as a P_Ω-module.

Thus we may assume that $V = F_\Omega^{(r)}(V)$. Now the unipotent radical U^Ω of P_Ω is generated by the root subgroups $\{U_{-\alpha} : \alpha \in \Phi^\Omega\}$. For $\alpha \in \Phi^\Omega$ we may choose an isomorphism of algebraic groups $\phi_\alpha : k^+ \to U_{-\alpha}$ from the additive group of the field to $U_{-\alpha}$ such that, for $t \in T$, $x \in k$, $t\phi_{-\alpha}(x)t^{-1} = \phi_{-\alpha}(x/\alpha(t))$ (see 26.3 Theorem (c) of [35]). It follows that, for $\lambda \in X$, $v \in V$, $x \in k$,

$$\phi_{-\alpha}(x)v \in v + \sum_{m=1}^{\infty} V^{\lambda - m\alpha} .$$

However, whenever λ and $\lambda - m\alpha$ are weights of V, $\delta_\Omega(\lambda - m\alpha) = r - m\delta_\Omega(\alpha) = r$. Hence $m\delta_\Omega(\alpha) = 0$ and, since $\alpha \in \Phi^\Omega$, $m = 0$. Thus $gv = v$ for all $g \in U_{-\alpha}$, and $v \in V^\lambda$ and so $gv = v$ for all $g \in U^\Omega$ and $v \in V$. Thus U^Ω acts trivially on V so that any good G_Ω filtration is also a good P_Ω filtration, in particular V has a good P_Ω filtration.

3.4 Good filtrations for semisimple groups

We begin with a proposition which leads to an improvement of Corollary 3.1.4.

Proposition 3.4.1 *Let* G *be a reductive group and* G_1 *its derived subgroup. Then* $\underline{H}_1(G)$ *(respectively* $\underline{H}_2(G)$*) is true if and only if* $\underline{H}_1(G_1)$ *(respectively* $\underline{H}_2(G_1)$*) is true.*

Proof We assume the usual notation for reductive groups, in particular T_1 is a maximal torus of G_1, T is a maximal torus of G containing T, $X_1 = X(T_1)$ and $X = X(T)$. Suppose first that $\underline{H}_1(G)$ is true. Let

$$V_1 = \mathrm{Ind}_{B_1}^{G_1}(\lambda_1) \otimes \mathrm{Ind}_{B_1}^{G_1}(\mu_1)$$

for $\lambda_1, \mu_1 \in X_1$ with $B_1 = B \cap G_1$ (B a Borel subgroup of G containing T). Then by Proposition 3.2.7 (i),

$V_1 = V|_{G_1}$ where $V = Ind_B^G(\lambda) \otimes Ind_B^G(\mu)$ for elements λ , $\mu \in X$ such that $\lambda|_{T_1} = \lambda_1$, $\mu|_{T_1} = \mu_1$. By the assumption V has a good filtration and so, by Proposition 3.2.7 (iii), $V|_{G_1} \cong V_1$ has a good filtration. Thus, by Proposition 3.1.1, $\underline{H}_1(G_1)$ is true. The proof that $\underline{H}_1(G_1)$ implies $\underline{H}_1(G)$ is similar.

Now suppose that $\underline{H}_2(G)$ is true. Let $M_1 = Ind_{B_1}^{G_1}(\lambda_1)$ for some $\lambda_1 \in X_1$. Thus, by Proposition 3.2.7 (i), $M_1 \cong M|_{G_1}$ where $M = Ind_B^G(\lambda)$ with $\lambda \in X$ such that $\lambda|_{T_1} = \lambda_1$. Let P_1 be a parabolic subgroup of G_1 containing B_1 . Then $P_1 = P_{\Omega_1}$ for some subset Ω_1 of $\Delta_1 = \{\alpha|_{T_1} : \alpha \in \Delta\}$. Let $\Omega = \{\alpha \in \Delta : \alpha|_{T_1} \in \Omega_1\}$ then it is not hard to see that the derived subgroup of G_Ω , say G'_Ω , is equal to the derived subgroup, say G'_{Ω_1} , of G_{Ω_1} . Now $M|_{P_\Omega}$ has a good filtration by the assumption and so, by Proposition 3.3.2, $M|_{G_\Omega}$ has a good filtration. Thus, by Proposition 3.2.7 (iii), $M|_{G'_\Omega} = M|_{G'_{\Omega_1}}$ has a good filtration. Again by Proposition 3.2.7 (iii) we have that $M|_{G_{\Omega_1}}$ has a good filtration and so, by Proposition 3.3.2 $M|_{P_1} = (M|_{G_1})|_{P_1}$ has a good filtration. We have shown that for any $\lambda_1 \in X_1$ and any parabolic subgroup P_1 of G_1 , $Ind_{B_1}^{G_1}(\lambda_1)|_{P_1}$ has a good filtration and so, by Proposition 3.1.1, $\underline{H}_2(G_1)$ is true. The proof that $\underline{H}_2(G_1)$ implies $\underline{H}_2(G_1)$ is similar.

We now obtain, from Corollary 3.1.4:

<u>Corollary 3.4.2</u> *For a root system* Φ *, the hypothesis* $\underline{H}_1(\Phi)$ *(respectively* $\underline{H}_2(\Phi)$*) is true if and only if* $\underline{H}_1(G)$ *(respectively* $\underline{H}_2(G)$*) is true for all semisimple groups type* Φ *.*

We now wish to make a further improvement by replacing semisimple by semisimple, simply connected in the above corollary. This is done, as one would expect, by considering for a semisimple group G_1 an epimorphism $\pi : G \to G_1$ from a semisimple, simply connected group G of the same type as G_1. However, some care must be taken in the choice of the epimorphism π. We require, in fact, that the natural map $\tilde{\pi} : X(T_1) \to X(T)$ $(\tilde{\pi}(\lambda)(t) = \lambda(\pi(t))$ for $\lambda \in X(T_1)$, $t \in T)$ induces a bijection from Φ_1 to Φ. Here T denotes a maximal torus of G, $T_1 = \pi(T)$ - a maximal torus of G_1 - Φ denotes the set of roots of G with respect to T and Φ_1 denotes the set of roots of G_1 with respect to T_1. An epimorphism $\pi : G \to G_1$ between semisimple groups such that the kernel of π is finite and such that π satisfies the above condition will be called a root regular epimorphism.

Proposition 3.4.3 *Let* $\pi : G \to G_1$ *be a root regular epimorphism between semisimple groups* G *and* G_1. *Then in the notation established above:*

(i) *for any* $\lambda \in X_1$, $\mathit{Ind}_{B_1}^{G_1}(\lambda)^{\pi} = \mathit{Ind}_B^G(\lambda^{\pi})$;

(ii) *if* V *is a rational* G_1 *module such that* $V^{\pi} \cong \mathit{Ind}_B^G(\tau)$ *for some* $\tau \in X$ *then* $V = \mathit{Ind}_{B_1}^{G_1}(\lambda)$ *for some* $\lambda \in X_1$;

(iii) *a finite dimensional rational* G_1*-module* V *has a good filtration if and only if* V^{π} *has a good filtration.*

Let B be a Borel subgroup of G and $B_1 = \pi(B)$. We show that the induced map $G/B \to G_1/B_1$ is an isomorphism of varieties. Proposition 3.4.3 then follows from the arguments of the proof of Proposition 3.2.7.

Let U be any maximal unipotent subgroup of G and $U_1 = \pi(U)$. We claim that the induced map $\pi : U \to U_1$ is an isomorphism of algebraic groups. Since all such U are conjugate we may assume U to be T invariant. The assertion is easy to check when G has rank one. Let $\alpha \in \Phi$ and $\alpha_1 \in \Phi_1$ be such that $\tilde{\pi}(\alpha_1) = \alpha$. By considering $\pi : G'_{\alpha} \to G'_{\alpha_1}$ between the derived subgroup of G_{α} and G_{α_1} (see section 1.7) we obtain that π induces an isomorphism $U_{\alpha} \to U_{\alpha_1}$ as root subgroups. It follows that, for any T invariant maximal unipotent subgroup $U_1, \pi: U \to U_1$ is a separable morphism and an isomorphism of groups. Hence $\pi: U \to U_1$ is an isomorphism of algebraic groups.

Now let $U = R_u(B)$, let U' be the opposite unipotent subgroup and let $U_1' = \pi(U')$. The map π induces an isomorphism $U'B/B \to U_1'B_1/B_1$ on the big cells and by homogeneity the induced map $G/B \to G_1/B_1$ is an isomorphism.

Let $\pi: G \to G_1$ be a root regular epimorphism between semisimple groups. The restriction $G_\Omega' \to G_{1,\Omega_1}'$, between the derived subgroup of G_Ω and G_{Ω_1} (for Ω a subset of the base and $\tilde{\pi}(\Omega_1) = \Omega$), is a root regular epimorphism. Thus we obtain by the argument of the proof of Proposition 3.4.1:

<u>Corollary 3.4.4</u> *Let* $\pi: G \to G_1$ *be a root regular epimorphism between semisimple groups. If* $\underline{H}_1(G)$ *(respectively* $\underline{H}_2(G)$*) is true then* $\underline{H}_1(G_1)$ *(respectively* $\underline{H}_2(G_1)$*) is true.*

Since, for any semisimple G_1 of type Φ, there is a root regular epimorphism $\pi: G \to G_1$ from the semisimple, simply connected group G of type Φ, by Proposition 1, Exp. n° 23 of [17], we obtain from the above and Corollary 3.4.2:

<u>Corollary 3.4.5</u> *For a root system* Φ *the hypothesis* $\underline{H}_1(\Phi)$ *(respectively* $\underline{H}_2(\Phi)$*) is true if and only if* $\underline{H}_1(G)$ *(respectively* $\underline{H}_2(G)$*) is true for the semisimple, simply connected group* G *of type* Φ.

This corollary together with the following generalities will allow us to reduce to irreducible root systems.

<u>Proposition 3.4.6</u> *Let* $G_1, G_2, \ldots, G_n$ *be connected algebraic groups. Then the hypothesis* $\underline{H}_1(G_1 \times G_2 \times \ldots \times G_n)$ *(respectively* $\underline{H}_2(G_1 \times G_2 \times \ldots \times G_n)$*) is true if and only if* $\underline{H}_1(G_i)$ *(respectively* $\underline{H}_2(G_i)$*) is true for all* i *satisfying* $1 \leq i \leq n$.

Proof By induction it is enough to consider the case $n = 2$. Let B_1 (respectively B_2) be a Borel subgroup of G_1 (respectively G_2), let T_1 (respectively T_2) be a maximal torus of B_1 (respectively B_2) and let X_1 (respectively X_2) be $X(T_1)$ (respectively $X(T_2)$). We set $G = G_1 \times G_2$, $B = B_1 \times B_2$, a Borel subgroup of G, and $T = T_1 \times T_2$, a maximal torus of G. We identify $X = X(T_1 \times T_2)$ with $X(T_1) \oplus X(T_2)$ in the obvious manner. For a parabolic subgroup P_1 of G_1 containing B_1, a parabolic subgroup P_2 of G_2 containing B_2, $V_1 \in \mathcal{M}_{P_1}$ and $V_2 \in \mathcal{M}_{P_2}$ we denote by $V_1 \boxtimes V_2$ the k-space $V_1 \otimes V_2$ viewed as a $P_1 \times P_2$ module with action satisfying

$(x,y)(v_1 \otimes v_2) = xv_1 \otimes yv_2$ for $x \in P_1, y \in P_2, v_1 \in V_1, v_2 \in V_2$.

It is not difficult to check, directly from the definitions, that for $\lambda_1 \in X_1$, $\lambda_2 \in X_2$

$$Ind_B^P(\lambda_1 + \lambda_2) \cong Ind_{B_1}^{P_1}(\lambda_1) \boxtimes Ind_{B_2}^{G_2}(\lambda_2) \qquad (1)$$

where $P = P_1 \times P_2$.

Suppose now that $\underline{H}_1(G_1)$ and $\underline{H}_1(G_2)$ are true. For $\lambda, \mu \in X$ we write $\lambda = \lambda_1 + \lambda_2$, $\mu = \mu_1 + \mu_2$ with $\lambda_1, \mu_1 \in X_1$, $\lambda_2, \mu_2 \in X_2$. Then we have, by (1),

$$\begin{aligned} &Ind_B^G(\lambda) \otimes Ind_B^G(\mu) \\ &\cong (Ind_{B_1}^{G_1}(\lambda_1) \boxtimes Ind_{B_2}^{G_2}(\lambda_2)) \otimes (Ind_{B_1}^{G_1}(\mu_1) \boxtimes Ind_{B_2}^{G_2}(\mu_2)) \\ &\cong (Ind_{B_1}^{G_1}(\lambda_1) \otimes Ind_{B_1}^{G_1}(\mu_1)) \boxtimes (Ind_{B_2}^{G_2}(\lambda_2) \otimes Ind_{B_2}^{G_2}(\mu_2)) \quad (2). \end{aligned}$$

By the assumption $M = Ind_{B_1}^{G_1}(\lambda_1) \otimes Ind_{B_1}^{G_1}(\mu_1)$ has a good filtration say $0 = M_0, M_1, \ldots$ and $N = Ind_{B_2}^{G_2}(\lambda_2) \otimes Ind_{B_2}^{G_2}(\mu_2)$ has a good filtration say $0 = N_0, N_1, \ldots$ Thus $M \boxtimes N$ has a filtration $0 = M_0 \boxtimes N$, $M_1 \boxtimes N, \ldots$ and moreover, for each $i > 0$, $(M_i \boxtimes N)/(M_{i-1} \boxtimes N) = (M_i/M_{i-1}) \boxtimes N$ which has a filtration $0 = (M_i/M_{i-1}) \boxtimes N_0, (M_i/M_{i-1}) \boxtimes N_1, \ldots$ A typical quotient in this filtration is isomorphic to $(M_i/M_{i-1}) \boxtimes (N_j/N_{j-1})$, $j > 0$, and so by (2) this is a good filtration. Hence, $M \boxtimes N$ has a good filtration and so, by Proposition 4.1.1, $\underline{H}_1(G)$ is true.

Now suppose that $\underline{H}_1(G)$ is true. Let λ_1, μ_1 be elements of X_1. Then by assumption

$$(Ind_{B_1}^{G_1}(\lambda_1) \otimes Ind_{B_1}^{G_1}(\mu_1)) \boxtimes k$$

$$= (Ind_{B_1}^{G_1}(\lambda_1) \boxtimes k) \otimes (Ind_{B_1}^{G_1}(\mu_1) \boxtimes k)$$

has a good filtration. However, any G submodule of $V = Z \boxtimes k$ where $Z = Ind_{B_1}^{G_1}(\lambda_1) \otimes Ind_{B_1}^{G_1}(\mu_1)$ has the form $L \boxtimes k$ for some G_1 submodule L of Z. Thus V has a good filtration $0 = Z_o \boxtimes k$, $Z_1 \boxtimes k \ldots$ where each Z_i $(i = 0,1,\ldots)$ is a G_1 submodule of Z. It is easy to see that $0 = Z_o, Z_1, \ldots$ is a good filtration of Z and so, by Proposition 4.1.1, $\underline{H}_1(G_1)$ is true. We obtain symmetrically that $\underline{H}_1(G_2)$ is true.

The proof of the assertions regarding $\underline{H}_2$ involve similar, but easier, manipulations and will therefore be left to the reader.

It is tempting to believe that for semisimple groups G_1, G_2 and V a rational $G = G_1 \times G_2$ module V has a good filtration if and only if $V|_{G_i}$ has a good filtration for both $i = 1$ and $i = 2$. However, this is false. To see this we take $G_1 = G_2 = SL_2(k)$ where the characteristic of k is 2 and let M be the three dimensional rational $SL_2(k)$ module induced from a one dimensional module for a Borel subgroup. Then there is a non-zero homomorphism $\theta: M \to k$. The kernel V of the map $\theta \boxtimes 1 \oplus 1 \boxtimes \theta : M \boxtimes k \oplus k \boxtimes M \to k \boxtimes k$ does not have a good filtration but $V|_{G_i}$ does for $i = 1,2$.

Corollary 3.4.7 *Let Φ be a root system and let $\Phi_1, \Phi_2, \ldots, \Phi_n$ be the indecomposable components of Φ. Then $\underline{H}_1(\Phi)$ (respectively $\underline{H}_2(\Phi)$) is true if and only if $\underline{H}_1(\Phi_i)$ (respectively $\underline{H}_2(\Phi_i)$) is true for all $1 \le i \le n$.*

Proof For $1 \le i \le n$, let G_i be the semisimple, simply connected group of type Φ_i. Then $G = G_1 \times G_2 \ldots \times G_n$ is the semisimple, simply connected group of type Φ. By Corollary 3.4.5, $\underline{H}_1(\Phi)$ (respectively $\underline{H}_2(\Phi)$) is true if and only if $\underline{H}_1(G)$ (respectively $\underline{H}_2(G)$) is true and, by Proposition 3.4.6, $\underline{H}_1(G)$ (respectively $\underline{H}_2(G)$) is true if and only if $\underline{H}_2(G_i)$ is true for all $1 \le i \le n$. A further application of Corollary 3.4.5 now yields the desired conclusion.

3.5 Good filtrations for semisimple, simply connected groups

For the next three results G denotes an arbitrary reductive group and the usual notation applies.

Proposition 3.5.1 Let Ω, Σ be subsets of Δ with $\Omega \subseteq \Sigma$. Suppose that $V \in M_{P_\Omega}$ has a good filtration $0 = V_0, V_1, \ldots$ such that

$\mathrm{RInd}_{P_\Omega}^{P_\Sigma}(V_i/V_{i-1}) = 0$ for all $i > 0$. Then, identifying $\mathrm{Ind}_{P_\Omega}^{P_\Sigma}(V_i)$ with a submodule of $\mathrm{Ind}_{P_\Omega}^{P_\Sigma}(V)$ in the natural manner,

$0 = \mathrm{Ind}_{P_\Omega}^{P_\Sigma}(V_0)$, $\mathrm{Ind}_{P_\Omega}^{P_\Sigma}(V_1)$, ... is a good filtration of $\mathrm{Ind}_{P_\Omega}^{P_\Sigma}(V)$.

This follows by induction on the filtration length and the long exact sequence of derived functors.

In the following proposition we see for the first time some interaction between $V|_{P_\Omega}$ having a good filtration and $V \otimes V'$ having a good filtration, for G modules V, V' which each have a good filtration.

Proposition 3.5.2 *Suppose that* V *is a rational* G-*module such that* $V|_{P_\Omega}$ *has a good filtration for all subsets* Ω *of* Δ *and* $(\tau, \alpha^\vee) \ge -2$ *for any weight* τ *of* V *and any* $\alpha \in \Delta$. *If* V' *is any rational* G-*module with a good filtration then* $V \otimes V'$ *has a good filtration.*

Proof Let $0 = V_0', V_1', \ldots$ be a good filtration of V'. Then we have a filtration $0 = V \otimes V_0', V \otimes V_1', \ldots$ of $V \otimes V'$ and by Proposition 4.4.1 it is enough to prove that each

$(V \otimes V_i')/(V \otimes V_{i-1}') \cong V \otimes (V_i'/V_{i-1}')$, for $i > 0$, has a good filtration. Thus we may assume that $V' = Y(\lambda)$ for some $\lambda \in X^+$. Now let $\Omega = \{\alpha \in \Delta : (\lambda, \alpha^\vee) = 0\}$. Then, for example by Weyl's Character Formula, $Y_\Omega(\lambda) \cong \lambda$ so, by (1.1.7),

$$V \otimes V' \cong Ind_{P_\Omega}^G (V \otimes \lambda) \qquad (1).$$

Let $0 = V_0, V_1, \ldots$ be a good filtration of $V|_{P_\Omega}$ then we have a filtration $0 = V_0 \otimes \lambda, V_1 \otimes \lambda, V_2 \otimes \lambda, \ldots$ of $V \otimes \lambda$. From (1) and Proposition 3.5.1 it suffices to prove that

$RInd_{P_\Omega}^G (V_i \otimes \lambda / V_{i-1} \otimes \lambda) \cong RInd_{P_\Omega}^G ((V_i/V_{i-1}) \otimes \lambda)$ is zero for all $i > 0$.

If $V_i/V_{i-1} = 0$ then certainly $RInd_P^G((V_i/V_{i-1}) \otimes \lambda)$ is zero. Now suppose that $V_i/V_{i-1} \cong Y_\Omega(\tau)$ for some $\tau \in X_\Omega$ and so $(V_i/V_{i-1}) \otimes \lambda = Y_\Omega(\tau) \otimes \lambda$ which is isomorphic to $Y_\Omega(\tau + \lambda)$, as one may see from (1.5.4). For $\alpha \in \Omega$ we have $(\tau + \lambda, \alpha^\vee) = (\tau, \alpha^\vee) \geq 0$ and for $\alpha \in \Delta \backslash \Omega$ we have $(\tau + \lambda, \alpha^\vee) = (\tau, \alpha^\vee) + (\lambda, \alpha^\vee) \geq 1 - 2 = -1$. Thus $\tau + \lambda \in X^+$ or there is some $\alpha \in \Delta \backslash \Omega$ such that $(\tau + \lambda, \alpha^\vee) = -1$. In the former case $RInd_{P_\Omega}^G Y_\Omega(\tau + \lambda) = 0$ by (2.1.4) and in the latter case $RInd_{P_\Omega}^G Y_\Omega(\tau + \lambda) = 0$ by (2.1.5).

For a subset Ω of Δ we denote, for short, by Φ_Ω the root system $(E_\Omega, \Phi_\Omega, (\ ,\ )|_{E_\Omega})$ discussed in section 1.7. For $\alpha \in \Delta$ we write $\Delta(\alpha)$ for $\Delta \backslash \{\alpha\}$. We put on record, for future use, the obvious:

<u>Proposition 3.5.3</u> *Suppose that, for* $V \in M_G$, $V|_{P_{\Delta(\alpha)}}$ *has a good filtration for all* $\alpha \in \Delta$ *and* $\underline{H}_2(\Phi_{\Delta(\alpha)})$ *is true for all* $\alpha \in \Delta$. *Then for any subset* $\Omega \neq \Delta$ *of* Δ , $V|_{P_\Omega}$ *has a good filtration.*

For a semisimple, simply connected group G the fundamental dominant weights $\lambda_\alpha (\alpha \in \Delta)$ are the elements of X determined by the conditions $(\lambda_\alpha, \beta^\vee) = \delta_{\alpha,\beta} (\beta \in \Delta)$ where $\delta_{\alpha,\beta}$ denotes the Kronecker delta.

The first part of the following proposition is a result of Wang Jian-pan (4.2 Theorem of [52]), also discovered independently by the author.

Proposition 3.5.4 *Let G be a semisimple, simply connected group.*

(i) *If* $Y(\lambda_\alpha) \otimes Y(\tau)$ *has a good filtration for all* $\alpha \in \Delta$, $\tau \in X^+$ *then* $\underline{H}_1(G)$ *is true.*

(ii) *If, for* $\Omega \subseteq \Delta$, $Y(\lambda_\alpha)|_{P_\Omega}$ *has a good filtration for all* $\alpha \in \Delta$, *and if* $\underline{H}_1(\Phi)$ *and* $\underline{H}_1(\Phi_\Omega)$ *are true then* $\underline{H}_2(G;P_\Omega)$ *is true.*

(iii) *If* $Y(\lambda_\alpha)|_{P_{\Delta(\beta)}}$ *has a good filtration for all* $\alpha, \beta \in \Delta$ *and if* $\underline{H}_1(\Phi_\Omega)$ *is true for all* $\Omega \subseteq \Delta$ *then* $\underline{H}_2(\Phi)$ *is true.*

Proof For this proof only we define a new partial ordering on X. We define $\lambda \geq \mu$ if $\lambda - \mu = \sum_{\alpha\in\Delta} m_\alpha \alpha$ for rational numbers m_α with $m_\alpha \geq 0$. By Exercise 8, section 13 of [34] each $\lambda_\alpha > 0$ and it follows that for any dominant weight λ, $\lambda \geq 0$ and there are only a finite number of dominant weights less that λ.

(ii) By Proposition 3.1.2 it is enough to prove that $Y(\lambda)|_{P_\Omega}$ has a good filtration for all $\lambda \in X^+$. Suppose for a contradiction that the proposition is false and that λ is a dominant weight as small as possible (with respect to the ordering introduced above) for which $Y(\lambda)|_{P_\Omega}$ fails to have a good filtration. Clearly $\lambda \neq 0$ and, by hypothesis, $\lambda \neq \lambda_\alpha$ for any $\alpha \in \Delta$. Hence there is some $\beta \in \Delta$ such that $(\lambda,\beta^\vee) > 0$ and $\lambda' = \lambda - \lambda_\beta$ is not zero. By $\underline{H}_1(\Phi)$ and Proposition 3.2.6 there is a G submodule M of $Y(\lambda') \otimes Y(\lambda_\beta)$ such that M has a good filtration and $(Y(\lambda') \otimes Y(\lambda_\beta))/M \cong Y(\lambda)$. By choice of λ, $Y(\lambda')|_{P_\Omega}$ has a good filtration and by hypothesis $Y(\lambda_\beta)|_{P_\Omega}$ has a good filtration. Thus, by $\underline{H}_1(\Phi_\Omega)$, $(Y(\lambda') \otimes Y(\lambda_\beta))|_{P_\Omega}$ has a good filtration. Let $0 = M_0, M_1, \ldots, M_r = M$ be a good filtration of M with $M_i/M_{i-1} \cong Y(\lambda_i)$ for $i > 0$, $\lambda_i \in X^+$. Each $\lambda_i < \lambda$ since λ is the unique maximal weight of $Y(\lambda') \otimes Y(\lambda_\beta)$ hence each $Y(\lambda_i)|_{P_\Omega}$ has a good filtration. Thus $M|_{P_\Omega}$ has a good filtration and so, by Proposition 3.2.4, $Y(\lambda)|_{P_\Omega}$ has a good filtration.

(iii) This is obtained from (ii) and Proposition 3.5.3.

3.6 The canonical filtration revisited

Let G be a reductive group. Let Ω be a maximal proper subset of Δ, that is $\Omega = \Delta\backslash\{\alpha\}$ for some $\alpha \in \Delta$. Suppose that $\tilde{\phi}:X \to X$ is a group homomorphism which takes Φ into Φ and determines an isomorphism of the root system. Suppose also that $\tilde{\phi}(\Omega) = \Omega$ and $\delta_\Omega(\tilde{\Phi}(\beta)) = -\delta_\Omega(\beta)$ for all $\beta \in \Phi$. Such a $\tilde{\phi}$ may not exist of course. However if $\Omega^\perp = \{\theta \in \Phi:(\theta,\gamma^\vee) = 0$ for all $\gamma \in \Omega\}$ is not empty then s_β for $\beta \in \Omega^\perp$ is such a $\tilde{\phi}$. This observation will be of crucial importance for types B_ℓ and C_ℓ. We also make use of a $\tilde{\phi}$ satisfying these conditions in type D_ℓ.

By Theorem 1, Exp. 23 of [13] there is an endomorphism $\phi:G \to G$ such that $\phi(T) = T$ and the induced map $X \to X$ is $\tilde{\phi}$. Since $\tilde{\phi}$ is an isomorphism, by Prop. 5, Exp. 18 of [13], ϕ is an isomorphism. For $\lambda \in X$ and any isomorphism $\psi:G \to G$, by (1.1.14), $Ind_B^G(\lambda)^\psi = Ind^G_{B^{\psi^{-1}}}(\lambda^\psi)$ and so we have:

(3.6.1) *For an isomorphism* $\psi:G \to G$ *a* G-*module* V *has a good filtration if and only if* V^ψ *has a good filtration.*

The restriction of ϕ from G_Ω to G_Ω will also be denoted by ϕ.

(3.6.2) *A* G_Ω-*module* V *has a good filtration if and only if* $F^O_\Omega(V)$ *and* $F^O_\Omega(V^\psi)$ *each has a good filtration.*

Suppose V has a good filtration. Then $F^{(r)}_\Omega(V)$ has a good filtration for each $r \in$ by Corollary 3.2.5. Since $F^O_\Omega(V)$ is the direct sum of those $F^{(r)}_\Omega(V)$ with $r \leq 0$, $F^O(V)$ has a good filtration by Corollary 3.2.5. V has a good filtration by (3.6.1) so the same argument shows that $F^O_\Omega(V^\phi)$ has a good filtration. Now suppose that $F^O_\Omega(V)$ and $F^O_\Omega(V\ )$ each has a good filtration. Since V is the direct sum of the $F^{(r)}_\Omega(V)$, by Corollary 3.2.5 it is only necessary to prove that each $F^{(r)}_\Omega(V)$ has a good filtration. If $r \leq 0$ then $F^{(r)}_\Omega(V)$ is a summand of $F^0_\Omega(V)$ and so has a good filtration by Corollary 3.2.5. For $v \in F^{(r)}_\Omega(V)$ it is clear from the definition that, when regarded as an element of V, v belongs to $F^{(-r)}_\Omega(V^\phi)$. Thus the subspace $F^{(r)}_\Omega(V)$ of V, when regarded as a subspace of V lies in $F^{(-r)}_\Omega(V^\phi)$. By dimensions the inclusion must be equality and we have shown that $F^{(r)}_\Omega(V)^\phi \cong F^{(-r)}_\Omega(V^\phi)$. Now for $r > 0$, $F^{(-r)}_\Omega(V^\phi)$ is a summand of $F^0_\Omega(V^\phi)$ and so has a good filtration by the hypotheses and Corollary 3.2.5. Hence, by (3.6.1) applied to the group G_Ω, $F^{(r)}_\Omega(V)$ has a good filtration and the proof is complete.

From Proposition 3.3.2 we deduce the following.

<u>Proposition 3.6.3</u> *Suppose that* V *is a* G*-module such that* $V^\phi \cong V$. *Then* $V|_{P_\Omega}$ *has a good filtration if and only if* $F^0_\Omega(V)$ *has a good filtration.*

<u>3.7 A useful lemma</u>

<u>Lemma 3.7.1</u> *Let* Ω, Σ *be subsets of* Δ *and* $\Gamma = \Omega \cap \Sigma$. *Suppose* $\underline{H}(P_\Omega)$ *is true and that* M *is a* P_Ω*-module with a good filtration such that any non-zero successive quotient has the form* $Y_\Omega(\mu)$ *for some* $\mu \in X_\Omega$ *satisfying* $(\mu,\alpha^\vee) \geq -1$ *for all* $\alpha \in \Sigma\backslash\Omega$. *Then* $(R^i Ind_{P_\Gamma}^{P_\Sigma})M$ *has a good filtration for* $i = 0$ *and is* 0 *for* $i > 0$. *If* M *is the restriction to* P_Ω *of a* G*-module* V *then* $V|_{P_\Sigma}$ *has a good filtration.*

<u>Proof</u> By the long exact sequence of derived functors we need only consider, for the first assertion, the case in which $M = Y_\Omega(\mu)$ where $\mu \in X_\Omega$ and $(\mu,\alpha^\vee) \geq -1$ for all $\alpha \in \Sigma\backslash\Omega$. Since $\underline{H}(P_\Omega)$ is true $M|_{P_\Gamma}$ has a good filtration $0 = M_0, M_1, \ldots$ If $M_r/M_{r-1} \neq 0$ then $M_r/M_{r-1} \cong Y_\Gamma(\tau)$ for some $\tau \in X_\Gamma$. Since τ is a weight of $Y_\Omega(\mu)$, by (1.5.2), τ has the form $\mu - \sum_{\beta\in\Omega} m_\beta\beta$ for $m_\beta \geq 0$ from which it is clear that $(\tau,\alpha^\vee) \geq -1$ for $\alpha \in \Sigma\backslash\Omega$. Thus either $\tau \in X_\Sigma$ or $(\tau,\alpha^\vee) = -1$ for some $\alpha \in \Sigma\backslash\Gamma$. In the former case $(R^i Ind_{P_\Gamma}^{P_\Sigma})Y_\Gamma(\tau)$

is $Y_\Sigma(\tau)$ for $i = 0$ and 0 for $i > 0$, by (2.1.2) and (1.7.4); in the latter case $(R^i Ind_{P_\Gamma}^{P_\Sigma}) Y_\Gamma(\tau) = 0$ for all $i \geq 0$ by (2.1.5). Thus in both cases $(R^i Ind_{P_\Gamma}^{P_\Sigma})(M_r/M_{r-1})$ has a good filtration for $i = 0$ and is 0 for $i > 0$ and so, by the long exact sequence of derived functors, $(R^i Ind_{P_\Gamma}^{P_\Sigma})M$ has a good filtration for $i = 0$ and is 0 for $i > 0$.

Of course if $M = V|_{P_\Omega}$ then

$(Ind_{P_\Gamma}^{P_\Sigma})(M|_{P_\Gamma}) = (Ind_{P_\Gamma}^{P_\Sigma})(V|_{P_\Gamma} \otimes k) = V|_{P_\Sigma} \otimes Ind_{P_\Gamma}^{P_\Sigma}(k) = V|_{P_\Sigma}$, by the tensor identity.

4. The Classical Groups

4.1 Exterior Powers

Let G be a reductive group; we assume the notation of section 1.6. We shall say that G is simply connected if the elements $\lambda_\alpha(\alpha \in \Delta)$ of E_Δ (see chapter 1) defined by $(\lambda_\alpha, \beta^\vee) = \delta_{\alpha,\beta}$ actually belong to X. (Of course for G semisimple this agrees with the usual definition). Suppose the root system of G is of type $A_\ell(\ell \geq 1)$, $B_\ell(\ell \geq 2)$, $C_\ell(\ell \geq 3)$ or $D_\ell(\ell \geq 4)$ and that G is simply connected. We list the elements $\alpha_1, \alpha_2, \ldots, \alpha_\ell$ of Δ as follows: for type $A_\ell(\ell \geq 1)$, α_1 is a terminal vertex of the Dynkin diagram and $(\alpha_{i+1}, \alpha_i^\vee) \neq 0$ for $1 \leq i < \ell$; for $B_\ell(\ell \geq 2)$, α_1 is a terminal vertex of the Dynkin diagram, $(\alpha_2, \alpha_1^\vee) = -1$ and $(\alpha_{i+1}, \alpha_i^\vee) \neq 0$ for $1 \leq i < \ell$; for $C_\ell(\ell \geq 3)$, α_1 is a terminal vertex of the Dynkin diagram attached by a single bond and $(\alpha_{i+1}, \alpha_i^\vee) \neq 0$ for $1 \leq i < \ell$; for $D_\ell(\ell \geq 4)$, $\alpha_{\ell-2}$ is the vertex of degree 3 of the Dynkin diagram and $(\alpha_{i+1}, \alpha_i^\vee) \neq 0$ for $1 \leq i \leq \ell-2$. We denote λ_{α_r} $(1 \leq r \leq \ell)$ simply by λ_r.

We record, for future use, some formulas for $\Lambda^r\chi(\lambda_1)$ (that is the character of the rth exterior power of $Y(\lambda_1)$).

(4.1.1) *For* G *of type* A_ℓ, $\Lambda^r\chi(\lambda_1) = \chi(\lambda_r)$ *for* $1 \leq r \leq \ell$.

For B *of type* B_ℓ, $\Lambda^r\chi(\lambda_1) = \chi(\lambda_r)$ *for* $1 \leq r < \ell$,

$$\Lambda^\ell\chi(\lambda_1) = \chi(2\lambda_\ell).$$

For G *of type* C_ℓ,

$$\Lambda^r\chi(\lambda_1) = \chi(\lambda_r) + \chi(\lambda_{r-2}) + \ldots \quad \textit{for } r \quad \ell.$$

For G *of type* D_ℓ, $\Lambda^r\chi(\lambda_1) = \chi(\lambda_r)$ *for* $r < \ell-1$,

$$\Lambda^{\ell-1}\chi(\lambda_1) = \chi(\lambda_{\ell-1} + \lambda_\ell).$$

$$\Lambda^\ell\chi(\lambda_1) = \chi(2\lambda_{\ell-1}) + \chi(2\lambda_\ell)$$

In interpreting the above for C_ℓ we set $\lambda_0 = 0$ and $\chi(\lambda_{-t}) = 0$ for $t > 0$.

Since the characters of the irreducible, finite dimensional modules for a complex, semisimple Lie algebra are also given by Weyl's character formula (4.1.1) follows from the corresponding results from the representation theory of complex, semisimple Lie algebras (see [13] or section 6, Chapter VII of [38]). They may also be obtained by use of (2.2.5) and "Brauer's Formula" (2.2.3).

4.2 Miniscule weights

Let $\Omega \subseteq \Delta$ and $\lambda \in X_\Omega$ be miniscule, i.e. $(\lambda,\alpha^\vee) \in \{-1,0,1\}$ for all $\alpha \in \Phi_\Omega$. It follows from (2.2.5) that $s_\Omega(\lambda) = \chi_\Omega(\lambda)$ and so $Y_\Omega(\lambda)$ is a simple G_Ω-module. This remark, together with Proposition 3.5.2 shows the following.

Proposition 4.2.1 *Suppose* $\lambda \in X^+$ *is a miniscule weight and* $\Omega \subseteq \Delta$. *Then any composition series of* $Y(\lambda)|_{P_\Omega}$ *is a good filtration. Moreover, for any* $\tau \in X^+$, $Y(\lambda) \otimes Y(\tau)$ *has a good filtration.*

Now suppose that G is semisimple, simply connected of type A_ℓ. It is well known that λ_r is miniscule for $1 \le r \le \ell$ and so the above and Proposition 3.5.4 implies that $\underline{H}_1(G)$ and $\underline{H}_2(G)$ are true. Thus by Corollary 3.4.5 we have the following.

Proposition 4.2.2 *For any root system* Φ *of type* A, $\underline{H}(\Phi)$ *is true.*

The validity of $\underline{H}_1(SL_{\ell+1}(k))$ is also demonstrated by Wang Jian-pan in [52], (4.2) Theorem.

4.3 Exceptional Weights

We now set out on the road leading to a proof that hypotheses 1 and 2 are satisfied by the other classical groups. The remainder of this section is devoted to proving:

Theorem 4.3.1 *For any root system* Φ *of classical type* (i.e. *the indecomposable components of* Φ *have type* A,B,C *or* D) $\underline{H}(\Phi)$ *is true.*

We prove this by induction on $|\Phi|$. If $|\Phi| = 0$ then $\underline{H}(\Phi)$ is trivially true. We now assume that $|\Phi| > 0$ and that $\underline{H}(\Phi')$ is true for classical root systems with $|\Phi'| < |\Phi|$. By Corollary 3.4.7 we may assume that Φ is indecomposable and so, by Proposition 4.2.2 we may assume that Φ is of type $B_\ell (\ell \geq 2)$, $C_\ell (\ell \geq 3)$ or $D_\ell (\ell \geq 4)$. We let G be semisimple, simply connected of type Φ and assume the notation of chapter 1 and section 4.1.

To simplify the calculations we define elements $\varepsilon_1, \varepsilon_2, \ldots, \varepsilon_\ell$ of X as follows: for B_ℓ, $\varepsilon_1 = \lambda_1$, $\varepsilon_i = \lambda_i - \lambda_{i-1}$ for $1 < i \leq \ell$, $\varepsilon_\ell = 2\lambda_\ell - \lambda_{\ell-1}$; for C_ℓ, $\varepsilon_1 = \lambda_1$, $\varepsilon_i = \lambda_i - \lambda_{i-1}$ for $1 < i \leq \ell$; for D_ℓ, $\varepsilon_1 = \lambda_1$, $\varepsilon_i = \lambda_i - \lambda_{i-1}$ for $1 < i \leq \ell-2$, $\varepsilon_{\ell-1} = \lambda_{\ell-1} + \lambda_\ell - \lambda_{\ell-2}$ and $\varepsilon_\ell = \lambda_\ell - \lambda_{\ell-1}$. It is not difficult to check that $\varepsilon_1, \varepsilon_2, \ldots, \varepsilon_\ell$ are conjugate under W and, adjusting (,) by a scalar factor if necessary, $\varepsilon_1, \varepsilon_2, \ldots, \varepsilon_\ell$ is an orthonormal basis of $E = \mathbb{R} \otimes_{\mathbb{Z}} X$.

What we must do, to prove that $\underline{H}(\Phi)$ is true, is to show that $Y(\lambda_r)|_{P_\Omega}$ has a good filtration for $1 \leq r \leq \ell$ and each subset Ω of Δ. For this purpose we divide the fundamental weights into two types, the exceptional weights and the non-exceptional weights. We decree that for B_ℓ, λ_ℓ is the only exceptional dominant weight, for C_ℓ there are no exceptional dominant weights and, for D_ℓ, $\lambda_{\ell-1}$ and λ_ℓ are the exceptional dominant weights. A fundamental dominant weight is called non-exceptional if it is not an exceptional dominant weight. Of course the exceptional weights are miniscule so we obtain the following from Proposition 4.2.1.

<u>Proposition 4.3.2</u> *Suppose that* G *is of type* $B_\ell (\ell \geq 2)$ *and* $\lambda = \lambda_\ell$ *or* G *is of type* $D_\ell (\ell \geq 4)$ *and* $\lambda = \lambda_{\ell-1}$ *or* λ_ℓ. *Then* $Y(\lambda)|_{P_\Omega}$ *has a good filtration for any* $\Omega \subseteq \Delta$ *and* $Y(\lambda) \otimes Y(\tau)$ *has a good filtration for any* $\tau \in X^+$.

<u>4.4 The first fundamental dominant weight</u>

Having disposed of the exceptional weights we must now show that $Y(\lambda_r)|_{P_\Omega}$ has a good filtration for λ_r non-exceptional and any subset

Ω of Δ. We make a start here with $r = 1$, proving the following result.

<u>Proposition 4.4.1</u> *For any subset* Ω *of* Δ, $Y(\lambda_1)|_{P_\Omega}$ *has a good filtration and, for any* $\tau \in X^+$, $Y(\lambda_1) \otimes Y(\lambda)$ *has a good filtration.*

<u>Proof</u> For G of type $C_\ell (\ell \geq 3)$ or $D_\ell (\ell \geq 4)$ λ_1 is miniscule so that the proposition follows from Proposition 4.2.1. We have

$$s(\lambda_1) = \sum_{i=1}^{\ell} e(\varepsilon_i) + \sum_{i=1}^{\ell} e(-\varepsilon_i) \ ;$$

$$\chi(\lambda_1) = s(\lambda_1) + s(0) \quad \text{for } G \text{ of type } B_\ell \ ;$$

$$\chi(\lambda_1) = s(\lambda_1) \quad \text{for } G \text{ of type } C_\ell \text{ or } D_\ell \qquad (1).$$

Moreover for $\Omega = \Delta \backslash \{\alpha_1\}$ we have:

$$\chi(\lambda_1) = \chi_\Omega(\varepsilon_1) + \chi_\Omega(\varepsilon_2) + \chi_\Omega(-\varepsilon_1) \qquad (2).$$

Now suppose that G has type B_ℓ, let $V = Y(\lambda_1)$, $V_1 = F_\Omega^{-1}(V)$ and $V_2 = F_\Omega^{0}(V)$. Then it is not hard to show, using (2), that $chV_1 = \chi_\Omega(-\varepsilon_1)$, $ch(V_2/V_1) = \chi_\Omega(\varepsilon_2)$ and $ch(V/V_2) = \chi_\Omega(\varepsilon_1)$. We have $V_1 \cong Y_\Omega(-\varepsilon_1)$, $V/V_2 \cong Y_\Omega(\varepsilon_1)$ so, in order to show that $V|_{P_\Omega}$ has a good filtration it is enough to prove that $V_2/V_1 \cong Y_\Omega(\varepsilon_2)$. We first note that $(V_2/V_1)^B = 0$. The short exact sequence

$$0 \to V_1 \to V_2 \to V_2/V_1 \to 0$$

gives rise to an exact sequence

$$0 \to {V_1}^B \to {V_2}^B \to (V_2/V_1)^B \to H^1(B,V_1) \qquad (3).$$

However $H^1(B,V_1) \cong H^1(B,-\varepsilon_1)$ and this is zero by Corollary 2.3.5 since $\varepsilon_1 = \alpha_1 + \alpha_2 + \ldots + \alpha_\ell$. Also ${V_2}^B \subseteq V^B$ which is zero by (1.5.2). Hence, from (3), $(V_2/V_1)^B = 0$. Now suppose that $(\lambda \otimes (V_2/V_1))^B \neq 0$ for some $\lambda \in X$. Then $-\lambda$ is a weight of V_2/V_1 and, by (1.1.2) and (1.1.7), $(Y_\Omega(\lambda) \otimes (V_2/V_1))^{P_\Omega} \neq 0$ so that $\lambda \in X_\Omega$. The weights of V_2/V_1 are $\{\pm\varepsilon_i : 2 \leq i \leq \ell\} \cup \{0\}$ each occurring with multiplicity one.

Thus $\lambda = 0$ or $\lambda = \varepsilon_2$ and $\dim(\lambda \otimes (V_2/V_1))^B \le 1$. We have already ruled out the possibility that $\lambda = 0$ so that the P_Ω-module V_2/V_1 has a simple B socle $-\varepsilon_2$. By the argument of the proof of Proposition 3.3.2, U^Ω acts trivially on V_2/V_1 so that the B-socle of V_2/V_1 is isomorphic to $-\varepsilon_2$ and, by (1.5.4), $V_2/V_1 \cong Y_\Omega(\varepsilon_2)$. This shows that $V|_{P_\Omega}$ has a good filtration and, by Lemma 3.7.1, $Y(\lambda_1)|_{P_\Gamma}$ has a good filtration for any $\Gamma \subseteq \Delta$. However $(\mu,\alpha^\vee) \ge -2$ for any weight μ of $Y(\lambda_1)$ and $\alpha \in \Delta$ so that $Y(\lambda_1) \otimes Y(\tau)$ has a good filtration for any $\tau \in X^+$ by Proposition 3.5.2.

4.5 Some module homomorphisms

Proposition 4.5.1 *For any non-exceptional weight* λ_r *with* $r > 1$ *we have*

$$Hom_G(Y(\lambda_{r-1}) , Y(\lambda_1) \otimes Y(\lambda_r)) \cong k .$$

Proof We claim that

$$\chi(\lambda_1)\ \chi(\lambda_r) = \chi(\lambda_{r-1}) + \sum_{i=1}^{m} \chi(\tau_i) \tag{1}$$

for some $m > 0$, and $\tau_i \in X^+$ $(1 \le i \le m)$ such that $\tau_i \le \lambda_{r-1}$. Proving (1) involves a case by case analysis using "Brauer's Formula" (2.2.3) and equation (1) of section 4.4. We record the results of the calculations, leaving the reader to check the details.

For B_ℓ, $1 < r < \ell-1$. For C_ℓ, $1 < r < \ell$ and for D_ℓ with $1 < r < \ell-2$

$$\chi(\lambda_1)\ \chi(\lambda_r) = \chi(\lambda_{r-1}) + \chi(\lambda_{r+1}) + \chi(\lambda_1 + \lambda_r)$$

and, if $1 = \ell-1$, for B_ℓ

$$\chi(\lambda_1)\ \chi(\lambda_{\ell-1}) = \chi(\lambda_{\ell-2}) + \chi(2\lambda_\ell) + \chi(\lambda_1 + \lambda_{\ell-1}) .$$

For D_ℓ, $r = \ell-2$, we have

$$\chi(\lambda_1)\ \chi(\lambda_{\ell-2}) = \chi(\lambda_{\ell-3}) + \chi(\lambda_{\ell-1} + \lambda_\ell) + \chi(\lambda_1 + \lambda_{\ell-2}) .$$

Thus (1) holds in all cases. By Lemma 3.2.2 we have a good filtration

$0 = V_0, V_1, \ldots, V_n = V = Y(\lambda_1) \otimes Y(\lambda_r)$, for some n , with $V_1 = Y(\lambda_{r-1})$ and, for $i > 1$, $(V_i/V_{i-1}) \cong Y(\eta_i)$ with $\eta_i \le \lambda_{r-1}$. Let $\theta \in \mathrm{Hom}_G(Y(\lambda_{r-1}), V)$, we claim that the image of θ lies in V_1 . Suppose for a contradiction that this is not the case and choose $m \le n$ minimal such that the image of θ lies in V_m . We denote by $\bar{\theta}$ the composite of θ with the natural map $V_m \to V_m/V_{m-1}$. By the choice of m this is non zero so that $\bar{\theta}$ **determines** a non zero map $Y(\lambda_{r-1}) \to Y(\eta_m) \cong (V_m/V_{m-1})$. However $\mathrm{Hom}_G(Y(\lambda_{r-1}), Y(\eta_m)) = 0$ by (1.5.3). This contradiction shows that $\theta(Y(\lambda_{r-1})) \subseteq V_1$ for any $\theta \in \mathrm{Hom}_G(Y(\lambda_{r-1}), V)$. Thus

$$\begin{aligned} Hom_G(Y(\lambda_{r-1}), V) &\cong Hom_G(Y(\lambda_{r-1}), V_1) \\ &\cong Hom_G(Y(\lambda_{r-1}), Y(\lambda_{r-1})) \cong k \end{aligned}$$

by (1.5.3), as required.

Corollary 4.5.2 *For any non-exceptional weight* λ_r *with* $r > 1$, *and* $\Omega = \Delta\backslash\{\alpha\ \}$,

$$\begin{aligned} Hom_B(Y(\lambda_{r-1}) \otimes -\varepsilon_1, Y(\lambda_r)) &\cong Hom_{P_\Omega}(Y(\lambda_{r-1}) \otimes -\varepsilon_1, Y(\lambda_r)) \\ &= k \ . \end{aligned}$$

Proof We have

$$\begin{aligned} \mathrm{Hom}_B(Y(\lambda_{r-1}) \otimes -\varepsilon_1, Y(\lambda_r)) &\cong \mathrm{Hom}_B(Y(\lambda_{r-1}), \varepsilon_1 \otimes Y(\lambda_r)) \\ &\cong Hom_G(Y(\lambda_{r-1}), Y(\lambda_1) \otimes Y(\lambda_r)) \cong k \end{aligned}$$

by (1.1.2). We may regard $-\varepsilon_1 \cong Y_\Omega(-\varepsilon_1)$ (by (1.5.2)) as a P_Ω-module.

$$Hom_B(Y(\lambda_{r-1}) \otimes -\varepsilon_1, Y(\lambda_r)) \cong Hom_{P_\Omega}(Y(\lambda_{r-1}) \otimes -\varepsilon_1, Y(\lambda_r))$$

by (2.1.3) and the proof is complete.

4.6 Character Formulas

In order to analyse $Y(\lambda_r)|_{P_\Omega}$ for $\Omega = \Delta\backslash\{\alpha_1\}$ we need to express $\chi(\lambda_r)$ as a sum of $\chi_\Omega(\tau)$'s for various $\tau \in X_\Omega$. Recall equation (2) of section 4.4:

$$\chi(\lambda_1) = \chi_\Omega(\varepsilon_1) + \chi_\Omega(-\varepsilon_1) + \chi_\Omega(\varepsilon_2) .$$

Applying Λ^r to this we obtain, for $r \geq 2$,

$$\Lambda^r\chi(\lambda_1) = \sum_{u+v=r} \Lambda^u(\chi_\Omega(\varepsilon_1) + \chi_\Omega(-\varepsilon_1))\Lambda^v\chi_\Omega(\varepsilon_2)$$

$$= \Lambda^r\chi_\Omega(\varepsilon_2) + (\chi_\Omega(\varepsilon_1) + \chi_\Omega(-\varepsilon_1))\Lambda^{r-1}\chi_\Omega(\varepsilon_2)$$

$$+ \Lambda^{r-2}\chi_\Omega(\varepsilon_2) \qquad (1).$$

If G has type B_ℓ and $r < \ell-1$ or G has type D_ℓ and $r < \ell-2$ we obtain from (4.1.1):

(4.6.1) $$\chi(\lambda_r) = \chi_\Omega(\lambda_{r+1} - \varepsilon_1) + (\chi_\Omega(\varepsilon_1) + \chi_\Omega(-\varepsilon_1))\chi_\Omega(\lambda_r - \varepsilon_1)$$

$$+ \chi_\Omega(\lambda_{r-1} - \varepsilon_1)$$

$$= \chi_\Omega(\lambda_r) + \chi_\Omega(\lambda_{r+1} - \varepsilon_1) + \chi_\Omega(\lambda_{r-1} - \varepsilon_1)$$

$$+ \chi_\Omega(\lambda_r - 2\varepsilon_1)$$

If G has type B_ℓ and $r = \ell-1$ we obtain, from (4.1.1),

(4.6.2) $$\chi(\lambda_{\ell-1}) = \chi_\Omega(2\lambda_\ell - \varepsilon_1) + (\chi_\Omega(\varepsilon_1) + \chi_\Omega(-\varepsilon_1))\chi_\Omega(\lambda_{\ell-1} - \varepsilon_1)$$

$$+ \chi_\Omega(\lambda_{\ell-2} - \varepsilon_1)$$

$$= \chi_\Omega(\lambda_{\ell-1}) + \chi_\Omega(2\lambda_\ell - \varepsilon_1) + \chi_\Omega(\lambda_{\ell-2} - \varepsilon_1)$$

$$+ \chi_\Omega(\lambda_{\ell-1} - 2\varepsilon_1)$$

If G has type D_ℓ and $r = \ell-2$ we obtain from (1) and (4.1.1):

(4.6.3) $$\chi(\lambda_{\ell-2}) = \chi_\Omega(\lambda_{\ell-1} + \lambda_\ell - \varepsilon_1)$$

$$+ (\chi_\Omega(\varepsilon_1) + \chi_\Omega(-\varepsilon_1))\chi_\Omega(\lambda_{\ell-2} - \varepsilon_1) + \chi_\Omega(\lambda_{\ell-3} - \varepsilon_1)$$

$$= \chi_\Omega(\lambda_{\ell-2}) + \chi_\Omega(\lambda_{\ell-1} + \lambda_\ell - \varepsilon_1) + \chi_\Omega(\lambda_{\ell-3} - \varepsilon_1)$$

$$+ \chi_\Omega(\lambda_{\ell-2} - 2\varepsilon_1)$$

We leave the reader to check that, for G of type C_ℓ , $1 < r < \ell$,

(4.6.4) $\quad \chi(\lambda_r) = \chi_\Omega(\lambda_r) + \chi_\Omega(\lambda_{r+1} - \varepsilon_1) + \chi_\Omega(\lambda_{r-1} - \varepsilon_1) + \chi_\Omega(\lambda_r - 2\varepsilon_1)$

and

(4.6.5) $\quad \chi(\lambda_\ell) = \chi_\Omega(\lambda_\ell) + \chi_\Omega(\lambda_{\ell-1} - \varepsilon_1) + \chi_\Omega(\lambda_\ell - 2\varepsilon_1)$

These formulas may be verified by using (1) above and (4.1.1).

The most striking feature of the formulas we have now assembled is that $\chi(\lambda_r)$ is a sum of at most four $\chi_\Omega(\tau)$'s for $\tau \in X_\Omega$. Thus there should be at most four terms in a good filtration of $Y(\lambda_r)|_{P_\Omega}$. It is this which makes the particular choice of Ω the correct one for the first part of our analysis of $Y(\lambda_r)|_{P_\Omega}$ for $\Omega \subseteq \Delta$. For other choices of Ω the number of terms in a good filtration of $Y(\lambda_r)|_{P_\Omega}$ increases with r.

4.7 The canonical filtration again

Proposition 4.7.1 *Let $\Omega = \Delta\backslash\{\alpha\ \}$. Suppose that G has type B_ℓ or C_ℓ and V is any G-module or G has type D_ℓ and $V = Y(\lambda_r)$ for λ_r non-exceptional. Then $V|_{P_\Omega}$ has a good filtration if and only if $F^0_\Omega(V)$ has a good filtration.*

Proof We deduce this from Proposition 3.6.3. For G of type B_ℓ let $w = s_{\varepsilon_1}$ and for G of type C_ℓ let $w = s_{2\varepsilon_1}$. Let $n_w \in N_G(T)$ be such that $n_w T = w$ and let $\phi : G \to G$ be conjugation by n_w. Then $\phi(T) = T$ and ϕ induces a map $\tilde{\phi} : X \to X$ satisfying the hypotheses of section 3.6. Moreover, since ϕ is conjugation, $V^\phi \cong V$ for any G-module V so that the proposition does indeed follow from Proposition 3.6.3.

Now suppose that G has type D_ℓ. Define $\tilde{\psi} : X \to X$ by ψ $\tilde{\psi}(\varepsilon_i) = \varepsilon_i$, $1 \le i < \ell$, $\tilde{\psi}(\varepsilon_\ell) = -\varepsilon_\ell$ and let $\psi : G \to G$ be an isomorphism preserving T and inducing $\tilde{\psi}$. Let $w = s_{\alpha_{\ell-1}} s_{\alpha_{\ell-2}} \cdots s_{\alpha_1}$ and $n_w \in N_G(T)$ be such that $n_w T = w$. We define $\phi = \theta\psi$ where θ is conjugation by n_w. Then $\tilde{\phi}(\varepsilon_1) = -\varepsilon_1$ and $\tilde{\phi}(\varepsilon_i) = \varepsilon_i$ for $i > 1$.

Thus $\tilde{\phi}$ satisfies the hypotheses of section 3.6. Now $\psi(B) = B$ so we have for $1 \le r < \ell-1$,

$$Y(\lambda_r)^{\phi} \cong (Y(\lambda_r)^{\psi}) \cong \mathit{Ind}_B^G(\tilde{\psi}(\lambda_r))^{\theta} \cong \mathit{Ind}_B^G(\lambda_r)^{\theta} = Y(\lambda_r)$$

from (1.1.14) and the fact that $M^{\theta} \cong M$ for all G-modules M. Thus Proposition 4.7.1 also follows from Proposition 3.6.3 for G of type D_ℓ.

4.8 The restriction of $Y(\lambda_r)$ to P_Ω

We shall show, in this section, that the restriction of $Y(\lambda_r)$ to P_Ω has a good filtration for a non-exceptional weight λ_r and $\Omega = \Delta\backslash\{\alpha_1\}$. It is convenient to deal first with some homological points which will be used in the proof.

Lemma 4.8.1 (i) $H^1(B, Y(\lambda_r) \otimes (\lambda_r - 2\varepsilon_1)) = 0$ *for* $1 \le r \le \ell$.

(ii) $H^r(B, \lambda_r - 2\varepsilon_1) = 0$ *for* $1 \le r \le \ell$.

(iii) $H^a(B, Y(\lambda_{r-a}) \otimes (\lambda_r - (a+1)\varepsilon_1)) = 0$ *for* $1 \le a < r$ *and* λ_r *non-exceptional.*

(iv) $H^{r-1}(B, Y(\lambda_1) \otimes (\lambda_r - (r+1)\varepsilon_1)) = 0$ *for* λ_r *non-exceptional.*

Proof (i) If $r = 1$, $H^1(B, Y(\lambda_r) \otimes (\lambda_r - 2\varepsilon_1))$ is 0 by (2.1.6). If $r \ne 1$ it is isomorphic to $(Y(\lambda_r) \otimes (\lambda_r - \lambda_2))^B$ by (2.3.3) which is 0 by (1.5.2).

(ii) If $r = 1$, $H^r(B, \lambda_r - 2\varepsilon_1)$ is clearly 0. If $r > 1$ it is isomorphic to $H^{r-1}(B, \lambda_r - \lambda_2)$ by (2.3.3) and hence 0 by (2.1.6) ($r \ne 2$) and (2.1.3), (2.1.4) ($r = 2$).

(iii) Let $\Sigma = \{\alpha_1, \alpha_2, \ldots, \alpha_a\}$. By the remark following (2.3.6) $R^n\mathit{Ind}_B^{P_\Sigma}(\lambda_r - (a+1)\varepsilon_1)$ is $\lambda_r - \lambda_{a+1}$ if $n = a$ and 0 otherwise. Thus, by the tensor identity and (1.1.9), $H^a(B, Y(\lambda_{r-a}) \otimes (\lambda_r - (a+1)\varepsilon_1))$ is isomorphic to $(Y(\lambda_{r-a}) \otimes (\lambda_r - \lambda_{a+1}))^{P_\Sigma}$ which is 0 by (1.5.1).

(iv) If $r = 1$ we have $H^0(B, Y(\lambda_1) \otimes -\lambda_1) = (Y(\lambda_1) \otimes -\lambda_1)^B$ which is 0, for example by (1.5.2). Now suppose that $r > 1$ and the result holds for $r' < r$. We let $\Sigma = \{\alpha_1, \alpha_2, \ldots, \alpha_{r-1}\}$. We have a short exact sequence of B-modules

$$0 \to -\varepsilon_1 \to Y_\Sigma(-\varepsilon_r) \to N \to 0$$

where N is the quotient of $Y_\Sigma(-\varepsilon_r)$ by its B-socle. Tensoring with $Y(\lambda_1) \otimes (\lambda_r - r\varepsilon_1)$ we obtain, by the long exact sequence of cohomology, an exact sequence

$$H^{r-2}(B, Y(\lambda_1) \otimes N \otimes (\lambda_r - r\varepsilon_1)) \to H^{r-1}(B, Y(\lambda_1) \otimes (\lambda_r - (r+1)\varepsilon_1))$$
$$\to H^{r-1}(B, Y(\lambda_1) \otimes Y_\Sigma(-\varepsilon_r) \otimes (\lambda_r - r\varepsilon_1)) .$$

By the remark following (2.3.6), $R^n Ind_B^{P_\Sigma}(\lambda_r - r\varepsilon_1)$ is isomorphic to the trivial module k if $n = r-1$ and is 0 otherwise. Hence by the tensor identity and (1.1.9),

$$H^{r-1}(B, Y(\lambda_1) \otimes Y_\Sigma(-\varepsilon_r) \otimes (\lambda_r - r\varepsilon_1)) = H^0(P_\Sigma, Y(\lambda_1) \otimes Y_\Sigma(-\varepsilon_r))$$

$= (Y(\lambda_1) \otimes Y_\Sigma(-\varepsilon_r))^{P_\Sigma}$. However, $(Y(\lambda_1) \otimes Y_\Sigma(-\varepsilon_r))^{P_\Sigma}$ is isomorphic to $(Y(\lambda_1) \otimes -\varepsilon_r)^B$ by (1.1.2) which is 0 , for example by (1.5.2). Thus $H^{r-1}(B, Y(\lambda_1) \otimes Y_\Sigma(-\varepsilon_r) \otimes (\lambda_r - r\varepsilon_1)) = 0$ and so, from the above sequence, we see that in order to complete the proof it is enough to show that

$$H^{r-2}(B, Y(\lambda_1) \otimes N \otimes (\lambda_r - r\varepsilon_1)) = 0 .$$

By the long exact sequence of cohomology it is enough to prove:

$$H^{r-2}(B, Y(\lambda_1) \otimes (\tau + \lambda_r - r\varepsilon_1)) = 0 \quad \textit{for each weight } \tau \textit{ of } N .$$

We obtain, by Weyl's character formula, that the weights of $Y_\Sigma(-\varepsilon_r)$ are $-\varepsilon_1, \ldots, -\varepsilon_r$ so the weights of N are $-\varepsilon_2, -\varepsilon_3, \ldots, -\varepsilon_r$. If $\tau = -\varepsilon_i$ for $2 \le i < r$ then $(\tau + \lambda_r - r\varepsilon_1, \alpha_i^\vee) = -1$ so that $H^{r-2}(B, Y(\lambda_1) \otimes (\tau + \lambda_r - r\varepsilon_1)) = 0$ by (2.1.6). If $\tau = -\varepsilon_r$ then $\tau + \lambda_r - r\varepsilon_1 = \lambda_{r-1} - r\varepsilon_1$ so that $H^{r-1}(B, Y(\lambda_1) \otimes (\tau + \lambda_r - r\varepsilon_1)) = 0$ by the inductive hypothesis. This completes the proof of (iv).

Lemma 4.8.2 *Suppose that* λ_r *is a non-exceptional weight and that* $Y(\lambda_a)|_{P_\Omega}$ *has a good filtration for all* $1 \le a < r$. *Then*

$$H^1(B, Y_\Omega(\lambda_r - 2\varepsilon_1) \otimes (\lambda_s - \varepsilon_1)) = 0 \quad \textit{for} \quad 1 \le s < r-1$$

and

$$H^1(B, Y_\Omega(\lambda_{r-1} - \varepsilon_1) \otimes (\lambda_s - \varepsilon_1)) = 0 \quad \textit{for} \quad 1 \le s \le r-1 .$$

Proof By (2.1.3),

$$H^1(B, Y_\Omega(\lambda_r - 2\varepsilon_1) \otimes (\lambda_s - \varepsilon_1)) = H^1(B, (\lambda_r - 2\varepsilon_1) \otimes Y_\Omega(\lambda_s - \varepsilon_1))$$

moreover, $Y_\Omega(\lambda_s - \varepsilon_1) \otimes -\varepsilon_1 = Y_\Omega(\lambda_s - 2\varepsilon_1)$ (by (1.5.2) since $Y_\Omega(\lambda_s - \varepsilon_1) \otimes -\varepsilon_1$ has the correct B-socle and dimension) so we must show that $H^1(B, (\lambda_r - \varepsilon_1) \otimes Y_\Omega(\lambda_s - 2\varepsilon_1)) = 0$. If $s = 1$, $H^1(B, (\lambda_r - \varepsilon_1) \otimes Y_\Omega(\lambda_1 - 2\varepsilon_1)) = H^1(B, \lambda_r - 2\varepsilon_1)$ which is 0 since by (2.1.6), $H^1(B, \lambda_r - 2\varepsilon_1) = (\lambda_r - \lambda_2)^B = 0$. For $s > 1$, we have, by Lemma 3.2.2, (4.6.1) and (4.6.4), a short exact sequence

$$0 \to Y_\Omega(\lambda_s - 2\varepsilon_1) \to Y(\lambda_s) \to A \to 0$$

where A has a good filtration with quotients $Y_\Omega(\lambda_{s-1} - \varepsilon_1)$, $Y_\Omega(\lambda_{s+1} - \varepsilon_1)$, $Y_\Omega(\varepsilon_s)$. Tensoring with $(\lambda_r - \varepsilon_1)$ and applying the long exact sequence of cohomology we obtain an exact sequence

$$((\lambda_r - \varepsilon_1) \otimes A)^B \to H^1(B, (\lambda_r - \varepsilon_1) \otimes Y_\Omega(\lambda_s - 2\varepsilon_1))$$
$$\to H^1(B, (\lambda_r - \varepsilon_1) \otimes Y(\lambda_s)) .$$

Now $((\lambda_s - \varepsilon_1) \otimes A)^B = 0$ since $\varepsilon_1 - \lambda_r$ does not occur in the socle of any successive quotient of the filtration of A described above. Moreover $H^1(B, (\lambda_r - \varepsilon_1) \otimes Y(\lambda_s)) = 0$ by (2.1.6) since $(\lambda_r - \varepsilon_1, \alpha_1^\vee) = -1$. Thus we have $H^1(B, (\lambda_r - \varepsilon_1) \otimes Y_\Omega(\lambda_s - 2\varepsilon_1)) = 0$, as required.

We now show that $H^1(B, Y_\Omega(\lambda_{r-1} - \varepsilon_1) \otimes (\lambda_s - \varepsilon_1)) = 0$. For $r = 2$ this is clear. Suppose now $r > 2$. By Lemma 3.2.2, (4.6.1) and (4.6.4) we have a short exact sequence of P_Ω-modules,

$$0 \to Y_\Omega(\lambda_{r-1} - 2\varepsilon_1) \to Y(\lambda_{r-1}) \to A' \to 0$$

where A' has a good filtration with quotients $Y_\Omega(\lambda_{r-2} - \varepsilon_1)$, $Y_\Omega(\lambda_r - \varepsilon_1)$, $Y_\Omega(\lambda_{r-1})$. Tensoring by λ_s and applying the long exact sequence of cohomology we obtain an exact sequence,

$$(A' \otimes \lambda_s)^B \to H^1(B, Y_\Omega(\lambda_{r-1} - 2\varepsilon_1) \otimes \lambda_s) \to H^1(B, Y(\lambda_{r-1}) \otimes \lambda_s) .$$

Now $(A' \otimes \lambda_s)^B = 0$ since $-\lambda_s$ does not occur in the B-socle of any successive quotient in the filtration, for A', described above. Moreover $H^1(B, Y(\lambda_{r-1}) \otimes \lambda_s) = 0$ by (2.1.3), (2.1.4) and so $H^1(B, Y_\Omega(\lambda_{r-1} - 2\varepsilon_1) \otimes \lambda_s) = 0$ and the proof is complete.

We define $Y(\lambda_0) = k$ and $M_0 = 0$. We define M_1 to be the $-\varepsilon_1$ weight space of $Y(\lambda_1)$ and, for $r > 1$, M_r is defined to be the image of a non-zero homomorphism $\theta_r : Y(\lambda_{r-1}) \otimes -\varepsilon_1 \to Y(\lambda_r)$. By Corollary 4.5.2 there is such a homomorphism and M_r does not depend on the choice of θ_r. We denote by Q_r the quotient module $Y(\lambda_r)/M_r$. The following proposition is really the crux of the whole analysis for classical groups.

Proposition 4.8.3 *For any non-exceptional weight* λ_r *or* $r = 0$:

(i) $M_r = Q_{r-1} \otimes - \varepsilon_1$ *as a* P_Ω*-module* ;

(ii) $$chM_r = \begin{cases} \chi_\Omega(\lambda_r - 2\varepsilon_1) + \chi_\Omega(\lambda_{r-1} - \varepsilon_1) & \textit{if } r > 1 \\ \chi_\Omega(-\varepsilon_1) & \textit{if } r = 1 ; \end{cases}$$

(iii) $Y(\lambda_r)|_{P_\Omega}$ *has a good filtration* ;

(iv) Q_r *has a good filtration* ;

(v) Q_r *has a simple* B*-socle* .

Proof We proceed by induction on r. If $r = 0$ then (i), (iii), (iv) and (v) are obvious and (ii) is empty.

Now assume that $r > 0$, λ_r is a non-exceptional weight and the Proposition holds for all λ_s with $s < r$. By (1.5.2) the B-socle of $Y(\lambda_r)$ is $-\lambda_r$ and, by (ii) for $r-1$, $-\lambda_r$ is not a weight of $M_{r-1} \otimes - \varepsilon_1$. Thus $M_{r-1} \otimes - \varepsilon_1$ is killed by any B-homomorphism from

$Y(\lambda_{r-1}) \otimes -\varepsilon_1$ to $Y(\lambda_r)$. Thus a non-zero B-map $Y(\lambda_{r-1}) \otimes -\varepsilon_1 \to Y(\lambda_r)$ gives rise to a non-zero map $\phi: Q_{r-1} \otimes -\varepsilon_1 \cong (Y(\lambda_{r-1}) \otimes -\varepsilon_1)/(M_{r-1} \otimes -\varepsilon_1) \to Y(\lambda_r)$. We have $chQ_{r-1} = chY(\lambda_{r-1}) - chM_{r-1}$ which is $\chi_\Omega(0)$ if $r = 1$ and, if $r > 1$, by (ii) for $r-1$, equation (2) of section 4.4, (4.6.1) and (4.6.4) is $\chi_\Omega(\lambda_{r-1}) + \chi_\Omega(\lambda_r - \delta_1)$. Thus, by Lemma 3.2.2, Q_{r-1} has a submodule isomorphic to $Y_\Omega(\lambda_r - \varepsilon_1)$. The B-socle of $Y_\Omega(\lambda_r - \varepsilon_1)$ is $\varepsilon_1 - \lambda_r$. But, by (v) for $r-1$, Q_{r-1} has a simple B-socle and so $\varepsilon_1 - \lambda_r$ is the B-socle of Q_{r-1}. Thus the B-socle, say L, of $Q_{r-1} \otimes -\varepsilon_1$, is isomorphic to $-\lambda_r$. The image of ϕ, being non-zero, contains the B-socle $Y(\lambda_r)^{-\lambda_r}$ of $Y(\lambda_r)$ and so the induced map

$$L = (Q_{r-1} \otimes -\varepsilon_1)^{-\lambda_r} \to Y(\lambda_r)^{-\lambda_r}$$

is non-zero. Thus ϕ is injective on the B-socle of $Q_{r-1} \otimes -\varepsilon_1$ and therefore is a monomorphism. Thus $M_r = \mathrm{Im}\theta_r$ is isomorphic to $Q_{r-1} \otimes -\varepsilon_1$ proving (i). Moreover, $chM_r = e(-\varepsilon_1)chQ_{r-1}$ which is $\chi_\Omega(-\varepsilon_1)$ if $r = 1$ and $\chi_\Omega(\lambda_{r-1} - \varepsilon_1) + \chi_\Omega(\lambda_r - 2\varepsilon_1)$ if $r > 1$, proving (ii).

We must now show that $Y(\lambda_r)|_{P_\Omega}$ has a good filtration. By Proposition 4.7.1, it is enough to prove that $F = F^0_\Omega(Y(\lambda_r))$ has a good filtration as a P_Ω-module. We see, from (ii), that $M_r \subseteq F$, moreover it follows, from (iv) for $r-1$ and (i), that M_r has a good filtration. Thus it suffices to prove that $V = F/M_r$ has a good filtration. We obtain from (ii) and (4.6.1) - (4.6.5):

$chV = \chi_\Omega(\lambda_{r+1} - \varepsilon_1)$ *for* G *of type* B_ℓ *with* $r < \ell-1$, G *of type* C_ℓ *with* $r < \ell$ *and* G *of type* D_ℓ *with* $r < \ell-2$;

$$chV = \begin{cases} \chi_\Omega(2\lambda_\ell - \varepsilon_1) & \textit{for type } B_\ell \textit{ with } r = \ell-1 \\ 0 & \textit{for type } C_\ell \textit{ with } r = \ell \\ \chi_\Omega(\lambda_{\ell-1} + \lambda_\ell - \varepsilon_1) & \textit{for type } D_\ell \textit{ with } r = \ell-2 . \end{cases}$$

Note that if G has type C_ℓ and $r = \ell$ our proof of (iii) is complete. We shall refer to the situation where G is of type B_ℓ and $r < \ell-1$, G is of type C_ℓ with $r < \ell$ or G is of type D_ℓ with $r < \ell-2$ as case I, we shall refer to the case on which G has type B_ℓ and $r = \ell-1$ as case II and the case in which G is of type D_ℓ and $r = \ell-1$ as case III.

Now, by (1.5.4), it suffices to prove that the B-socle (which is the B_Ω-socle since U^Ω acts trivially on V) is the following:

$$\varepsilon_1 - \lambda_{r+1} \quad \textit{in case I} \ ; \quad \varepsilon_1 - 2\lambda_\ell \quad \textit{in case II and}$$

$$\varepsilon_1 - \lambda_{\ell-1} - \lambda_\ell \quad \textit{in case III} \ .$$

Note that $\varepsilon_1 - \lambda_{r+1}$ is W-conjugate to λ_r in case I, that $\varepsilon_1 - 2\lambda_\ell$ is conjugate to λ_r in case II and $\varepsilon_1 - \lambda_{\ell-1} - \lambda_\ell$ is conjugate to $\lambda_{\ell-2}$ in case III. Thus the multiplicity of these weights in $Y(\lambda_r)$ is at most one and so the multiplicity of these weights in the B-socle is at most one. Thus it suffices to show that no other weights occur in the B-socle of V, that is $(\lambda \otimes V)^B = 0$ for $\lambda \neq \lambda_{r+1} - \varepsilon_1$ in case I, for $\lambda \neq 2\lambda_\ell - \varepsilon_1$ in case II and for $\lambda \neq \lambda_{\ell-1} + \lambda_\ell - \varepsilon_1$ in case III. Now $(V \otimes \lambda)^B \cong (V \otimes Y_\Omega(\lambda))^P$, by (1.1.7) so that if $(V \otimes \lambda)^B \neq 0$, $-\lambda$ is a weight of V and $\lambda \in X_\Omega$. However, if $-\lambda$ is a weight of V then λ is a weight of the module $Y_\Omega(\tau)^*$ dual to $Y_\Omega(\tau)$, where $\tau = \lambda_{r+1} - \varepsilon_1$ in case I, $\tau = 2\lambda_\ell - \varepsilon_1$ in case II and $\tau = \lambda_{\ell-1} + \lambda_\ell - \varepsilon_1$ in case III. It is not difficult to deduce from Weyl's Character Formula that $ch Y_\Omega(\tau)^* = \chi_\Omega(-w_\Omega\tau)$ where w_Ω is the longest element of W_Ω. But $w_\Omega\tau = -\tau$ in all the above cases so that τ is a weight of V. Now it follows from (4.1.1) that

$$\lambda \in \{\lambda_s - \varepsilon_1 : 1 \leq s \leq r\} \cup \{\tau\}$$

and, by the above remarks, to prove (iii) it is only necessary to show that

$$(V \otimes (\lambda_s - \lambda_1))^B = 0 \quad \textit{for} \quad 1 \leq s \leq r \qquad (1).$$

However V is a submodule of $Q_r = Y(\lambda_r)/M_r$ so that (1) follows from:

$$(Q_r \otimes (\lambda_s - \varepsilon_1))^B = 0 \quad for \quad 1 \le s \le r \qquad (2).$$

Now the short exact sequence

$$0 \to M_r \otimes (\lambda_s - \varepsilon_1) \to Y(\lambda_r) \otimes (\lambda_s - \varepsilon_1) \to Q_r \otimes (\lambda_s - \varepsilon_1) \to 0$$

gives rise to an exact sequence

$$(Y(\lambda_r) \otimes (\lambda_s - \varepsilon_1))^B \to (Q_r \otimes (\lambda_s - \varepsilon_1))^B \to H^1(B, M_r \otimes (\lambda_s - \varepsilon_1)) .$$

Moreover $(Y(\lambda_r) \otimes (\lambda_s - \varepsilon_1))^B = 0$ since the B-socle of $Y(\lambda_r)$ is, by (1.5.2), $-\lambda_r$. Thus to prove (2) it suffices to show:

$$H^1(B, M_r \otimes (\lambda_s - \varepsilon_1)) = 0 \quad for \quad 1 \le s \le r \qquad (3).$$

Suppose first that $s < r-1$. Thus r is at least 2. As remarked above M_r has a good filtration so, by (ii) and Proposition 3.2.6, M_r has a submodule, A say, with $A \cong Y_\Omega(\lambda_r - 2\varepsilon_1)$ and $M_r/A \cong Y_\Omega(\lambda_{r-1} - \varepsilon_1)$. Thus (3) follows from Lemma 4.8.2 and the long exact sequence of cohomology.

Now suppose that $s = r-1$. By Lemma 4.8.2 the short exact sequence

$$0 \to Y_\Omega(\lambda_r - 2\varepsilon_1) \otimes (\lambda_{r-1} - \varepsilon_1) \to M_r \otimes (\lambda_{r-1} - \varepsilon_1)$$
$$\to Y_\Omega(\lambda_{r-1} - \varepsilon_1) \otimes (\lambda_{r-1} - \varepsilon_1) \to 0$$

gives rise to the exact sequence

$$(M_r \otimes (\lambda_{r-1} - \varepsilon_1))^B \to (Y_\Omega(\lambda_{r-1} - \varepsilon_1) \otimes (\lambda_{r-1} - \varepsilon_1))^B \to$$
$$H^1(B, Y_\Omega(\lambda_r - 2\varepsilon_1) \otimes (\lambda_{r-1} - \varepsilon_1)) \to H^1(B, M_r \otimes (\lambda_{r-1} - \varepsilon_1)$$
$$\to 0 \qquad (4).$$

Since M_r is a submodule of $Y(\lambda_r)$ its B socle is $-\lambda_r$ and $(M_r \otimes (\lambda_{r-1} - \varepsilon_1))^B = 0$. The B socle of $Y_\Omega(\lambda_{r-1} - \varepsilon_1)$ however is $-\lambda_{r-1} + \varepsilon_1$ so that $(Y_\Omega(\lambda_{r-1} - \varepsilon_1) \otimes (\lambda_{r-1} - \varepsilon_1))^B \cong k$. Thus, from (4), we see that the statement $H^1(B, M_r \otimes (\lambda_{r-1} - \varepsilon_1)) = 0$ is equivalent to:

$$H^1(B, Y_\Omega(\lambda_r - 2\varepsilon_1) \otimes (\lambda_{r-1} - \varepsilon_1)) \cong k \qquad (5).$$

Now $Y_\Omega(\lambda_r - 2\varepsilon_1) \otimes (\lambda_{r-1} - \varepsilon_1) = Y_\Omega(\lambda_r - \varepsilon_1) \otimes (\lambda_{r-1} - 2\varepsilon_1)$ and, by (2.1.3)

$H^1(B, Y_\Omega(\lambda_r - \varepsilon_1) \otimes (\lambda_{r-1} - 2\varepsilon_1)) \cong H^1(B, (\lambda_r - \varepsilon_1) \otimes Y_\Omega(\lambda_{r-1} - 2\varepsilon_1))$ so it is enough to prove

$$H^1(B, Y_\Omega(\lambda_{r-1} - 2\varepsilon_1) \otimes (\lambda_r - \varepsilon_1)) \cong k \qquad (6).$$

Now by (iii) for $r-1$, Lemma 3.2.3, Equation (2) of section 4.4, (4.6.1) and (4.6.4) we have a short exact sequence

$$0 \to Y_\Omega(\lambda_{r-1} - 2\varepsilon_1) \to Y(\lambda_{r-1}) \to A'' \to 0$$

where A'' has a filtration with successive quotients isomorphic to $Y_\Omega(\lambda_{r-2} - \varepsilon_1)$, $Y_\Omega(\lambda_r - \varepsilon_1)$, $Y_\Omega(\lambda_{r-1})$ for $r > 2$ and $Y_\Omega(\lambda_2 - \varepsilon_1)$, $Y_\Omega(\lambda_1)$ for $r = 2$. Tensoring (7) with $\lambda_r - \varepsilon_1$ and applying the long exact sequence of cohomology we obtain

$$0 \to (A'' \otimes (\lambda_r - \varepsilon_1))^B \to H^1(B, Y_\Omega(\lambda_{r-1} - 2\varepsilon_1) \otimes (\lambda_r - \varepsilon_1))$$
$$\to H^1(B, Y(\lambda_{r-1}) \otimes (\lambda_r - \varepsilon_1)) \qquad (8).$$

However $H^1(B, Y(\lambda_{r-1}) \otimes (\lambda_r - \varepsilon_1)) = 0$ by (2.1.6) and $\dim(A'' \otimes (\lambda_r - \varepsilon_1))^B \leq 1$ since A'' has a good filtration in which $Y_\Omega(\lambda_r - \varepsilon_1)$ occurs exactly once. Hence, from (8), we see that

$$\dim H^1(B, Y_\Omega(\lambda_{r-1} - 2\varepsilon_1) \otimes (\lambda_r - \varepsilon_1)) \leq 1 .$$

However it is clear from the above observations on (4) that

$$\dim H^1(B, Y_\Omega(\lambda_r - 2\varepsilon_1) \otimes (\lambda_{r-1} - \varepsilon_1))$$
$$= \dim H^1(B, (\lambda_r - \varepsilon_1) \otimes Y_\Omega(\lambda_{r-1} - 2\varepsilon_1)) \geq 1 .$$

Thus we have proved (5) and therefore that $H^1(B, M_r \otimes (\lambda_{r-1} - \varepsilon_1)) = 0$. To conclude our proof of (iii) we must show that $H^1(B, M_r \otimes (\lambda_r - \varepsilon_1)) = 0$. This is a special case of the assertion:

$$H^a(B, M_{r+1-a} \otimes (\lambda_r - a\varepsilon_1)) = 0 \quad for \quad 1 \leq a \leq r \qquad (9)$$

which we prove by downward induction on a . We have

$$H^{r}(B,M_1 \otimes (\lambda_r - r\varepsilon_1)) \cong H^{r}(B,\lambda_r - (r+1)\varepsilon_1)$$

which is zero by Corollary 2.3.5 since $\lambda_r - (r+1)\varepsilon_1$ does not belong to $\mathbb{Z}\Sigma$ for any subset Σ of Δ of size r . Suppose that (9) is true for $a+1$ with $1 < a+1 \le r$ we must show that it is true for a . By (ii) for $r+1-a$ (which is less than r) we have $M_{r+1-a} \cong Q_{r-a} \otimes -\varepsilon_1$ and so a short exact sequence

$$0 \to M_{r-a} \otimes (\lambda_r - (a+1)\varepsilon_1) \to Y(\lambda_{r-a}) \otimes (\lambda_r - (a+1)\varepsilon_1) \to$$
$$M_{r+1-a} \otimes (\lambda_r - a\varepsilon_1) \to 0 .$$

By the long exact sequence of cohomology we have an exact sequence

$$H^{a}(B,Y(\lambda_{r-a}) \otimes (\lambda_r - (a+1)\varepsilon_1)) \to H^{a}(B,M_{r+1-a} \otimes (\lambda_r - a\varepsilon_1))$$
$$\to H^{a+1}(B,M_{r-a} \otimes (\lambda_r - (a+1)\varepsilon_1)) .$$

The first term is zero by the inductive hypothesis and the third term is zero by Lemma 4.8.1. This proves (9) and thus completes the proof of (iii), that $Y(\lambda_r)|_{P_\Omega}$ has a good filtration.

We now prove (iv). We know, by (iv) for $r-1$, that Q_{r-1} has a good filtration. Thus, by (i), M_r has a good filtration and so, by (iii) and Proposition 3.2.4, Q_r has a good filtration.

To complete the proof of the proposition we must prove (v), that Q_r has a simple B socle. Now $chQ_r = chY(\lambda_r) - chM_r$ so that

$$chQ_r = \chi_\Omega(\lambda_r) + \chi_\Omega(\lambda_{r+1} - \varepsilon_1) \quad \textit{in case I ,}$$

$$chQ_r = \chi_\Omega(\lambda_{\ell-1}) + \chi_\Omega(2\lambda_\ell - \varepsilon_1) \quad \textit{in case II ,}$$

$$chQ_r = \chi_\Omega(\lambda_{\ell-2}) + \chi_\Omega(\lambda_{\ell-1} + \lambda_\ell - \varepsilon_1) \quad \textit{in case III}$$

and

$$chQ_r = \chi_\Omega(\lambda_\ell) \quad \textit{if } G \textit{ has type } C_\ell \textit{ and } r = \ell .$$

Thus, by (iv), in case I, Q_r has a filtration in which successive quotients are $Y_\Omega(\lambda_{r+1} - \varepsilon_1)$ and $Y_\Omega(\lambda_r)$ each with multiplicity one.

Therefore, by (1.5.2), the B-socle of Q_r is either simple or $(-\lambda_{r+1}+\varepsilon_1)\oplus(-\lambda_r+2\varepsilon_1)$ $(-\lambda_{r+1}+\varepsilon_1$ is the smallest weight of $Y_\Omega(\lambda_{r+1}-\varepsilon_1)$ and $-\lambda_r+2\varepsilon_1$ is the smallest weight of $Y_\Omega(\lambda_r))$. Thus it suffices to prove that $-\lambda_r+2\varepsilon_1$ does not occur in the B socle of Q_r, that is

$$(Q_r\otimes(\lambda_r-2\varepsilon_1))^B=0 \tag{10}$$

Similar considerations apply to the other cases so it is enough to prove (10) in cases I, II and III. If G has type C_ℓ and $r=\ell$ then there is only one non-zero term in a good filtration of Q_r (by the above character formula) so that Q_r must have a simple B socle. We shall now prove (10) when G does not have type C_ℓ or $r\neq\ell$.

By Lemma 4.8.1 the short exact sequence

$$0\to M_r\otimes(\lambda_r-2\varepsilon_1)\to Y(\lambda_r)\otimes(\lambda_r-2\varepsilon_1)$$
$$\to Q_r\otimes(\lambda_r-2\varepsilon_1)\to 0$$

gives rise, via the long exact sequence of cohomology, to an isomorphism $(Q_r\otimes(\lambda_r-2\varepsilon_1))^B\to H^1(B,M_r\otimes(\lambda_r-2\varepsilon_1))$. Thus it suffices to prove:

$$H^1(B,M_r\otimes(\lambda_r-2\varepsilon_1)=0 \tag{11}$$

We claim that

$$H^1(B,M_r\otimes(\lambda_r-2\varepsilon_1))\cong H^a(B,M_{r+1-}\otimes(\lambda_r-(a+1)\varepsilon_1))$$
$$\textit{for}\quad 1\le a\le r-1 \tag{12}$$

This may be proved by induction on a as in the proof of (9). Thus, for $a=r-1$,

$$H^1(B,M_r\otimes(\lambda_r-2\varepsilon_1))\cong H^{r-1}(B,M_2\otimes(\lambda_r-r\varepsilon_1))$$
$$\cong H^{r-1}(B,Q_1\otimes(\lambda_r-(r+1)\varepsilon_1)) \tag{13}$$

using (12) and (i). Now the short exact sequence

$$0\to-\varepsilon_1\otimes(\lambda_r-(r+1)\varepsilon_1)\to Y(\lambda_1)\otimes(\lambda_r-(r+1)\varepsilon_1)$$
$$\to Q_1\otimes(\lambda_r-(r+1)\varepsilon_1)\to 0$$

gives rise to an exact sequence

$$H^{r-1}(B,Y(\lambda_1)\otimes(\lambda_r-(r+1)\varepsilon_1)) \to H^{r-1}(B,Q_1\otimes(\lambda_r-(r+1)\varepsilon_1))$$
$$\to H^r(B,\lambda_r-(r+2)\varepsilon_1)\ .$$

However, $H^r(B,\lambda_r-(r+2)\varepsilon_1)=0$ by Corollary 2.3.5 (we leave to the reader the task of expressing $\lambda_r-(r+2)\varepsilon_1$ as a linear combination of simple roots for type B , C and D) . Also $H^{r-1}(B,Y(\lambda_1)\otimes(\lambda_r-(r+1)\varepsilon_1))$ is zero by Lemma 4.8.1 (iv) so that $H^{r-1}(B,Q_1\otimes(\lambda_r-(r+1)\varepsilon_1))=0$. By (13) we now have $H^1(B,M_r\otimes(\lambda_r-2\varepsilon_1))=0$. This completes the proof of (11) and therefore the proof of the proposition.

4.9 The restriction of $Y(\lambda_r)$ to maximal parabolic subgroups

I am grateful to a referee for considerably tidying up this section.

In section 4.8 we showed that for λ_r non-exceptional and $\Omega=\Delta\backslash\{\alpha_1\}$, $Y(\lambda_r)|_{P_\Omega}$ has a good filtration. We must now analyse $Y(\lambda_r)|_{P_\Sigma}$ for $\Sigma=\Delta\backslash\{\alpha_t\}$ with $1<t\le \ell$. We write simply F for $Ind_B^{P_\Sigma}$.

Proposition 4.9.1 *For all $b>0$ and $j\ge 0$ we have*

(i) $R^jF(-b\varepsilon_1)=0$ *if* $j\ge t$;

(ii) $R^jF(-b\varepsilon_1)=0$ *if* $b<t$;

(iii) $$R^jF(-b\varepsilon_1)=\begin{cases}-\lambda_b & \text{if } j=b-1 \text{ and } b=t\\[2ex] 0 & \text{if } j\ne b-1 \text{ and } b=t\end{cases}$$

Proof (i) follows from (2.3.7). In order to obtain (ii) and (iii) we consider $\pi=\{\alpha_1,\ldots,\alpha_{b-1}\}\subset\Sigma$. By (2.3.6) we have that $R^jInd_B^{P_\pi}(-b\varepsilon_1)$ is $-\lambda_b$ if $j>b-1$ and 0 otherwise. By (1.1.6) we obtain that $R^jF(-b\varepsilon_1)$ is 0 if $j<b-1$ and $R^{j+1-b}Ind_{P_\pi}^{P_\Sigma}(-\lambda_b)$ if

$j \geq b-1$. Now (ii) follows from (2.1.5) and (iii) from (2.1.2).

Corollary 4.9.2 *For all* $j \geq b > 0$ *we have* $R^j F(-b\varepsilon_1) = 0$.

Proof This follows from Proposition 4.9.1 (i) for $j \geq t$ and from Proposition 4.9.1 (ii) for $j < t$.

Proposition 4.9.3 *We have for all* $b \geq 0$ *and* $r \geq 0$:

(i) $R^j F(M_r \otimes -b\varepsilon_1) = 0$ *if* $j \geq t$;

(ii) $R^j F(M_r \otimes -b\varepsilon_1) = 0$ *if* $j > b$

(iii) $$R^b F(M_r \otimes -b\varepsilon_1) = \begin{cases} 0 & \text{if } r+b < t \\ -\lambda_t & \text{if } r+b = t,\ r \neq 0 . \end{cases}$$

(iv) *For* $r+b > t$ *there is an epimorphism of* P_Σ*-modules*

$$Y(\lambda_{r+b-t}) \otimes -\lambda_t \to R^b F(M_r \otimes -b\varepsilon_1) .$$

Proof The proof is by induction on r . All is clear for $r = 0$ (since $M_0 = 0$) . Now suppose that $r > 0$ and the result holds for $r-1$. Tensoring the short exact sequence $0 \to M_{r-1} \to Y(\lambda_{r-1}) \to Q_{r-1} \to 0$ with $-(b+1)\varepsilon_1$ and using Proposition 4.8.3 (i) we obtain a short exact sequence $0 \to M_{r-1} \otimes -(b+1)\varepsilon_1 \to Y(\lambda_{r-1}) \otimes -(b+1)\varepsilon_1 \to M_r \otimes -b\varepsilon_1 \to 0$. We obtain, via the long exact sequence of cohomology, an exact sequence

$$Y(\lambda_{r-1}) \otimes R^j F(-(b+1)\varepsilon_1) \to R^j F(M_r \otimes -b\varepsilon_1)$$
$$\to R^{j+1} F(M_{r-1} \otimes -(b+1)\varepsilon_1) \to Y(\lambda_{r-1}) \otimes R^{j+1} F(-(b+1)\varepsilon_1) \qquad (1).$$

In the case of (i) (resp. (ii)) the first and last terms are 0 by Proposition 4.9.1 (i) (resp. Corollary 4.9.2). Thus the middle terms are equal and the assertion follows from the inductive hypothesis.

Now consider the case $j = b$. For $r = 1$, $R^b F(M_1 \otimes -b\varepsilon_1) = R^b F(-(b+1)\varepsilon_1)$ which is 0 for $r+b = b+1 < t$ by Proposition 4.9.1 (ii) and $-\lambda_{b+1}$ for $r+b = b+1 = t$ by Proposition 4.9.1 (iii). This proves (iii) for $r = 1$. Now suppose $r > 1$. The last term of (1) is 0 by Proposition 4.9.1 (i). Also if $r+b < t$ or $r+b = t$ then $b+1 < t$ and the first term in (1) is 0 by Proposition 4.9.1 (ii). Hence the middle terms are equal and (iii) follows by induction.

In dealing with (iv) we may assume that $b < t$ because otherwise (i) implies the result. If $b < t-1$ we obtain, from Proposition 4.9.1 (ii) applied to (1), an isomorphism $R^b F(M_r \otimes -b\varepsilon_1) \to R^{b+1}F(M_{r-1} \otimes -(b+1)\varepsilon_1)$ and the result follows by induction. If $b = t-1$ the epimorphism is provided by (1) since the first term is $Y(\lambda_{r+b-t}) \otimes -\lambda_t$ by Proposition 4.9.1 (ii) and the third term is 0 by (i).

Taking $b = 0$ in Proposition 4.9.3 (ii), (iii) we obtain the following.

Corollary 4.9.4 *We have for all* r :

(i) $R^j FM_r = 0$ *for all* $j > 0$

(ii) $$FM_r = \begin{cases} 0 & \text{if } r < t \\ -\lambda_t & \text{if } r = t \end{cases}$$

Proposition 4.9.5 (i) *For all* r *there is an exact sequence*

$$0 \to FM_r \to Y(\lambda_r) \to FQ_r \to 0 .$$

(ii) *If* $r \le t$ *then* FM_r *has a good filtration.*

(iii) FQ_r *has a good filtration for all* r .

Proof (i) This is obtained by applying F to $0 \to \mathbf{M}_r \to Y(\lambda_r) \to Q_r \to 0$ and using Corollary 4.9.4 (i).

(ii) This is clear from Corollary 4.9.4.

(iii) By Proposition 4.8.1 (i), (ii), (iv), Q_r satisfies the hypotheses of Lemma 3.7.1 so that $Ind_{P_\Gamma}^{P_\Sigma} Q_r$ has a good filtration where $\Gamma = \Omega \cap \Sigma$. But $(R^i Ind_B^{P_\Gamma})Q_r \cong Q_r \otimes (R^i Ind_B^{P_\Gamma})k$, by the tensor identity, which is Q_r if $i = 0$ and 0 otherwise. Hence by (1.1.6), $FQ_r \cong Ind_{P_\Gamma}^{P_\Sigma} Q_r$.

We are now ready to establish the highly desirable:

Proposition 4.9.6 *For any* λ_r *non-exceptional and any* $\Sigma = \Delta \backslash \{\alpha\ \}$

for $1 < t \leq \ell$, $Y(\lambda_r)|_{P_\Sigma}$ *has a good filtration.*'

Proof Suppose, for a contradiction, that the result is false and r is as small as possible for which it fails. By Proposition 4.9.5, $r > t$; we let $a = r-t$. By Proposition 4.9.3 (iv) there is an epimorphism of P_Σ-modules $\theta: Y(\lambda_a) \otimes -\lambda_t \to FM_r$. By the minimality of r , $Y(\lambda_a)|_{P_\Sigma}$ has a good filtration. We define $\delta: X \to$ by $\delta(\Sigma_i b_i \varepsilon_i) = \Sigma_i b_i$ and note that if $\lambda \leq \mu$ then $\delta(\lambda) \leq \delta(\mu)$. Thus, if r is a weight of $Y(\lambda_a)$, $\delta(r) \leq \delta(\lambda_a) = a$. By repeated application of Lemma 3.2.2, there is a P_Σ-submodule, N say, of $Y(\lambda_a)$ such that N and $Y(\lambda_a)/N$ each has a good filtration, if $Y_\Sigma(\tau)$ is a non-zero successive quotient in a good filtration of N then $\delta(\tau) < a$ and if $Y_\Sigma(\tau)$ is a non-zero successive quotient in a good filtration of $Y(\lambda_a)/N$ then $\delta(\tau) = a$. Now $\mu = -\varepsilon_{a+1}-\ldots-\varepsilon_r$ is W conjugate to $\varepsilon_{t+1}+\ldots+\varepsilon_r$ so that, if μ occurs as a weight of a successive quotient $Y_\Sigma(\tau)$ in a good filtration of $Y(\lambda_a)|_{P_\Sigma}$, $\tau \geq \varepsilon_{t+1}+\ldots+\varepsilon_r$ hence $\delta(\tau) \geq a$. Thus $\delta(\tau) = a$ and $Y_\Sigma(\tau)$ occurs as a successive quotient of $Y(\lambda_a)/N$. Hence $(Y(\lambda_a)/N)^\mu \neq 0$ so that $N^\mu = 0$ and therefore $(N \otimes -\lambda_t)^{-\lambda_r} = 0$. Hence $\theta(N \otimes -\lambda_t)$ has zero intersection with the socle $-\lambda_r$ of $Y(\lambda_r)$ and so $\theta(N \otimes -\lambda_t) = 0$. Thus:

θ *induces an epimorphism* $\bar{\theta}: (Y(\lambda_a)/N) \otimes -\lambda_t \to FM_r$ (1)

By minimality and Proposition 4.9.5, $\bar{\theta}$ is non-zero. We claim that $\bar{\theta}$ is an isomorphism, from which it immediately follows that FM_r has a good filtration and so, by Proposition 4.9.5, $Y(\lambda_r)|_{P_\Sigma}$ has a good filtration. In view of (1) the claim follows from equality of the dimension of $(Y(\lambda_a)/N) \otimes -\lambda_t$ and FM_r and so, a fortiori, from the assertion $ch((Y(\lambda_a)/N) \otimes -\lambda_t) = chFM_r$. However $chFM_r = \sum_{i\geq 0} (-1)^i chR^i FM_r$ by Proposition 4.9.3 (ii) so that the assertion is independent of the characteristic. We may, therefore, (and do) take the characteristic of k to be zero.

Now $Hom_{P_\Sigma}(Y(\lambda_a) \otimes -\lambda_t , Y(\lambda_r)) \cong Hom_{P_\Sigma}(Y(\lambda_a), \lambda_t \otimes Y(\lambda_r))$ which, by reciprocity and the tensor identity, is isomorphic to $Hom_G(Y(\lambda_a), Y(\lambda_t)\otimes Y(\lambda_r))$. Since $Y(\lambda_t)$ is self dual (in characteristic zero) this is isomorphic to $Hom_G(Y(\lambda_a)\otimes Y(\lambda_t), Y(\lambda_r))$. Now $Y(\lambda_a)\otimes Y(\lambda_t)$ has highest weight λ_r with multiplicity one and so $Y(\lambda_r)$ occurs exactly once in $Y(\lambda_a)\otimes Y(\lambda_r)$. Thus $Hom_G(Y(\lambda_a)\otimes Y(\lambda_t), Y(\lambda_r))$, and hence $Hom_{P_\Sigma}(Y(\lambda_a)\otimes -\lambda_t, Y(\lambda_r))$ is one dimensional. Since M_r is a submodule of $Y(\lambda_r)$, FM_r may be identified with a submodule of $Y(\lambda_r)$ and so $Hom_{P_\Sigma}(Y(\lambda_a)\otimes -\lambda_t, FM_r)$ is one dimensional and θ is determined upto a scalar multiple.

Let e_i (resp. e_{-i}) be a non zero vector of weight ε_i (resp. $-\varepsilon_i$) in $Y(\lambda_1)$, for $1\le i\le \ell$. We identify $Y(\lambda_a)$ (resp. $Y(\lambda_r)$) with the G-submodule of $\Lambda^a Y(\lambda_1)$ (resp. $\Lambda^r Y(\lambda_1)$) generated by $e_1\wedge e_2\wedge\ldots\wedge e_a$ (resp. $e_1\wedge e_2\wedge\ldots\wedge e_r$). Let V_o be a non-zero vector in the one dimensional P_Σ-module $-\lambda_t$ and let $\phi:\Lambda^a Y(\lambda_1)\otimes -\lambda_t\to\Lambda^r Y(\lambda_1)$ be the P_Σ-module homomorphism which takes $M\otimes V_o$ to $M\wedge e_{-1}\wedge e_{-2}\wedge\ldots\wedge e_{-t}$ for $M\in\Lambda^a Y(\lambda_1)$. By restriction ϕ determines a non-zero map $Y(\lambda_a)\otimes -\lambda_t\to Y(\lambda_r)$ which we may take to be θ. If $\bar\theta$ is not injective then $\bar\theta$ kills some copy of $Y_\Sigma(\tau)\otimes -\lambda_t$ where $\tau\in X_\Sigma$ and $\delta(\tau)=a$. Thus $\tau=\varepsilon_{i_1}+\varepsilon_{i_2}+\ldots+\varepsilon_{i_a}$ for some $i_1<i_2<\ldots<i_a$. However, for such τ, $Y(\lambda_a)^\tau$ is one dimensional and spanned by $e_{i_1}\wedge e_{i_2}\wedge\ldots\wedge e_{i_a}$. But $e_{i_1}\wedge e_{i_2}\wedge\ldots\wedge e_{i_a}\wedge e_{-1}\wedge\ldots\wedge e_{-t}$ is not zero so that $\theta(Y(\lambda_a)^\tau\otimes -\lambda_t)\neq 0$ and $\bar\theta$ is injective. Hence $\bar\theta$ is an isomorphism in characteristic zero, $ch(Y(\lambda_a)/N)\otimes -\lambda_t = chFM_r$ in all characteristics and so $\bar\theta$ is an isomorphism in all characteristics. This completes the proof of the proposition.

We now have, by Proposition 4.3.2, Proposition 4.8.3 and Proposition 4.9.6 and the inductive hypothesis, that $Y(\lambda_\alpha)|_{P_\Sigma}$ has a good filtration for every $\alpha\in\Delta$ and $\Sigma\subseteq\Delta$. Hence, by Proposition 3.5.2, $Y(\lambda_\alpha)\otimes Y(\tau)$ has a good filtration for every $\alpha\in\Delta$, $\tau\in X^+$ and so, by Proposition 3.5.4 (i), $\underline{H}_1(\Phi)$ is true and, by Proposition 3.5.4 (iii), $\underline{H}_2(\Phi)$ is true. This completes the proof of Theorem 4.3.1.

Remark It is shown, in section 4.4 of the recent preprint "Cohomology of Induced Representations for Algebraic Groups" by H.H. Andersen and J.C. Jantzen, that if G has type B_ℓ or D_ℓ and the characteristic of k is not 2 or G has type C_ℓ then $\Lambda^i Y(\lambda_1)$ has a good filtration for each i. This can be used to provide an alternative proof that $\underline{H}(G)$ is true for such G. For G of type B_ℓ or D_ℓ and the characteristic of k equal to 2 one has $(\Lambda^2 Y(\lambda_1))^G \cong k$ and $ch\Lambda^2 Y(\lambda_1) = \chi(\lambda_2)$ so that $\Lambda^2 Y(\lambda_1)$ cannot have a good filtration by (1.5.2).

5. Homological algebra revisited

The homological properties developed in chapters 1 and 2 were sufficient to prove our hypotheses for the classical groups. To make progress with the exceptional groups we shall use the new properties developed in the present chapter.

As usual, G denotes a (connected) reductive group and the notation of chapter 1 is in force. For subsets Γ, Ω of Δ with $\Gamma \subseteq \Omega$ we shall sometimes write simply I_{Γ}^{Ω} for $Ind_{P_{\Gamma}}^{P_{\Omega}}$.

(5.1) Lemma *Let* Ω, Σ *be subsets of* Δ *with* $\Omega \subseteq \Sigma$. *Suppose that* V *is a* P_{Ω}*-module with a filtration* $0 = V_0, V_1, \ldots, V_n = V$. *If* $I_{\Omega}^{\Sigma}(V_i/V_{i-1})$ *has a good filtration for* $0 < i \leq n$ *and* $RI_{\Omega}^{\Sigma}(V_i/V_{i-1}) = 0$ *for* $0 < i < n$ *then* $I_{\Omega}^{\Sigma}(V)$ *has a good filtration.*

This follows by induction on n and the long exact sequence of derived functors.

(5.2) Proposition *Let* Ω, Σ *be subsets of* Δ, *let* $\Gamma = \Omega \cap \Sigma$, *let* $F = I_{\Gamma}^{\Sigma}$ *and suppose that* $\underline{H}(P_{\Omega})$ *is true. Suppose that* V *is a* P_{Ω}*-module with a good filtration* $0 = V_0, V_1,$ *where each* V_i/V_{i-1} *is either* 0 *or isomorphic to* $Y_{\Omega}(\mu_i)$ *for* $\mu_i \in X_{\Omega}$ *where* $(\mu_i, \alpha^{\vee}) \geq -1$ *for all* $\alpha \in \Sigma \backslash \Omega$. *Then* $F(V|_{P_{\Gamma}})$ *has a good filtration and* $R^j F(V) = 0$ *for all* $j > 0$.

Proof By the long exact sequence of derived functors and (5.1) it suffices to show that if $\mu \in X_{\Omega}$ and $(\mu, \alpha^{\vee}) \geq -1$ for all $\alpha \in \Sigma \backslash \Gamma$ then $F(Y_{\Omega}(\mu)|_{P_{\Gamma}})$ has a good filtration and, for $j > 0$ $R^j F(Y_{\Omega}(\mu)|_{P_{\Gamma}}) = 0$. Let $M = Y_{\Omega}(\mu)$ and let $0 = M_0, M_1, \ldots$ be a good filtration of $M|_{P_{\Gamma}}$ (using $\underline{H}(P_{\Omega})$). Again by (5.1) and the long exact sequence of derived functors it suffices to prove that, for each $i > 0$, $F(M_i/M_{i-1})$ has a good filtration and $R^j F(M_i/M_{i-1}) = 0$ for $j > 0$. Now M_i/M_{i-1} is either 0 or isomorphic to $Y_{\Gamma}(\tau)$ for some $\tau \in X$ so that $F(M_i/M_{i-1})$, if non-zero, is isomorphic to $Y_{\Sigma}(\tau)$ which certainly has a good filtration. Also, if $M_i/M_{i-1} \cong Y_{\Gamma}(\tau)$ for $\tau \in X_{\Gamma}$

then τ is a weight of $Y_\Omega(\mu)$ and so $\mu - \tau = \sum_{\beta\in\Omega} m_\beta \beta$ for non-negative integers m_β. Thus, for $\alpha \in \Sigma\backslash\Omega$, $(\mu - \tau,\alpha^\vee) \le 0$ and so $(\tau,\alpha^\vee) \ge (\mu,\alpha^\vee) \ge -1$. Hence $\tau \in X_\Sigma$ or $(\tau,\alpha^\vee) = -1$ for some $\alpha \in \Sigma\backslash\Omega$ and $R^j F(Y_\Gamma(\tau)) = 0$, for $j > 0$, by (2.1.2), (1.7.4) and (2.1.5).

(5.3) Lemma *Suppose that* $\Gamma \subseteq \Omega \subseteq \Sigma \subseteq \Delta$, V *is a* P_Ω*-module and* $\lambda \in X_\Omega$. *Then we have*

$$R^i I_\Omega^\Sigma(V \otimes Y_\Omega(\lambda)) \cong R^i I_\Omega^\Sigma(V \otimes Y_\Gamma(\lambda))$$

for all $i \ge 0$, *in particular,* $R^i I_\Omega^\Sigma(V) \cong R^i I_\Gamma^\Sigma(V)$.

Proof By the tensor identity, (2.1.2) and (1.7.4) $R^i I_\Gamma^\Omega(V \otimes Y_\Gamma(\lambda)) = 0$ for all $i > 0$ so the result follows from (1.1.6).

(5.4) Lemma *Suppose that* $H^i(B,\lambda) \ne 0$ *for some* $\lambda \in X$, $i > 0$. *Then there is an* $\alpha \in \Delta$ *such that* $(\lambda,\alpha^\vee) < 0$ *and, for any such* α, *there is an integer* m *satisfying* $0 < m < -(\lambda,\alpha^\vee)$ *and* $H^{i-1}(B,\lambda + m\alpha) \ne 0$.

Proof By a result of Cline, Parshall, Scott and van der Kallen, (see (2.1.3) and (2.1.4)) λ is not dominant so there is some $\alpha \in \Delta$ such that $(\lambda,\alpha^\vee) < 0$. Let $\Omega = \{\alpha\}$. We have, by (1.7.1), (1.1.15) and (1.6.1), that $(R^i Ind_B^{P_\Omega})\lambda = 0$ for $i \ne 1$. Hence, by (1.1.9) $H^i(B,\lambda) \cong H^{i-1}(P_\Omega,(RInd_B^{P_\Omega})\lambda)$ which, by (2.1.3), is $H^{i-1}(B,(RInd_B^{P_\Omega})\lambda)$.

The character of $(RInd_B^{P_\Omega})\lambda$ is, by (1.1.15), $ch(RInd_{B_\Omega}^{G_\Omega})\lambda$. Hence $ch(RInd_B^{P_\Omega})\lambda$ is $-\chi_\Omega(\lambda)$, where $\chi_\Omega(\lambda)$ is the Euler character of λ with respect to $Ind_{B_\Omega}^{G_\Omega}$. However $\chi_\Omega(\lambda) = -\chi_\Omega(s_\alpha.\lambda)$ by (2.2.5). Thus the character of $(RInd_B^{P_\Omega})\lambda$ is, by Weyl's Character Formula (see (2.2.6) $\sum_{m=1}^{n-1} e(\lambda + m\alpha)$, where $n = -(\lambda,\alpha^\vee)$. Since $H^{i-1}(B,(RInd_B^P)\lambda) \ne 0$ we must have $H^{i-1}(B,\tau) \ne 0$ for some weight τ of $(RInd_B^{P_\Omega})\lambda$, i.e. for $\tau = \lambda + m\alpha$ for some $0 < m < n$.

The following result and its corollary are proved in the same way as (2.3.1), (2.3.3). However the result can not be generalised to the obvious assertion about $R^n Ind_B^{P_\Sigma}(\lambda)$ for $(\lambda,\alpha^\vee) = -m$, m an arbitrary positive integer, since the module $Y_{\{\alpha\}}(\lambda + (m-1)\alpha)$ may not be simple.

(5.5) Proposition *Suppose that* $\alpha \in \Sigma \subseteq \Delta$, $\lambda \in X$ *and* $(\lambda,\alpha^\vee) = -3$. *Then* $Ind_B^{P_\Sigma}(\lambda) = 0$ *and, for* $n > 0$, $R^n Ind_B^P(\lambda) \cong R^{n-1} Ind_B^{P_\Sigma}(\lambda + 2\alpha)$.

(5.6) Corollary *Suppose* $\alpha \in \Sigma \subseteq \Delta$, V *is a* P_Σ*-module and* $\lambda \in X$ *with* $(\lambda,\alpha^\vee) = -3$. *Then* $(V \otimes \lambda)^B = 0$ *and* $H^n(B,V \otimes \lambda) \cong H^{n-1}(B,V \otimes (\lambda + 2\alpha)$ *for* $n > 0$.

We need (5.5) to prove the next result in the case of G_2 (which isn't actually used in the sequel).

(5.7) Proposition *Suppose that* $\Gamma \subseteq \Delta$, $\alpha \in \Delta\setminus\Gamma$ *and* $\Sigma = \Gamma \cup \{\alpha\}$. *Suppose that* V *is a rational* P_Γ*-module with a good filtration in which each successive quotient has the form* $Y_\Gamma(\tau)$ *for some* $\tau \in X_\Gamma$ *(not necessarily the same for each quotient) satisfying* $-1 \geq (\tau,\alpha^\vee) \geq -2$. *Then* $(R^i I_\Gamma^\Sigma)V = 0$ *for* $i \neq 1$ *and* $RI_\Gamma^\Sigma(V)$ *has a good filtration.*

Proof By the long exact sequence of derived functors it suffices to consider the case in which $V = Y_\Gamma(\tau)$ for $\tau \in X_\Gamma$ with $-1 \geq (\tau,\alpha^\vee) \geq -2$. By (2.1.2), $(R^i I_\Gamma^\Sigma)V \cong (R^i Ind_B^{P_\Sigma})\tau$ for all i . Now if $(\tau,\alpha^\vee) = -1$ we have $(R^i Ind_B^{P_\Sigma})\tau = 0$ for all i by (2.1.5) and so we may assume that $(\tau,\alpha^\vee) = -2$. Thus, by (2.3.1), $Ind_B^{P_\Sigma}(\tau) = 0$ and $(R^i Ind_B^{P_\Sigma})\tau \cong (R^{i-1} Ind_B^{P_\Sigma})(\tau + \alpha)$ for $i > 0$. Now $Ind_B^{P_\Sigma}(\tau + \alpha)$ is either 0 or $Y_\Sigma(\tau + \alpha)$ so it suffices to show that $(R^i Ind_B^{P_\Sigma})(\tau + \alpha) = 0$ for all $i > 0$. If $\tau + \alpha \in X_\Sigma$ then this is true by Kempf's Vanishing Theorem (see (1.7.4)) so we may assume that $(\tau + \alpha,\beta^\vee) < 0$ for some $\beta \in \Gamma$. Also, if $(\tau + \alpha,\beta^\vee) = -1$ then

$(R^{i}Ind_B^{P_\Sigma})(\tau + \alpha) = 0$ for all i by (2.1.5) so we may assume that $(\tau + \alpha, \beta^\vee) \le -2$. Since $(\tau + \alpha, \beta^\vee) \ge (\alpha, \beta^\vee)$ we must have $(\tau + \alpha, \beta^\vee) = -2$ or -3. Suppose first that $(\tau + \alpha, \beta^\vee) = -2$. Since $(\alpha, \beta^\vee) < -1$ we must have $(\beta, \alpha^\vee) = -1$. By (2.3.1), $(R^{i}Ind_B^{P_\Sigma})(\tau + \alpha) \cong (R^{i-1}Ind_B^{P_\Sigma})(\tau + \alpha + \beta)$ for $i > 0$ and since $(\tau + \alpha + \beta, \alpha^\vee) = (\beta, \alpha^\vee) = -1$ we have $(R^{i}Ind_B^{P_\Sigma})(\tau + \alpha) = 0$ for all $i > 0$ by (2.1.5). Thus we are left with the case $(\tau + \alpha, \beta^\vee) = -3$. In this case $(\tau, \beta^\vee) = 0$, $(\alpha, \beta^\vee) = -3$ and $(\beta, \alpha^\vee) = -1$. By (5.5) we have $(R^{i}Ind_B^{P_\Sigma})(\tau + \alpha) \cong (R^{i-1}Ind_B^{P_\Sigma})(\tau + \alpha + 2\beta)$, for $i > 0$. However $(\tau + \alpha + 2\beta, \alpha^\vee) = -2$ so that $Ind_B^{P_\Sigma}(\tau + \alpha + 2\beta) = 0$, by (1.7.1) and, by (2.3.1), $(R^{i-1}Ind_B^{P_\Sigma})(\tau + \alpha + 2\beta) \cong (R^{i-2}Ind_B^{P_\Sigma})(\tau + 2\alpha + 2\beta)$ for $i > 1$. Thus it suffices to show that $(R^{i}Ind_B^{P_\Sigma})(\tau + 2\alpha + 2\beta) = 0$ for all $i \ge 0$. Now $(\tau + 2\alpha + 2\beta, \beta^\vee) = 0 - 6 + 4 = -2$ so that $Ind_B^{P_\Sigma}(\tau + 2\alpha + 2\beta) = 0$ and $(R^{i}Ind_B^{P_\Sigma})(\tau + 2\alpha + 2\beta) \cong (R^{i-1}Ind_B^{P})(\tau + 2\alpha + 3\beta)$ for $i \ge 0$. Thus it suffices to show that $(R^{i}Ind_B^{P_\Sigma})(\tau + 2\alpha + 3\beta) = 0$ for all $i \ge 0$. However this is true by (2.1.5) since $(\tau + 2\alpha + 3\beta, \alpha^\vee) = -1$.

(5.8) <u>Proposition</u> *Suppose that* $\Gamma \subseteq \Omega \subseteq \Delta$, $\lambda \in X_\Omega$ *and* $\underline{H}(P_\Omega)$ *is true. Let* $M_\Gamma(\lambda)$ *be the largest* P_Γ*-submodule of* $Y_\Omega(\lambda)$ *of which* λ *is not a weight. Then* $M_\Gamma(\lambda)$ *has a good filtration and* $Y_\Omega(\lambda)/M_\Gamma(\lambda)$ *is isomorphic to* $Y_\Gamma(\lambda)$.

<u>Proof</u> Of course if M', M'' are P_Γ-submodules of $Y_\Omega(\lambda)$ of which λ is not a weight then λ is not a weight of $M' + M''$ so that $M_\Gamma(\lambda)$ is well defined. By Proposition 3.2.6, there is a P_Γ-submodule, say M, of $Y_\Omega(\lambda)$ such that M has a good filtration and $Y_\Omega(\lambda)/M$ is

isomorphic to $Y_\Gamma(\lambda)$. Since λ occurs with multiplicity one as a weight of $Y_\Omega(\lambda)$ and $Y_\Gamma(\lambda)$, $M \subseteq M_\Gamma(\lambda)$. If $M \neq M_\Gamma(\lambda)$ then, by (1.5.2), $s_\Gamma\lambda$ and hence λ is a weight of $M_\Gamma(\lambda)/M$. Hence $M = M_\Gamma(\lambda)$.

(5.9) <u>Lemma</u> *Suppose that* $\Omega \subseteq \Delta$, $\underline{H}(P_\Omega)$ *is true,* $\lambda , \mu \in X_\Omega$ *and* $\lambda - \mu \in \mathbb{Z}\Sigma$ *for some* $\Sigma \subseteq \Delta$. *Then if* $\Gamma = \Omega \cap \Sigma$ *we have* $Ext^i_{P_\Gamma}(M_\Gamma(\mu), Y_\Gamma(\lambda)) = 0$ *for all* $i \geq 0$.

<u>Proof</u> Let S be the set of weights ξ of $Y_\Omega(\mu)$ such that $\mu - \xi \in \mathbb{Z}\Gamma$ and let N be the k-span of the weight spaces of weight $\xi \in S$. By p.230 of [39] the character of N is $\chi_\Gamma(\mu)$. Since $Y_\Omega(\mu)/M_\Gamma(\mu)$ is isomorphic to $Y_\Gamma(\mu)$, for any weight τ of $M_\Gamma(\mu)$, $\tau \notin S$. Now suppose that $Ext^i_{P_\Gamma}(M_\Gamma(\mu), Y_\Gamma(\lambda)) \neq 0$ for some i . Thus $Ext^i_{P_\Gamma}(Y_\Gamma(\tau), Y_\Gamma(\lambda)) \neq 0$ for some $\tau \in X_\Gamma$ such that τ is a weight of $M_\Gamma(\mu)$. By Lemma 3.2.1, $\tau \geq \lambda$ and so $\mu - \lambda = (\mu - \tau) + (\tau - \lambda) \in \mathbb{Z}$ and $\mu - \tau$ non-negative implies that $\mu - \tau \in \mathbb{Z}\Sigma \cap \mathbb{Z}\Omega = \mathbb{Z}\Gamma$. Hence $\tau \in S$ contradicting the initial remark.

(5.10) <u>Proposition</u> *Under the same hypotheses as* (5.9), $Ext^i_{P_\Omega}(Y_\Omega(\mu), Y_\Omega(\lambda)) \cong Ext^i_{P_\Gamma}(Y_\Gamma(\mu), Y_\Gamma(\lambda))$ *for all* $i \geq 0$.

This follows from (5.9), (2.1.3) and the long exact sequence of cohomology.

<u>Remark</u> The proposition is similar to the theorem of [27] where we showed that for $\lambda , \mu \in X^+$ with $\lambda - \mu \in \mathbb{Z}\Omega$, the composition multiplicity of the simple G-module $L(\lambda)$ in $Y(\mu)$ is equal to the multiplicity of the simple G-module $L_\Omega(\lambda)$ in $Y_\Omega(\mu)$.

The above proposition is useful in conjunction with the following well known generality.

(5.11) <u>Lemma</u> *Let* Y, Z *be* P_Ω*-modules where* Ω *is a subset of* Δ . *If* $\mathrm{Ext}^1_{P_\Omega}(Z,Y) \cong k$ *then there is, up to isomorphism, precisely one* P_Ω*-module which is a non-split extension of* Y *by* Z .

(5.12) <u>Proposition</u> *Assume the hypotheses of* (5.9). *Let* $M = M_\Gamma(\lambda)$. *If*

$$0 \to Y_\Omega(\lambda) \overset{i}{\to} K \to Y_\Omega(\mu) \to 0$$

is any non-split short exact sequence of P_Ω*-modules then the induced short exact sequence of* P_Γ*-modules*

$$0 \to Y_\Gamma(\lambda) \to K/i(M) \to Y_\Omega(\mu) \to 0$$

is also non-split.

<u>Proof</u> Let $F = I_\Gamma^\Omega$ and identify $Y_\Omega(\lambda)$ with a P_Ω-submodule of K . Then if K/M is a direct sum of $Y_\Gamma(\lambda)$ and $Y_\Omega(\mu)$, applying F , we see that $F(K/M)$ is a direct sum of $Y_\Omega(\lambda)$ and $Y_\Omega(\mu)$. However, the short exact sequence

$$0 \to M \to K \to K/M \to 0$$

gives rise to an exact sequence

$$0 \to F(M) \to K \to F(K/M) \to RF(M) \qquad (1)$$

(since $F(K) \cong K \otimes F(k) \cong K$) . Also, the short exact sequence

$$0 \to M \to Y_\Omega(\lambda) \to Y_\Gamma(\lambda) \to 0$$

gives rise to an exact sequence

$$0 \to F(M) \to Y_\Omega(\lambda) \overset{j}{\to} Y_\Omega(\lambda) \to RF(M) \to 0 \qquad (2)$$

(using transitivity of induction and (1.7.4). By (1.5.3), j is either 0 or an isomorphism. Suppose that j is 0 . Then we have an isomorphism $F(M) \to Y_\Omega(\lambda)$ and so λ is a weight of $F(M)$. However M has a filtration with quotients $Y_\Gamma(\tau)$, where $\tau \in X_\Gamma$ and $\tau < \lambda$.

By the left exactness of F , λ must be a weight of $F(Y_\Gamma(\tau)) \cong Y_\Omega(\tau)$ for some such τ . We obtain $\lambda \le \tau < \lambda$ - a contradiction. Thus j is an isomorphism and, from (2), $F(M) = RF(M) = 0$. From (1) we thus obtain $K \cong F(K/M)$, the direct sum of $Y_\Omega(\lambda)$ and $Y_\Omega(\mu)$ contrary to hypothesis. Thus K/M is a non-split extension of $Y_\Gamma(\lambda)$ by $Y_\Omega(\mu)$.

Let Γ , Ω be subsets of Δ with $\Gamma \subseteq \Omega$. For λ , $\mu \in X_\Omega$ we regard $E_\Omega(\mu,\lambda) = Ext^1_{P_\Omega}(Y_\Omega(\mu),Y_\Omega(\lambda))$ as the set of equivalence classes of extensions, $0 \to Y_\Omega(\lambda) \to E \to Y_\Omega(\mu) \to 0$, of $Y_\Omega(\lambda)$ by $Y_\Omega(\mu)$, made into a k-vector space in the usual manner. We define $\phi: E_\Gamma(\mu,\lambda) \to E_\Omega(\mu,\lambda)$ to be the map taking the equivalence class containing $0 \to Y(\lambda) \xrightarrow{i} F \xrightarrow{\pi} Y_\Gamma(\mu) \to 0$ to the equivalence class containing $0 \to Y_\Omega(\lambda) \xrightarrow{i^*} I^\Omega_\Gamma(F) \xrightarrow{\pi^*} Y_\Omega(\mu) \to 0$ (exact by (2.1.2) and (1.7.4)) where $\theta^* = I^\Omega_\Gamma(\theta)$ for a morphism θ of P_Γ-modules. It is easy to check that ϕ is a linear map.

(5.13) <u>Proposition</u> *Suppose that* Ω , Σ *are subsets of* Δ *and that* $\underline{H}(\Phi_\Omega)$ *is true. Suppose that* λ , $\mu \in X_\Omega$ *and that* $\lambda - \mu \in \mathbb{Z}\Sigma$. *Then, for* $\Gamma = \Omega \cap \Sigma$, *the map* $\phi: E_\Gamma(\mu,\lambda) \to E_\Omega(\mu,\lambda)$, *defined above, is an isomorphism.*

<u>Proof</u> By (5.10), it suffices to show that ϕ is surjective. Let $0 \xrightarrow{i} Y_\Omega(\lambda) \to E \xrightarrow{\pi} Y_\Omega(\mu) \to 0$ be a P_Ω extension of $Y_\Omega(\lambda)$ by $Y_\Omega(\mu)$. Let $K = i(Y_\Omega(\lambda))$, $M = i(M_\Gamma(\lambda))$ and $L = \pi^{-1}(M_\Gamma(\mu))$. We have $Ext^1_{P_\Gamma}(L/K,K/M) = 0$ by (5.9) so that $L/M = K/M \oplus N/M$ for some P_Γ submodule N of L containing M . We define $F = E/N$. The map $i: Y_\Omega(\lambda) \to E$ takes $M_\Gamma(\lambda)$ into $M \subseteq N$ and so induces a map $i_0: Y_\Gamma(\lambda) \to F$. Also the map $\pi: E \to Y_\Omega(\mu)$ takes L , and hence N , into $M_\Gamma(\mu)$ and so induces a map $\pi_0: F \to Y_\Gamma(\mu)$. We now have a commutative diagram

$$\begin{array}{ccccccccc}
0 & \to & Y_\Omega(\lambda) & \xrightarrow{i} & E & \xrightarrow{\pi} & Y_\Omega(\mu) & \to & 0 \\
 & & \downarrow & & \downarrow & & \downarrow & & \\
0 & \to & Y_\Gamma(\lambda) & \xrightarrow{i_0} & F & \xrightarrow{\pi_0} & Y_\Gamma(\mu) & \to & 0
\end{array}$$

with rows exact, where $E \to F$ is the quotient map and the remaining vertical maps are the natural ones. Applying I_Γ^Ω to the above diagram shows that the equivalence class containing (E,i,π) is the image, under ϕ, of the equivalence class containing (F,i_o,π_o). In particular we have:

(5.14) Corollary *Under the hypothesis of* (5.13), *any extension* E *of* $Y_\Omega(\lambda)$ *by* $Y_\Omega(\mu)$ *is isomorphic to* $I_\Gamma^\Omega(F)$, *where* F *is an extension of* $Y_\Gamma(\lambda)$ *by* $Y_\Gamma(\mu)$. *Moreover* E *is split if and only if* F *is split.*

(5.15) Proposition *Let* $\mathcal{F}$ *be the set of* G-*modules* V *such that, for every* G-*module* V' *with a good filtration,* $V \otimes V'$ *has a good filtration and let* S *be the set of* G-*modules* V *in* $\mathcal{F}$ *such that* $V|_{P_\Sigma}$ *has a good filtration for every* $\Sigma \subseteq \Delta$. *Assume that* $\underline{H}(P_\Sigma)$ *is true for every proper subset* Σ *of* Δ. *If* $0 \to V' \to V \to V'' \to 0$ *is a short exact sequence of* G-*modules and* V', $V \in \mathcal{F}$ *(respectively* S*) then* $V'' \in \mathcal{F}$ *(respectively* S*). For rational* G-*modules* V', V'' *the direct sum* $V' \oplus V''$ *belongs to* $\mathcal{F}$ *(respecitvely* S*) if and only if* V' *and* V'' *belong to* $\mathcal{F}$ *(respectively* S*). If* V', V'' *belong to* $\mathcal{F}$ *(respectively* S*) then* $V' \otimes V''$ *belongs to* $\mathcal{F}$ *(respectively* S*).*

Proof The first two assertions follow from Proposition 3.2.4 and Corollary 3.2.5. If V', V'' belong to $\mathcal{F}$ and V is a G-module with a good filtration then $V'' \otimes V$ has a good filtration and so $(V' \otimes V'') \otimes V = V' \otimes (V'' \otimes V)$ has a good filtration. Thus $V' \otimes V''$ belongs to $\mathcal{F}$. If V', V'' belong to S then $V' \otimes V''$ has a good filtration since V' belongs to $\mathcal{F}$ and V'' has a good filtration. If $\Sigma \subseteq \Delta$, $\Sigma \neq \Sigma$ then $V'|_{P_\Sigma}$ and $V''|_{P_\Sigma}$ each has a good filtration and so $(V' \otimes V'')|_{P_\Sigma}$ has a good filtration by $\underline{H}(P_\Sigma)$.

(5.16) Lemma *Let* V *be a* G-*module,* $\Omega \subseteq \Delta$ *and suppose that* $V|_{P_\Omega}$ *has a good filtration. Suppose that* $\tau \in X^+$, $(\tau,\alpha^\vee) = 0$ *for all* $\alpha \in \Omega$ *and* $(\tau + \eta,\alpha^\vee) \geq -1$ *for all* $\alpha \in \Delta$ *and* $\eta \in X_\Omega$ *such that* $Y_\Omega(\eta)$ *is a successive quotient in a good filtration of* $V|_{P_\Omega}$. *Then*

$V \otimes Y(\tau)$ *has a good filtration.*

This follows by the proof of Proposition 3.5.2.

(5.17) Proposition *Suppose that* $\Omega \subseteq \Delta$ *and that* $Y(\lambda_\alpha) \otimes Y(\tau)$ *has a good filtration for all* $\alpha \in \Omega$ *and* $\tau \in X^+$. *Suppose that* $\beta \in \Delta$ *and that* $Y(\lambda_\beta) \otimes Y(\tau)$ *has a good filtration for all* $\tau \in X^+$ *such that* $(\tau, \alpha^\vee) = 0$ *for all* $\alpha \in \Omega$. *Then* $Y(\lambda_\beta) \otimes Y(\tau)$ *has a good filtration for all* $\tau \in X^+$.

Proof For this proof we define a new partial ordering on X . We define $\lambda \geq \mu$ if $\lambda - \mu = \sum_{\alpha \in \Delta} m_\alpha \alpha$ for rational numbers m_α with $m_\alpha \geq 0$. By Exercise 8, section 13 of [34] each $\lambda_\alpha > 0$ and it follows that for any non-zero dominant weight λ , $\lambda \geq 0$ and there are only a finite number of dominant weights less than λ . Suppose, for a contradiction, the proposition is false and choose $\tau \in X^+$ minimal such that $Y(\lambda_\beta) \otimes Y(\tau)$ does not have a good filtration. By the hypotheses $(\tau, \alpha^\vee) > 0$ for some $\alpha \in \Omega$. Let $\tau' = \tau - \lambda_\alpha$. Now $Y(\lambda_\alpha) \otimes Y(\tau')$ has a good filtration so by Proposition 3.2.6, there is a submodule M with a good filtration such that $(Y(\lambda_\alpha) \otimes Y(\tau'))/M$ is isomorphic to $Y(\tau)$. By the minimality of τ , $Y(\lambda_\beta) \otimes Y(\tau')$ has a good filtration and so by the hypotheses, $Y(\lambda_\alpha) \otimes Y(\lambda_\beta) \otimes Y(\tau')$ has a good filtration. Also by minimality $Y(\lambda_\beta) \otimes M$ has a good filtration and so by Proposition 3.2.4, $Y(\lambda_\beta) \otimes Y(\tau) \cong (Y(\lambda_\beta) \otimes Y(\lambda_\alpha) \otimes Y(\tau'))/(Y(\lambda_\beta) \otimes M)$ has a good filtration. This contradiction completes the proof of the proposition.

The final result of this chapter will be of use to us in producing good filtrations for various modules for parabolic subgroups.

(5.18) Proposition *Suppose that* $\Gamma \subseteq \Omega \subseteq \Delta$ *with* $\Gamma = \Omega \backslash \{\alpha\}$ *for some* $\alpha \in \Omega$ *and* $\underline{H}(P_\Omega)$ *is true. Suppose* $\lambda \in X$ *is such that* $(\lambda, \alpha^\vee) = -2$ *and* $\lambda + \alpha \in X_\Omega$. *If* K *is a* P_Γ*-module which is a non-split extension of* $Y_\Gamma(\lambda)$ *by* $Y_\Gamma(\lambda + \alpha)$ *then* $(R^i Ind_B^{P_\Omega})K = 0$ *for all* $i \geq 0$.

Proof We have a short exact sequence of P_Γ-modules $0 \to Y_\Gamma(\lambda) \to K \to Y_\Gamma(\lambda + \alpha) \to 0$. For $F = Ind_B^{P_\Omega}$ we have $FY_\Gamma(\lambda) = 0$

and $R^{i}FY_{\Gamma}(\lambda) \cong R^{i-1}FY_{\Gamma}(\lambda + \alpha)$ for $i > 0$ moreover $R^{i}FY_{\Gamma}(\lambda + \alpha)$ is $Y_{\Omega}(\lambda + \alpha)$ for $j = 0$ and 0 for $j > 0$, by (1.7.1), (2.3.1) and the Vanishing Theorem. Thus the long exact sequence of derived functors gives $(R^{i}F)K = 0$ for $i > 1$ and an exact sequence

$$0 \to FK \to Y_{\Omega}(\lambda + \alpha) \to Y_{\Omega}(\lambda + \alpha) \to (RF)K \to 0 .$$

Hence it suffices to show that $FK = 0$. If this is not the case then the map $Y_{\Omega}(\lambda + \alpha) \to Y_{\Omega}(\lambda + \alpha)$ is not an isomorphism and so by (1.5.3) is the zero map. This implies that FK is isomorphic to $Y_{\Omega}(\lambda + \alpha)$. Thus $Hom_{P_{\Omega}}(Y_{\Omega}(\lambda + \alpha), FK)$ is non-zero and by reciprocity there exists a non-zero P_{Γ}-map say $\theta: Y_{\Omega}(\lambda + \alpha) \to K$. If the image of θ intersects $Y_{\Gamma}(\lambda)$ then λ is a weight of $Y_{\Omega}(\lambda + \alpha)$. But then $s_{\alpha}\lambda = \lambda + 2\alpha$ is also a weight of $Y_{\Omega}(\lambda + \alpha)$ contrary to (1.5.2). Hence $Im\theta \cap Y_{\Gamma}(\lambda) = 0$. This implies that the induced map $\bar{\theta}: Y_{\Omega}(\lambda + \alpha) \to Y_{\Gamma}(\lambda + \alpha)$ is not zero. Now $\bar{\theta}(M_{\Gamma}(\lambda + \alpha)) = 0$ since $\lambda + \alpha$ is not a weight of $M_{\Gamma}(\lambda + \alpha)$ so that $\bar{\theta}$ induces a non-zero map $\bar{\bar{\theta}}: Y_{\Gamma}(\lambda + \alpha) \to Y_{\Gamma}(\lambda + \alpha)$ which, by (1.5.3), is surjective. Thus $\bar{\theta}$ is surjective and $Im\theta$ is isomorphic to $Y_{\Gamma}(\lambda + \alpha)$ and so $Im\theta$ is a complement to $Y_{\Gamma}(\lambda)$ in K. But K is a non-split extension. This contradiction shows that $FK = 0$ and thus completes the proof of the proposition.

6. G_2

We begin our analysis of the exceptional groups with G_2. Use of the homological properties developed in the previous chapter is not essential here and it would have been an easy matter to treat G_2 along with the classical groups. Most of the results in this chapter could also be deduced from p.123-127 of [40].

Let G be a semisimple, simply connected group of type G_2. Let $\Delta = \{\alpha, \beta\}$ with $(\alpha, \beta^\vee) = -1$, $(\beta, \alpha^\vee) = -3$. We denote by (a,b), for integers a,b, the element $a\lambda_\alpha + b\lambda_\beta$ of X. Thus $\alpha = (2,-1)$ snd $\beta = (-3,2)$. We let $\Omega = \{\alpha\}$, $\Sigma = \{\beta\}$. Since there is a root, $(0,1)$, orthogonal to α by the argument of Proposition 4.7.1 we have the following.

(6.1) *Let* V *be a* G-*module. Then* $V|_{P_\Omega}$ *has a good filtration if and only if* $F^0_\Omega(V)$ *has a good filtration.*

It is easy to verify:

(6.2) $\chi(1,0) = \chi_\Omega(1,0) + \chi_\Omega(2,-1) + \chi_\Omega(1,-1)$

and

(6.3) $\chi(0,1) = \chi_\Omega(0,1) + \chi_\Omega(3,-1) + \chi_\Omega(2,-1) + \chi_\Omega(0)$

$$+ \chi_\Omega(3,-2) + \chi_\Omega(0,-1) .$$

Let $V = Y(1,0)$. Then $F^{-1}_\Omega V \cong Y_\Omega(1,-1)$. To show that $V|_{P_\Omega}$ has a good filtration, by (6.1), it is enough to prove that $F^0_\Omega V/F^{-1}_\Omega V$ is isomorphic to $Y_\Omega(2,-1)$. By (6.2) the character of $F^0_\Omega V/F^{-1}_\Omega V$ is $\chi_\Omega(2,-1)$ and so, by (1.5.4), if this section is not isomorphic to $Y_\Omega(2,-1)$ then $(F^0_\Omega V/F^{-1}_\Omega V)^B \neq 0$ and so $Q^B \neq 0$ where $Q = V/F^{-1}_\Omega V$. However $V^B = 0$, by (1.5.2) and $H^1(B, Y_\Omega(1,-1)) = 0$ by (2.1.6) so that $Q^B = 0$ by the long exact sequence of cohomology. We thus have that $Y(1,0)|_{P_\Omega}$ has a good filtration. By Lemma 3.7.1 and the truth of $\underline{H}(\Phi)$ for Φ of type A_1 we have the following.

(6.4) $Y(1,0)|_{P_\Gamma} \otimes Y_\Gamma(\tau)$ *has a good filtration for all* $\Gamma \subseteq \Delta$, $\tau \in X_\Gamma$.

We now turn our attention to $Y(0,1)$. Let $Y = Y(0,1)$, $K = F_\Omega^{-2}Y$, $M = F_\Omega^{-1}Y$ and $N = F_\Omega^{0}Y$. By (5.3), the character of M/K is $\chi_\Omega(3,-2)$ so that, by (1.5.4), if M/K is not isomorphic to $Y_\Omega(3,-2)$, $(-1,0)$ occurs in the B socle of M/K . However $(Y \otimes (1,0))^B = 0$ by (1.5.2) and $H^1(B,K \otimes (1,0)) \cong H^1(B,(1,-1)) = 0$ by (2.1.6). Hence, by the long exact sequence of cohomology, $((Y/K) \otimes (1,0))^B = 0$ and M/K is isomorphic to $Y_\Omega(3,-2)$. Let $U/M = (Y/M)^B$. Then $U \subseteq F_\Omega^0 Y$ so that $U/M = (N/M)^B$. We have $H^i(B,Y) = 0$ for all i , by (1.5.2) and (2.1.4) so that $U/M \cong H^1(B,M)$, by the long exact sequence of cohomology. Now $H^i(B,Y_\Omega(0,-1)) = 0$ for all i by (2.1.6) so that $H^1(B,M) \cong H^1(B,Y_\Omega(3,-2))$. The latter is k by (2.1.3) and (1.4.7). Hence U/M is isomorphic, as a k-space, to k . Also $(Y/M)^B = (Y/M)^{P_\Omega}$ by (2.1.3) so that U/M is a trivial, one dimensional P_Ω-submodule of Y/M . By (5.3), the character of N/U is $Y_\Omega(2,-1)$ so that, if N/U is not isomorphic to $Y_\Omega(2,-1)$, by (1.5.4), we must have that $(N/U)^B$ 0 . Let $U'/U = (N/U)^B$. Then U'/M is a trivial B-module by Lemma 3.2.1 so that $U'/M = U/M$. Thus $(N/U)^B = 0$ and $N/U = Y_\Omega(2,-1)$. Now by (6.1), $Y(0,1)|_{P_\Omega}$ has a good filtration.

We now show that $Y|_{P_\Sigma}$ has a good filtration. It follows from the tensor identity that $Y|_{P_\Sigma} \cong \mathrm{Ind}_B^{P_\Sigma}(Y|_B)$ so by the long exact sequence of derived functors we need only show that $R^i\mathit{Ind}_B^{P_\Sigma}(Z)$ has a good filtration for $i = 0$ and $Z = K$, U/K , Y/U and is 0 for $i = 1$ and $Z = K$, U/K . By Lemma 3.7.1 it suffices to prove that $R^i\mathit{Ind}_B^{P_\Sigma}(U/K)$ has a good filtration for $i = 0$ and is 0 for $i = 1$. We claim that U/K is a non-split extension of M/K by U/M . If not we obtain $(Y/K)^B \neq 0$. However $Y^B = 0$ and, by (2.1.6), $H^1(B,K) = 0$. Thus $(Y/K)^B = 0$ and U/K is a non-split extension. Let $\bar{U} = U/K$ and let $\bar{U}'$ be the sum of the weight spaces of weight $(1,-1)$ and $(-1,0)$. $\bar{U}'$ is a B-submodule of $\bar{U}$, by (1.4.3), and by Lemma

3.7.1, $R^i Ind_B^{P_\Sigma}(\overline{U}')$ has a good filtration for $i = 0$ and is 0 for $i > 0$. Thus it suffices, by the long exact sequence of derived functors, to show that $R^i Ind_B^{P_\Sigma}(Z)$ has a good filtration for $i = 0$ and is 0 for $i = 1$, where $Z = \overline{U}/\overline{U}'$. By (5.12) Z is a non-split extension of (3,-2) by k. By (1.4.7) and (5.11) it is the unique non-split B-module extension and hence isomorphic to $Y_\Sigma(0,1) \otimes (0,-1)$ (by (1.5.2) applied to G_Σ). Thus $R^i Ind_B^{P_\Sigma}(Z) = 0$ for all i, by the tensor identity and (2.1.5). This completes the proof that $Y(0,1)|_{P_\Sigma}$ has a good filtration. We have now established the following.

(6.5) $Y(0,1)|_{P_\Gamma}$ *has a good filtration for all* $\Gamma \subseteq \Delta$.

It is now easy to prove the following.

(6.6) $Y(0,1) \otimes Y(\tau)$ *has a good filtration for all* $\tau \in X^+$.

By (5.16), (6.3) and (6.5), $Y(0,1) \otimes Y(\tau)$ has a good filtration when $\tau = (0,m)$, $m \geq 0$. Thus by (5.17) and (6.4) (with $\Gamma = \Delta$) $Y(0,1) \otimes Y(\tau)$ has a good filtration for all $\tau \in X^+$. By (6.4), (6.6) above and Propsoition 3.5.4(i) $\underline{H}_1(G)$ is true. Thus, by (6.4), (6.5) and Proposition 3.5.4, $\underline{H}_2(G)$ is true. We have now proved the following.

(6.7) *Let* G *be a semisimple, simply connected group of type* G_2. *Then* $\underline{H}(G)$ *is true.*

7. F_4

7.1 Introduction

The analysis of filtrations for groups of type F_4 is the most complicated of those exceptional groups for which we have a complete proof of our hypotheses. The proof of the hypotheses given for the classical groups and G_2 is characteristic free. This is not so with the other exceptional groups. Our proof for F_4 is obtained by analysing the problem in odd and even characteristics separately. In characteristic not equal to 2 we need only analyse in detail the modules $Y(\lambda_1)$ and $Y(\lambda_4)$ corresponding to the end nodes of the Dynkin-Diagram and obtain information about $Y(\lambda_2)$ and $Y(\lambda_3)$ by considering the exterior squares $\Lambda^2 Y(\lambda_1)$ and $\Lambda^2 Y(\lambda_3)$. In characteristic 2 a detailed analysis of $Y(\lambda_2)$ is also needed.

7.2 $Y(\lambda_1)$

For the rest of the chapter we let G be a semisimple group of type F_4. We label the elements of Δ, $\alpha_1,\alpha_2,\alpha_3,\alpha_4$ so that

$$(\alpha_1,\alpha_2^\vee) = (\alpha_2,\alpha_3^\vee) = (\alpha_3,\alpha_4^\vee) = -1 .$$

For $1 \le i \le 4$, λ_i denotes the fundamental dominant weight dual to α_i, thus $(\lambda_i,\alpha_j^\vee) = \delta_{i,j}$ - the Kronecker delta - for $1 \le j \le 4$. For $a,b,c,d \in \mathbb{Z}$, (a,b,c,d) denotes $a\lambda_1 + b\lambda_2 + c\lambda_3 + d\lambda_4$ and $\langle a,b,c,d\rangle$ denotes $a\alpha_1 + b\alpha_2 + c\alpha_3 + d\alpha_4$. Thus we have:

(7.2.1) $\alpha_1 = (2,-1,0,0)$, $\alpha_2 = (-1,2,-1,0)$, $\alpha_3 = (0,-2,2,-1)$,

$\alpha_4 = (0,0,-1,2)$

and

(7.2.2) $\lambda_1 = \langle 2,3,2,1\rangle$, $\lambda_2 = \langle 3,6,4,2\rangle$, $\lambda_3 = \langle 4,8,6,3\rangle$,

$\lambda_4 = \langle 2,4,3,2\rangle$.

Moreover it is easy to see that

(7.2.3) *the* W-*orbit of* λ_1 *consists of the following elements:*

(1,0,0,0), (-1,1,0,0), (0,-1,1,0), (0,1,-1,1), (1,-1,0,0), (0,1,0,-1), (-1,0,0,1), (1,-1,1,-1), (-1,0,1,-1), (1,1,-1,0), (-1,2,-1,0), (2,-1,0,0), (1,-2,1,0), (-2,1,0,0), (-1,-1,1,0), (1,0,-1,1), (-1,1,-1,1), (1,0,0,-1), (0,-1,0,1), (-1,1,0,-1), (0,-1,1,-1), (0,1,-1,0), (1,-1,0,0), (-1,0,0,0)

and

(7.2.4) *the* W *orbit of* λ_4 *consists of the following elements*:

(0,0,0,1), (0,0,1,-1), (0,2,-1,0), (2,-2,1,0), (-2,0,1,0), (2,0,-1,1), (-2,2,-1,1), (2,0,0,-1), (0,-2,1,1), (-2,2,0,-1), (0,0,-1,2), (0,-2,2,-1), (0,0,1,-2), (0,2,-2,1), (2,-2,0,1), (0,2,-1,-1), (2,-1,0,1), (2,-2,1,-1), (-2,0,0,1), (-2,0,1,-1), (-2,2,-1,0), (0,-2,1,0), (0,0,-1,1), (0,0,0,-1) .

Thus, using "Brauer's Formula", (2.2.3) and (2.2.5), $s(\lambda_1) = \chi(\lambda_1) - 2\chi(0)$ and so

(7.2.5) $\chi(\lambda_1) = s(\lambda_1) + 2s(0)$.

Let $\Omega = \Delta_4 = \Delta\backslash\{\alpha_4\}$. We wish to show that $Y(\lambda_1)|_{P_\Omega}$ has a good filtration. By definition λ_4 is orthogonal to Ω moreover, by (7.2.4), λ_4 is conjugate to α_4 and is therefore a root. Thus

(7.2.6) *there is a root orthogonal to* Ω

and so, by the argument of Proposition 4.7.1,

(7.2.7) $Y(\lambda_1)|_{P_\Omega}$ *has a good filtration if and only if* $F_\Omega^0 Y(\lambda_1)$ *has a good filtration.*

For a G module V and $1 \le i \le 4$, we shall write simply $F_i^r(V)$ (respectively $F_i^{(r)}(V)$) for $F_{\Delta_i}^r(V)$ (respectively $F_{\Delta_i}^{(r)}(V)$) - where $\Delta_i = \Delta\backslash\{\alpha_i\}$. (We hope this is not confusing when $i = 4$). Thus we have $F_4^0 Y(\lambda_1) = F_\Omega^0 Y(\lambda_1)$. Moreover we shall write $\chi_i(\tau)$ (respectively $Y_i(\tau)$) for $\chi_{\Delta_i}(\tau)$ (respectively $Y_{\Delta_i}(\tau)$) for $\tau \in X$.

Using "Brauer's Formula", (2.2.3) and (2.2.5) for G_Ω together with (7.2.5) and (7.2.4) we obtain

(7.2.8) $\chi(\lambda_1) = \chi_4(1,0,0,0) + \chi_4(0,1,0,-1) + \chi_4(1,0,0,-1)$.

Hence, we have, $\mathit{ch}F_4^{-1}Y(\lambda_1) = \chi_4(1,0,0,-1)$. However, U^{Ω} acts trivially on $F_4^{-1}Y(\lambda_1)$ and $Y(\lambda_1)$ has, by (1.5.2), a simple B socle $-\lambda_1$. It follows, by (1.5.4), that $F_4^{-1}Y(\lambda_1)$ is isomorphic to $Y_4(1,0,0,-1)$. Also, from (7.2.8) we see that $\mathit{ch}F_4^{0}Y(\lambda_1)/F_4^{-1}Y(\lambda_1) = \chi_4(0,1,0,-1)$ and so, by (1.5.4), to prove that $F_4^{0}Y(\lambda_1)/F_4^{-1}Y(\lambda_1)$ is isomorphic to $Y_4(0,1,0,-1)$ it suffices to show that $V = F_4^{0}Y(\lambda_1)/F_4^{-1}Y(\lambda_1)$ has a simple B socle $(0,-1,0,1)$.

We denote by W_i , $(1 \le i \le 4)$ the subgroup of W generated by $\{s_{\alpha_j} : j \ne i\}$ and by $s_i(\lambda)$ the sum of the $e(\tau)$ such that τ is W_i conjugate to λ , for $\lambda \in X$. It is easy to check that

(7.2.9) *the* W_4 *orbit of* $(0,1,0,-1)$ *consists of the following elements*: $(0,1,0,-1),(1,-1,1,-1),(1,1,-1,0),(-1,0,1,-1),(2,-1,0,0),$ $(-1,2,-1,0),(-2,1,0,0),(1,-2,1,0),(-1,-1,1,0),(1,0,-1,1),(-1,1,-1,1),$ $(0,-1,0,1)$.

Thus, by "Brauer's Formula" and (2.2.5),
$s_4(0,1,0,-1) = \chi_4(0,1,0,-1) - 2\chi_4(0)$ and

(7.2.10) $\chi_4(0,1,0,-1) = s_4(0,1,0,-1) + 2s_4(0)$.

Now if, for $\tau \in X$, $(V \otimes \tau)^B \ne 0$ we have, by reciprocity and the tensor identity, $Y_\Omega(\tau) \ne 0$. Thus, by (1.7.1), $\tau \in X_\Omega$. Moreover $-\tau$ is a weight of V and so, by (7.2.10) and (7.2.9) $\tau = 0$ or $(0,-1,0,1)$. Thus to show that V has a simple socle $(0,-1,0,1)$ it is enough to prove that $V^B = 0$. Now V embeds in $Y(\lambda_1)/F_4^{-1}Y(\lambda_1)$ so that it is enough to prove that

(7.2.11) $(Y(\lambda_1)/F_4^{-1}Y(\lambda_1))^B = 0$.

From the short exact sequence

$$0 \to Y_4(1,0,0,-1) \to Y(\lambda_1) \to Y(\lambda_1)/F_4^{-1}Y(\lambda_1) \to 0$$

we obtain, via the long exact sequence of cohomology, an exact sequence,

$$Y(\lambda_1)^B \to (Y(\lambda_1)/F_4^{-1}Y(\lambda_1))^B \to H^1(B,Y_4(1,0,0,-1)) .$$

But $Y(\lambda_1)^B = 0$ by (1.5.2) and $H^1(B,Y_4(1,0,0,-1) = 0$ by (2.1.6). Hence $V^B = 0$ and $V = Y_4(0,1,0,-1)$ so that $F_4^0Y(\lambda_1)$ has a good filtration and therefore

(7.2.12) $Y(\lambda_1)_{P_\Omega}$ *has a good filtration.*

Hence, by (7.2.8) and Lemma 3.7.1,

(7.2.13) $Y(\lambda_1)|_{P_\Sigma}$ *has a good filtration for any* $\Sigma \subseteq \Delta$.

It follows from (7.2.3), (7.2.5) above and Proposition 3.5.2 that

(7.2.14) *if* V *is any rational* G-*module with a good filtration then* $Y(\lambda_1) \otimes V$ *has a good filtration.*

7.3 $Y(\lambda_4)$

We now analyse $Y(\lambda_4)|_{P_\Sigma}$ for subsets Σ of Δ . By "Brauer's Formula" (2.2.3) and also (2.2.5) we obtain from (7.2.4)

$$s(\lambda_4) = \chi(\lambda_4) - \chi(\lambda_1) - 2\chi(0)$$

so that

(7.3.1) $\chi(\lambda_4) = s(\lambda_4) + s(\lambda_1) + 4s(0)$.

Using (7.3.1), (7.2.4), (2.2.3) and (2.2.5) one can now show:

(7.3.2) $\chi(\lambda_4) = \chi_4(0,0,0,1) + \chi_4(0,0,1,-1) + \chi_4(2,0,0,-1)$
$+ \chi_4(0) + \chi_4(0,0,1,-2) + \chi_4(0,0,0,-1)$.

Thus $chF_4^{-2}Y(\lambda_4) = \chi_4(0,0,0,-1)$, which is $s_4(0,0,0,-1) = e(0,0,0,-1)$ and so $F_4^{-2}Y(\lambda_4) \cong Y_4(0,0,0,-1)$ has a good filtration.

We have:

(7.3.3) *the* W_4 *orbit of* $(0,0,1,-2)$ *consists of the following elements:*

(0,0,1,-2),(0,2,-1,-1),(2,-2,1,-1),(-2,0,1,-1),(2,0,-1,0),(-2,2,-1,0),
(0,-2,1,0),(0,0,-1,1) .

Hence

$$s_4(0,0,1,-2) = \chi_4(0,0,1,-2) - \chi_4(1,0,0,-1)$$

and so we have

(7.3.4) $\chi_4(0,0,1,-2) = s_4(0,0,1,-2) + \chi_4(1,0,0,-1)$.

Now, by (7.3.2), $chV = \chi_4(0,0,1,-2)$ where $V = F_4^{-1}Y(\lambda_4)$. Moreover U^{Ω} acts trivially on V and so, by (1.5.4), to prove that $V/F_4^{-2}Y(\lambda_4)$ is isomorphic to $Y_4(0,0,1,-2)$ it suffices to show that the B socle of V is $(0,0,-1,1)$. The analysis here is similar to the analysis of $F_4^{0}Y(\lambda_1)/F_4^{-1}Y(\lambda_1)$ in section 2 - it suffices to show that $((Y(\lambda_4)/F_4^{-2}Y(\lambda_4)) \otimes \tau)^B = 0$ for $\tau \in X_{\Omega}$ such that $-\tau$ is a weight of $Y_4(0,0,1,-2)$ except for the case in which $-\tau$ is the lowest weight of $Y_4(0,0,1,-2)$. Thus we see from (7.3.4) that it suffices to show that $((Y(\lambda_4)/F_4^{-2}Y(\lambda_4)) \otimes \lambda_1)^B = 0$. Tensoring the short exact sequence

$$0 \to -\lambda_4 \to Y(\lambda_4) \to Y(\lambda_4)/F_4^{-2}Y(\lambda_4) \to 0$$

with λ_1 and taking B cohomology we obtain an exact sequence

$$(Y(\lambda_4) \otimes \lambda_1)^B \to ((Y(\lambda_4)/F_4^{-2}Y(\lambda_4)) \otimes \lambda_1)^B$$
$$\to H^1(B,\lambda_1 - \lambda_4) .$$

However $(Y(\lambda_4) \otimes \lambda_1)^B = 0$ since the B socle of $Y(\lambda_4)$ is $-\lambda_4$ and $H^1(B,\lambda_1 - \lambda_4) = 0$ by (2.1.6) hence $((Y(\lambda_4)/F_4^{-2}Y(\lambda_4)) \otimes \lambda_1)^B = 0$ and

(7.3.5) $F_4^{-1}Y(\lambda_4)/F_4^{-2}Y(\lambda_4) \cong Y_4(0,0,1,-2)$.

Let $C = F_4^{-1}Y(\lambda_4)$. The short exact sequence

$$0 \to C \to Y(\lambda_4) \to Y(\lambda_4)/C \to 0$$

gives rise to an exact sequence

$$Y(\lambda_4)^B \to (Y(\lambda_4)/C)^B \to H^1(B,C) \to H^1(B,Y(\lambda_4)) \ .$$

Moreover $Y(\lambda_4)^B = 0$ since the B socle of $Y(\lambda_4)$ is $-\lambda_4$ and $H^1(B,Y(\lambda_4)) = 0$ by the result, Corollary 3.3 of [20] of Cline, Parshall, Scott and van der Kallen (see also (2.1.4) hence

(7.3.6) $(Y(\lambda_4)/C)^B \cong H^1(B,C)$.

Now by (7.3.5) we have a short exact sequence

$$0 \to -\lambda_4 \to C \to Y_4(0,0,1,-2) \to 0$$

giving an exact sequence

$$H^1(B,-\lambda_4) \to H^1(B,C) \to H^1(B,Y_4(0,0,1,-2)) \to H^2(B,-\lambda_4) \ .$$

By (2.1.6), the first and last terms are zero so we obtain

$$H^1(B,C) \cong H^1(B,Y_4(0,0,1,-2)) \ .$$

By (2.1.3) and (1.4.7), the latter is isomorphic to k so we obtain

(7.3.7) $H^1(B,C) \cong k$.

Thus, by (7.3.6) there is a submodule D of $Y(\lambda_4)$ containing C with $D/C = (Y(\lambda_4)/C^B \cong k$. Moreover $D/C = (Y(\lambda_4)/C)^{P_\Omega}$ by (2.1.3) and D is a P_Ω submodule of $Y(\lambda_4)$. Now let $E = F_4^0 Y(\lambda_4)$. To prove that $F_4^0 Y(\lambda_4)$, and hence $Y(\lambda_4)|_{P_\Omega}$, has a good filtration it suffices to show that E/D has a good filtration. From (7.3.2) we see that $\mathit{ch}E/D = \chi_4(2,0,0,-1)$ so we must prove that $E/D \cong Y_4(2,0,0,-1)$. The lowest weight of $Y_4(2,0,0,-1)$ is $(-2,0,0,1)$ so that it is enough to prove that $(E/D \otimes \tau)^B$ is zero for $\tau \neq (2,0,0,-1)$ such that $\tau \in X_\Omega$ and $-\tau$ is a weight of $Y_4(2,0,0,-1)$. It is not difficult to show that $\chi_4(2,0,0,-1) = s_4(2,0,0,-1) + s_4(0,1,0,-1) + 3s_4(0)$ so that it suffices to show that $(E/D \otimes \tau)^B = 0$ for $\tau = 0$ and $\tau = (0,1,0,-1)$.

Let $(E/D)^B = F/D$. Then, by (2.1.3), F is a P_Ω submodule of $Y(\lambda_4)$ and, by Lemma 4.2.1, F/C is completely reducible hence

$F/C \subseteq (Y(\lambda_1)/C)^B$ which, as we have already seen, is D/C . Hence $F = D$ and $(E/D)^B = 0$.

It remains to show that $(E/D \otimes (0,1,0,-1))^B = 0$. However, E/D is a submodule of $Y(\lambda_4)/D$ so that it suffices to show that $(Y(\lambda_4)/D \otimes (0,1,0,-1))^B = 0$. Tensoring the short exact sequence

$$0 \to D \to Y(\lambda_4) \to Y(\lambda_4)/D \to 0$$

by $(0,1,0,-1)$ and taking B cohomology we obtain an exact sequence

$$(Y(\lambda_4) \otimes (0,1,0,-1))^B \to (Y(\lambda_4)/D \otimes (0,1,0,-1))^B$$
$$\to H^1(B,D \otimes (0,1,0,-1)) \ .$$

Moreover $(Y(\lambda_4) \otimes (0,1,0,-1))^B$ is isomorphic to $Y(\lambda_4) \otimes \mathit{Ind}_B^G(0,1,0,-1)$, by reciprocity and the tensor identity for induction, and so is zero by (1.5.1). Thus it suffices to prove that $H^1(B,D \otimes (0,1,0,-1))$ is zero. By construction D has a filtration in which the successive factors are $-\lambda_4, Y_4(0,0,1,-2)$ and k . However

$$H^1(B,(0,1,0,-1) \otimes -\lambda_4) = H^1(B,(0,1,0,-2))$$
$$= H^0(B,(0,1,-1.0)) = Y(0,1,-1.0)^G = 0$$

using (2.3.3), reciprocity and (1.5.1). Also $H^1(B,(0,1,0,-1)) = 0$ by (2.1.6) and so it suffices to prove that $H^1(B,Y_4(0,0,1,-2) \otimes (0,1,0,-1)) = 0$. It is easy to check that, for any weight τ of $Y_4(0,0,1,-2)$, $\tau + (0,1,0,-1)$ is not an integral multiple of a simple root so that $H^1(B,\tau + (0,1,0,-1)) = 0$ by (2.3.5) and therefore $H^1(B,Y_4(0,0,1,-2) \otimes (0,1,0,-1)) = 0$ as required. This completes the proof that $F_4^0 Y(\lambda_4)$ has a good filtration and so by the argument of Proposition 4.7.1 and (7.2.6) above

(7.3.8) $Y(\lambda_4)|_{P_\Omega}$ *has a good filtration.*

Let $K = D/F_4^{-2}Y(\lambda_4)$. We next claim:

(7.3.9) *The short exact sequence*

$$0 \to C/F_4^{-2}Y(\lambda_4) \to K \to D/C \to 0$$

is not split.

If the sequence were split, since $D/C \cong k$, we would have $(C/F_4^{-2}Y(\lambda_4))^B \neq 0$ and hence $(Y(\lambda_4)/F_4^{-2}Y(\lambda_4))^B = 0$. However, from the short exact sequence

$$0 \to -\lambda_4 \to Y(\lambda_4) \to Y(\lambda_4)/F_4^{-2}Y(\lambda_4) \to 0$$

we obtain an exact sequence

$$Y(\lambda_4)^B \to (Y(\lambda_4)/F_4^{-2}Y(\lambda_4))^B \to H^1(B,-\lambda_4) .$$

Also $Y(\lambda_4)^B = 0$ since $Y(\lambda_4)$ has B-socle $-\lambda_4$ and $H^1(B,-\lambda_4) = 0$ by, for example, (2.1.6). Thus $(Y(\lambda_4)/F_4^{-2}Y(\lambda_4))^B = 0$ and (7.3.9) is proven.

Now let $\Gamma = \Delta_i \cap \Omega$, for $1 \le i < 4$, and $F = I_\Gamma^{\Delta_i}$. We shall now prove:

(7.3.10) $Y(\lambda_4)|_{P_{\Delta_i}}$ *has a good filtration.*

We have $F(Y(\lambda_4)) \cong Y(\lambda_4) \otimes F(k) = Y(\lambda_4)$ so, by (5.1) it is enough to prove that $F(F_4^{-2}Y(\lambda_4))$, $F(K)$ and $F(Y(\lambda_4)/D)$ have a good filtration and that $RF(F_4^{-2}Y(\lambda_4)) = RF(K) = 0$. By (5.2), since $F_4^{-2}Y(\lambda_4) \cong -\lambda_4, F(F_4^{-2}Y(\lambda_4))$ has a good filtration and $RF(F_4^{-2}Y(\lambda_4)) = 0$. Now by (7.3.8) and Proposition 3.2.4, $Y(\lambda_4)/D$ has a good filtration as a P_Ω module. Moreover, $ch Y(\lambda_4)/D = ch Y(\lambda_4) - ch D = \chi_4(0,0,0,1) + \chi_4(0,0,1,-1) + \chi_4(2,0,0,-1)$ (by (7.3.2) and the construction of D) so that $F(Y(\lambda_4)/D)$ has a good filtration and $RF(Y(\lambda_4)/D) = 0$ by (5.2). Thus it remains to show that $F(K)$ has a good filtration and $RF(K) = 0$. By (5.8) we have a P_Γ submodule M of $C/F_4^{-2}Y(\lambda_4)$ such that M has a good filtration and $(C/F_4^{-2}Y(\lambda_4))/M = Y_\Gamma(0,0,1,-2)$. Let $K' = K/M$. By (5.12) and (7.3.9):

(7.3.11) *the short exact sequence* $0 \to Y_\Gamma(0,0,1,-2) \to K' \to k \to 0$ *is not split.*

For any weight τ of $Y_\Omega(0,0,1,-2)$ other than $(0,0,1,-2)$, $(\tau,\alpha^\vee) \geq -1$ thus if $Y_\Gamma(\xi)$ occurs as a successive quotient in a good filtration of M then $(\xi,\alpha^\vee) \geq -1$. Hence by (5.2), $F(M)$ has a good filtration and $R^j F(M) = 0$ for all $j > 0$. The short exact sequence

$$0 \to M \to K \to K' \to 0$$

thus gives rise to an exact sequence

$$0 \to F(M) \to F(K) \to F(K') \to 0$$

and an isomorphism $RF(K) \to RF(K')$. Hence it suffices to prove that $F(K')$ has a good filtration and $RF(K') = 0$. Let $G = Ind_B^{P_{\Delta_i}}$, by (5.3) it suffices to show that $G(K')$ has a good filtration and that $RG(K') = 0$. Let M' be the B submodule of K' spanned by the weight vectors of weight less that $(0,0,1,-2)$ and let $K'' = K'/M'$. By (5.12) and (7.3.11) the sequence

$$0 \to (0,0,1,-2) \to K'' \to k \to 0$$

is non-split. We leave it to the reader to check that if τ is any weight of M'' then $(\tau,\alpha_j^\vee) = -1$ for some $\alpha_j \in \Delta_i$ from which it follows, from (2.1.5) and the long exact sequence of derived functors, that $R^j G(K') = R^j G(K'')$ for all $j \geq 0$. Thus it suffices to prove that $G(K'')$ has a good filtration and $RG(K'') = 0$. By (5.11) and (1.4.7), K'' is the unique non-split B-module extension of $(0,0,1,-2)$ by k.

However $Y_{\{4\}}(0,0,0,1)$ is a two dimensional B module with simple B-socle $(0,0,1,-1)$ (for example by Weyl's Character Formula and (1.5.2). Hence we have $K'' \cong Y_{\{4\}}(0,0,0,1) \otimes (0,0,1,-1)$.

Thus by the tensor identity, Kempf's vanishing theorem and (2.1.5), $G(K'')$ has a good filtration and $RG(K'') = 0$. This completes the proof of (7.3.10).

We now have, by (7.3.8), (7,3.10) above and Proposition 3.5.3:

(7.3.12) *For any subset* Σ *of* Δ, $Y(\lambda_4)|_{P_\Sigma}$ *has a good filtration.*

From (7,3.1), (7.2.2) and (7.2.4) we see that if τ is any weight of $Y(\lambda_4)$ then $(\tau,\alpha_i^\vee) \geq -2$ for $1 \leq i \leq 4$ and so, by (7.3.12) above and

Proposition 3.5.2:

(7.3.13) *If* V *is any* G-*module with a good filtration then* $V \otimes Y(\lambda_4)$ *has a good filtration.*

7.4 Odd characteristic

Using (7.2.5) and (7.2.1) together with (2.2.3) and (2.2.5) we obtain:

(7.4.1) $\chi(\lambda_1)^2 = \chi(2\lambda_1) + \chi(\lambda_2) + \chi(\lambda_1) + \chi(\lambda_4) + \chi(0)$

and similarly

(7.4.2) $\chi(\lambda_4)^2 = \chi(2\lambda_4) + \chi(2\lambda_1) + \chi(\lambda_3) + \chi(\lambda_4) + \chi(0)$.

Now in characteristic zero all rational G modules are completely reducible and $\{Y(\lambda) : \lambda \in X^+\}$ is a full set of simple modules. Thus $\Lambda^2 Y(\lambda_1)$ and $\mathrm{Symm}^2(Y(\lambda_1))$ have a good filtration and $Y(\lambda_1) \otimes Y(\lambda_1)$ is the sum of its alternating and symmetric parts. Thus $ch\Lambda^2 Y(\lambda_1)$ and $ch\mathrm{Symm}^2(Y(\lambda_1))$ have an expression as the sum of $\chi(\tau)$'s with non-negative coefficients and the sum of these characters is $\chi(\lambda_1)^2$. However these characters are independent of the characteristics so these facts remain true in any characteristic. Also $\lambda_2 = 2\lambda_1 - \alpha_1$ is the unique highest weight of $\Lambda^2 Y(\lambda_1)$ and $\dim Y(\lambda_2) = 273$ (see p.43 of [51]) so we must have, by (7.4.1) and dimensions

(7.4.3) $\Lambda^2\chi(\lambda_1) = \chi(\lambda_2) + \chi(\lambda_4)$

(where $\Lambda^2\chi(\lambda_1) = ch\Lambda^2 Y(\lambda_1)$)

(7.4.4) $\Lambda^2\chi(\lambda_4) = \chi(\lambda_3) + \chi(\lambda_4)$.

Now suppose that the characteristic of k is not 2 and set $D_2 = \Lambda^2 Y(\lambda_1)$, $D_3 = \Lambda^2 Y(\lambda_4)$. Then D_2 is a direct summand of $Y(\lambda_1) \otimes Y(\lambda_1)$ and so has a good filtration by (7.2.14) above and Corollary 3.2.5. Similarly D_3 has a good filtration and so by Proposition 3.2.6, we have a short exact sequence

(7.4.5) $\quad 0 \to Y(\lambda_4) \to D_i \to Y(\lambda_i) \to 0$

for $i = 2,3$. Now if $\Sigma \subseteq \Delta$, $\Sigma \neq \Delta$, $(Y(\lambda_1) \otimes Y(\lambda_1))|_{P_\Sigma}$ has a good filtration by (7.2.14) and $\underline{H}_1(P_\Sigma)$ (proved in chapter 4). Thus $D_1|_{P_\Sigma}$, and similarly $D_2|_{P_\Sigma}$ has a good filtration by Corollary 3.2.5. We have already shown that $Y(\lambda_4)|_{P_\Sigma}$ has a good filtration, (7.3.12), so that by (7.4.5) and Proposition 3.2.4:

(7.4.6) $Y(\lambda_i)|_{P_\Sigma}$ *has a good filtration for* $i = 2,3$ *and any* $\Sigma \subseteq \Delta$.

We now show:

(7.4.7) *If* V *is any* G-*module with a good filtration then* $V \otimes Y(\lambda_i)$ *has a good filtration* ($i = 2,3$) .

We have, from (7.4.5), a short exact sequence

$$0 \to V \otimes Y(\lambda_4) \to V \otimes D \to V \otimes Y(\lambda_i) \to 0$$

moreover $V \otimes Y(\lambda_4)$ has a good filtration by (7.3.13) so, by Proposition 3.2.4, it suffices to show that $V \otimes D_i$ has a good filtration. However, $V \otimes Y(\lambda_1) \otimes Y(\lambda_1)$ has a good filtration by repeated application of (7.2.15) so that $V \otimes D_2$ has a good filtration by Corollary 3.2.5. Similarly $V \otimes D_3$ has a good filtration. We now obtain from (7.2.14), (7.2.13), (7.3.12), (7.3.13), (7.4.6), (7.4.7) and Proposition 3.5.4:

(7.4.8) *If the characteristic of* k *is not* 2 *then* $\underline{H}_1(G)$ *and* $\underline{H}_2(G)$ *are true.*

7.5 Exact sequences

By writing down various W_4 orbits and using (2.2.3), (2.2.5) together with (7.2.8), (7.4.3), (7.3.2) above it is not difficult to obtain:

(7.5.1) $\chi(\lambda_2) = \chi_4(0,1,0,0) + \chi_4(1,1,0,-1) + \chi_4(1,0,0,0)$

$+ \chi_4(1,0,1,-2) + \chi_4(2,0,0,-1) + \chi_4(0,1,0,-1)$

$+ \chi_4(1,1,0,-2) + \chi_4(1,0,0,-1) + \chi_4(0,1,0,-2)$.

We have, from (2.2.3), (2.2.5) and (7.2.3), (7.2.5):

(7.5.2) $\chi(\lambda_1) = \chi_1(1,0,0,0) + \chi_1(-1,1,0,0) + \chi_1(-1,0,0,1)$

$+ \chi_1(0) + \chi_1(-2,1,0,0) + \chi_1(-1,0,0,0)$.

We are going to prove that $Y(\lambda_2)|_{P_\Omega}$ has a good filtration where $\Omega = \Delta\backslash\{\alpha_4\}$. We have $(Y(\lambda_2) \otimes Y(\lambda_4))^G = 0$, for example by Lemma 2.1 of [25]. Thus $\mathrm{Hom}_G(k,Y(\lambda_2) \otimes Y(\lambda_4)) = 0$ and the dimension, say a , of $\mathrm{Hom}_G(Y(\lambda_1),Y(\lambda_2) \otimes Y(\lambda_4))$ (by (7.3.13), Proposition 3.2.6 and (1.5.3)) is equal to the coefficient of $\chi(\lambda_1)$ in an expression of $\chi(\lambda_2)\chi(\lambda_4)$ as a linear combination of $\chi(\tau)$'s , for $\tau \in X^+$. This coefficient is independent of the characteristic of k so to determine a we may assume that the characteristic of k is zero. In this case $Y(\lambda_1),Y(\lambda_2)$ are self dual so that
$\mathrm{Hom}_G(Y(\lambda_1),Y(\lambda_2) \otimes Y(\lambda_4)) = \mathrm{Hom}_G(Y(\lambda_2),Y(\lambda_1) \otimes Y(\lambda_4))$. However, by (7.2.3), (7.2.5) and (2.2.3), (2.2.5),

$$\chi(\lambda_1)\chi(\lambda_4) = \chi(\lambda_1) + \chi(\lambda_2) + \chi(\lambda_1 + \lambda_4) .$$

Thus $a = 1$ and we have, using the reciprocity of induction,

(7.5.3) $\mathrm{Hom}_{P_\Omega}(Y(\lambda_1) \otimes -\lambda_4,Y(\lambda_2)) \cong k$.

We choose a non-zero P_Ω map $\theta:Y(\lambda_1) \otimes -\lambda_4 \to Y(\lambda_2)$ and let M denote the image of θ . To determine the character of M we need to return to the study (begun in section 7.2) of the structure of $Y(\lambda_1)|_{P_\Omega}$. We claim that the B socle of $Y(\lambda_1)/F_\Omega^{-1}Y(\lambda_1)$ is $(0,-1,0,1)$. By (7.2.12), $Y(\lambda_1)|_{P_\Omega}$ has a good filtration and, for example by the construction of a good filtration given in section 7.2, $Y(\lambda_1)/F_\Omega^{-1}Y(\lambda_1)$ has a good filtration with successive quotients

$Y_4(0,1,0,-1)$ and $Y_4(1,0,0,0)$. The B socle of $Y_4(1,0,0,0)$ is $(-1,0,0,1)$ and the B socle of $Y_4(0,1,0,-1)$ is $(0,-1,0,1)$ thus it suffices to prove that $((Y(\lambda_1)/F_\Omega^{-1}Y(\lambda_1)) \otimes (1,0,0,-1))^B = 0$. However, $H^i(B,Y(\lambda_1) \otimes (1,0,0,-1)) = 0$ for all $i \geq 0$ by (2.1.6) so that, by the long exact sequence of cohomology, it suffices to prove that $H^1(B,Y_4(1,0,0,-1) \otimes (1,0,0,-1)) = 0$. Writing down the W_4 orbit of $(1,0,0,-1)$ and using (2.2.3), (2.2.5) we see that

$$\chi_4(1,0,0,-1)^2 = \chi_4(2,0,0,-2) + \chi_4(0,1,0,-2) + \chi_4(0,0,0,-1) .$$

Hence $Y_4(1,0,0,-1) \otimes Y_4(1,0,0,-1)$ has a good filtration with successive quotients $Y_4(2,0,0,-2)$, $Y_4(0,1,0,-2)$, $Y_4(0,0,0,-1)$. Using the long exact sequence of cohomology and (2.1.3), (2.1.6) and (2.3.3) we now see that

(7.5.4) $H^i(B,Y_4(1,0,0,-1) \otimes Y_4(1,0,0,-1)) = 0$ *for all* $i \geq 0$.

Thus $Y(\lambda_1)/F_\Omega^{-1}Y(\lambda_1)$ has a simple B-socle $(0,-1,0,1)$. Now $-\lambda_2$ is not a weight of $F_\Omega^{-1}Y(\lambda_1) \otimes -\lambda_4$ so that θ is zero on $F_\Omega^{-1}Y(\lambda_1) \otimes -\lambda_4$ (since $-\lambda_2$ is the B socle of $Y(\lambda_2)$) and so θ induces a map $\bar{\theta}: (Y(\lambda_1)/F_\Omega^{-1}Y(\lambda_1)) \otimes -\lambda_4 \to Y(\lambda_2)$. But the B-socle of $(Y(\lambda_1)/F_\Omega^{-1}Y(\lambda_1)) \otimes -\lambda_4$ is $-\lambda_2$ and $-\lambda_2$ occurs precisely once as a weight of $(Y(\lambda_1)/F_\Omega^{-1}Y(\lambda_1)) \otimes -\lambda_4$. Thus $\bar{\theta}$ is a monomorphism on the socle and therefore is a monomorphism. We have shown:

(7.2.5) *there is a short exact sequence*
$0 \to Y_4(1,0,0,-2) \to Y(\lambda_1) \otimes -\lambda_4 \to M \to 0$, *of* P_Ω*-modules.*

Thus M has a good filtration with successive quotients $Y_4(0,1,0,-2)$ and $Y_4(1,0,0,-1)$.

7.6 The restriction of $Y(\lambda_2)$ to P_Ω

Let $F = Ind_B^{P_\Omega}$. We next claim:

(7.6.1) $Hom_{P_\Omega}(F(Y(\lambda_1) \otimes (\lambda_1 - \lambda_4)), Y/M) \cong k$

where $Y = Y(\lambda_2)$. We have that $F(Y(\lambda_1) \otimes (\lambda_1 - \lambda_4))$ is isomorphic to $Y(\lambda_1) \otimes Y_4(\lambda_1 - \lambda_4)$ and the dual $Y_4(\lambda_1 - \lambda_4)^*$ is isomorphic to $Y_4(\lambda_1)$ ($Y_4(\lambda_1 - \lambda_4)$ is a simple P_Ω module since $\chi_4(\lambda_1 - \lambda_4) = s_4(\lambda_1 - \lambda_4)$) . Thus (7.6.1) is equivalent to the assertion $Hom_{P_\Omega}(Y(\lambda_1), Y(\lambda_1) \otimes Y/M) \cong k$ (using the reciprocity of induction). Now from (2.2.3), (2.2.5) and (7.2.4), (7.2.5) we have

$$\chi(\lambda_1)\chi(\lambda_2) = \chi(\lambda_1 + \lambda_2) + \chi(\lambda_3) + \chi(\lambda_2) + \chi(\lambda_1 + \lambda_4) + \chi(2\lambda_1) + \chi(\lambda_4) + \chi(\lambda_1) .$$

Thus, by (7.2.15) and Proposition 3.2.6, $Y(\lambda_1)|_{P_\Omega}$ has a good filtration which the successive quotients, in ascending order, are $Y(\lambda_1)$, $Y(\lambda_4)$, $Y(\lambda_2)$, $Y(2\lambda_1)$, $Y(\lambda_1 + \lambda_4)$, $Y(\lambda_3)$, $Y(\lambda_1 + \lambda_2)$. Thus, by (1.5.3) and (2.1.3), $\mathrm{Hom}_{P_\Omega}(Y(\lambda_1), Y(\lambda_1) \otimes Y) \cong \mathrm{Hom}_G(Y(\lambda_1), Y(\lambda_1) \otimes Y(\lambda_2))$ which is isomorphic to k . Hence, by the long exact sequence of cohomology, it suffices to show that $Hom_{P_\Omega}(Y(\lambda_1), Y(\lambda_1) \otimes M) = Ext^1_{P_\Omega}(Y(\lambda_1), Y(\lambda_1) \otimes M) = 0$. However, by (2.1.6) $Ext^i_{P_\Omega}(Y(\lambda_1), Y(\lambda_1) \otimes Y(\lambda_1) \otimes - \lambda_4) = 0$ for all $i \geq 0$, so by (7.5.5) and the long exact sequence of cohomology, $Ext^i_{P_\Omega}(Y(\lambda_1), Y(\lambda_1) \otimes M) \cong Ext^{i+1}_{P_\Omega}(Y(\lambda_1), Y(\lambda_1) \otimes Y_4(1,0,0,-2))$ for all $i \geq 0$. Thus, by (1.1.10), (1.1.7) and (1.7.4), it suffices to show that $Ext^i_B(Y(\lambda_1), Y(\lambda_1) \otimes (1,0,0,-2)) = 0$ for all $i \geq 0$. However, this follows from (2.3.3) and (2.1.6). Thus the proof of (7.6.1) is complete.

Let K be a P_1 submodule of $Y(\lambda_1)$ with a good filtration such that $Q = Y(\lambda_1)/K$ has a good filtration with successive quotients $Y_1(-1,1,0,0)$ and $Y_1(1,0,0,0)$ (K exists by (7.5.2), (7.2.13) and Proposition 3.2.6). We have $R^iF(K \otimes (\lambda_1 - \lambda_4)) = R^iF(K \otimes \lambda_1) \otimes - \lambda_4$, by the tensor identity, which is zero, for $i > 0$, by (5.2). Moreover

$R^i F(Y(\lambda_1) \otimes (\lambda_1 - \lambda_4)) = 0$ for $i > 0$ by the tensor identity and (2.1.5). Thus we have:

(7.6.2) *the sequence*

$$0 \to F(K \otimes (\lambda_1 - \lambda_4)) \to F(Y(\lambda_1) \otimes (\lambda_1 - \lambda_4) \to F(Q \otimes (\lambda_1 - \lambda_4)) \to 0$$

is exact and $R^i F(Q \otimes (\lambda_1 - \lambda_4)) = 0$ *for all* $i > 0$.

Moreover $F(K \otimes (\lambda_1 - \lambda_4))$ has a filtration in which the successive quotients are $F(-\lambda_4)$, $F(Y_1(-1,1,0,0) \otimes - \lambda_4)$, $F(\lambda_1 - \lambda_4)$, $F(Y_1(0,0,0,1) \otimes - \lambda_4)$. Thus restricting to P_Γ for $\Gamma = \{\alpha_2,\alpha_3\}$ then applying F we see that $F(K \otimes (\lambda_1 - \lambda_4))$ has a good filtration with successive quotients $Y_4(0,0,0,-1)$, $Y_4(0,1,0,-2)$, $Y_4(1,0,0,-1)$, $Y_4(0)$, $Y_4(0,0,1,-2)$, $Y_4(2,0,0,-2)$. Let $0 \neq \theta \in \mathrm{Hom}_{P_\Omega}(F(Y(\lambda_1) \otimes (\lambda_1 - \lambda_4)),Y/M)$ (such a θ exists by (7.6.1). If $\theta(F(K \otimes (\lambda_1 - \lambda_4)) \neq 0$ then Y/M contains one of the following P_Ω modules in its socle: $L_4(-\lambda_4)$, $L_4(0,1,0,-2)$, $L_4(1,0,0,-1)$, $L_4(0)$, $L_4(0,0,1,-2)$, $L_4(2,0,0,-2)$. It follows that Y/M contains τ in the B socle for some $\tau \in \{-\lambda_4,-\lambda_2,-\lambda_1,0,-\lambda_3 + \lambda_4,-2\lambda_1\}$. Hence $((Y/M) \otimes \tau)^B \neq 0$ for some $\tau \in S = \{\lambda_4,\lambda_2,\lambda_1,0,\lambda_3 - \lambda_4,2\lambda_1\}$.

Now if $\tau \in S$ and $\tau \neq \lambda_2$ then $H^i(B,Y \otimes \tau) = 0$ for all $i \geq 0$, by the fact that the B-socle of Y is $-\lambda_2$, (2.1.6) and, for example, (3.3') Corollary of [12]. Also $(M \otimes \lambda_2)^B = (Y\ )\ \lambda_2)^B = k$ and $H^i(B,Y \otimes \lambda_2) = 0$ for all $i > 0$ (by (3.3') Corollary of [12]). It follows, from the long exact sequence of cohomology, that $((Y/M) \otimes \tau)^B = H^1(B,M \otimes \tau)$ for all $\tau \in S$. Also $H^i(B,Y(\lambda_1) \otimes \tau - \lambda_4) = 0$ for all $i > 0$ and $\tau \in S$ by (3.3') Corollary of [12], (2.1.6) and (2.3.3). Hence $((Y/M) \otimes \tau)^B \cong H^2(B,Y_4(1,0,0,-2) \otimes \tau)$ by (7.4.5) and the long exact sequence of cohomology. For $\tau = 0$, $H^2(B,Y_4(1,0,0,-2) \otimes \tau) = H^2(B,(1,0,0,-2)) = 0$, by (2.1.3), (2.3.3) and (2.1.6). For $\tau = \lambda_4$, $Y_4(1,0,0,-2) \otimes \tau \cong Y_4(1,0,0,-1)$ so that

$H^2(B,Y_4(1,0,0,-2) \otimes \tau) = 0$ by (2.1.6) and (2.1.3). Also $((Y/M) \otimes 2\lambda_1)^B = 0$ since $-2\lambda_1$ is not a weight of $Y(\lambda_2)$. Thus, by (2.1.3), we must have $H^2(B,Y_4(1,0,0,-2) \otimes Y_4(\tau)) \neq 0$ for some $\tau = \lambda_2, \lambda_1, \lambda_3 - \lambda_4$. Now by (2.2.3), (2.2.5), $\chi_4(1,0,0,-2)\chi_4(\chi_2) = \chi_4(1,1,0,-2) + \chi_4(0,0,1,-2) + \chi_4(1,0,0,-1)$. Hence $Y_4(1,0,0,-2) \otimes Y_4(\lambda_2)$ has a good filtration with successive quotients $Y_4(1,1,0,-2)$, $Y_4(0,0,1,-2)$, $Y_4(1,0,0,-1)$. It now follows by (2.1.3), (2.3.3) and (2.1.6) that $H^2(B,Y_4(1,0,0,-2)) \otimes Y_4(\lambda_2)) = 0$. Similar arguments show that $H^2(B,Y_4(1,0,0,-2) \otimes Y_4(\tau)) = 0$ for $\tau = \lambda_1$ and $\tau = \lambda_3 - \lambda_4$. This is a contradiction and so we must have $\theta(F(K \otimes (\lambda_1 - \lambda_4)) = 0$. Thus by (7.6.2):

(7.6.3) θ *induces a non-zero* P_Ω *map* $\bar{\theta}: F(Q \otimes (\lambda_1 - \lambda_4)) \to Y/M$.

As a P_1 module Q has a filtration with quotients $Y_1(-1,1,0,0)$ and $Y_1(1,0,0,0)$. Thus as a P_Γ module, where $\Gamma = \{\alpha_2,\alpha_3\}$, Q has a filtration with successive quotients $Y_\Gamma(0,1,0,-1)$, $Y_\Gamma(-1,1,0,0)$, $Y_\Gamma(1,0,0,0)$. Thus, by (2.1.2) and (1.7.4),

(7.6.4) $R^i F(Q \otimes (\lambda_1 - \lambda_4))$ *has a good filtration with successive quotients* $Y_4(1,1,0,-2)$, $Y_4(0,1,0,-1)$, $Y_4(2,0,0,-1)$ *if* $i = 0$ *and is* 0 *if* $i > 0$.

We claim:

(7.6.5) *the* B-*socle of* $F(Q \otimes (\lambda_1 - \lambda_4))$ *is* $(-1,-1,0,1)$.

The B-socle of $Y_4(1,1,0,-2)$ is $(-1,1,0,1)$, that of $Y_4(0,1,0,-1)$ is $(0,-1,0,1)$ and that of $Y_4(2,0,0,-1)$ is $(-2,0,0,1)$ so that it suffices to prove that $(F(Q \otimes (\lambda_1 - \lambda_4)) \otimes \tau)^B = 0$ for $\tau \in T = \{(0,1,0,-1), (2,0,0,-1)\}$. Now, by the reciprocity and tensor identity for induction and (2.1.3), $(F(Q \otimes (\lambda_1 - \lambda_4) \otimes \tau)^B \cong (Q \otimes (\lambda_1 - \lambda_4) \otimes Y_4(\tau))^B$ so it suffices to prove that the latter is zero. Hence, by the tensor identity and reciprocity of induction and (2.1.3), it suffices to prove that $(Q \otimes G(Y_4(\tau) \otimes (\lambda_1 - \lambda_4)))^B = 0$ where $G = \mathrm{Ind}_B^{P_1}$. Let $\Gamma = \{\alpha_2\ \alpha_3\}$. As a P_Γ module $Y_4(0,1,0,-1)$ has a filtration with successive quotients $Y_\Gamma(-2,1,0,0)$, $Y_\Gamma(0)$, $Y_\Gamma(-1,0,1,-1)$, $Y_\Gamma(0,1,0,-1)$ and so $Y_4(0,1,0,-1) \otimes (\lambda_1 - \lambda_4)$ has a filtration with successive quotients $Y_\Gamma(-1,1,0,-1)$, $Y_\Gamma(1,0,0,-1)$, $Y_\Gamma(0,0,1,-2)$, $Y_\Gamma(1,1,0,-2)$. Hence by (1.7.1), (2.1.5), (2.3.1) and the long exact sequence of cohomology:

(7.6.6) *$R^iG(Y_4(0,1,0,-1) \otimes (\lambda_1 - \lambda_4))$ is 0 for $i \neq 1$ and k for $i = 1$.*

Now $Y_4(2,0,0,-1) \otimes (\lambda_1 - \lambda_4)$ as a P_Γ module has a filtration with successive quotients $Y_\Gamma(0,0,0,-1)$, $Y_\Gamma(-1,1,0,-1)$, $Y_\Gamma(1,1,0,-2)$, $Y_\Gamma(0)$, $Y_\Gamma(-1,2,0,-2)$, $Y_\Gamma(3,0,0,-2)$. It follows that

(7.6.7) *$R^iG(Y_4(2,0,0,-1) \otimes (\lambda_1 - \lambda_4))$ is k for $i = 0$ and 0 for $i > 0$.*

Thus, by (7.3.6) and 7.3.7), to prove that $(Q \otimes G(Y_4(\tau) \otimes (\lambda_1 - \lambda_4)))^B = 0$ for $\tau \in T$ it suffices to show that $Q^B = 0$. However this is true by (1.5.2) since Q has a filtration with successive quotients $Y_1(-1,1,0,0)$ and $Y_1(1,0,0,0)$, as a P_1 module. This completes the proof of (7.6.5).

We record:

(7.6.8) *the W_4 orbit of $(1,1,0,-2)$ is:* (1,1,0,-2),(-1,2,0,-2), (2,-1,1,-2),(1,-2,2,-2),(-2,1,1,-2),(2,1,-1,-1),(-1,-1,2,-1),(1,2,-2,0), (-2,3,-1,-1),(3,-1,0,-1),(-1,3,-2.0),(3,-2,0,0),(1,-3,2,-1),(-3,2,0,-1), (2,-3,1,0),(-3,1,0,0),(-1,-2,2,-1),(1,1,-2,1),(-2,-1,1,0),(2,-1,-1,1), (-1,2,-2,1),(-2,1,-1,1),(1,-2,0,1),(-1,-1,0,1) .

Hence, by (2.2.3) and (2.2.5),

$$s_4(1,1,0,-2) = \chi_4(1,1,0,-2) - 2\chi_4(0,0,1,-2) - 2\chi_4(1,0,0,-1)$$

and so

(7.6.9) $\quad \chi_4(1,0,0,-2) = s_4(1,1,0,-2) + 2\chi_4(0,0,1,-2) + 2\chi_4(1,0,0,-1)$.

We now claim:

(7.6.10) $\quad \overline{\theta}: F(Q \otimes (\lambda_1 - \lambda_4)) \to Y/M$ *is a monomorphism.*

If this is not so then, from (7.6.5), $\overline{\theta}$ is zero on the P_4-socle $L_4(1,0,0,-2)$ of $F(Q \otimes (\lambda_1 - \lambda_4))$. Thus some composition factor $L(\tau)$, say, of $F(Q \otimes (\lambda_1 - \lambda_4))$ other than $L_4(1,1,0,-2)$ occurs in the P_4 socle of Y/M . However, if $L_4(\tau)$ is a composition factor of $Y_4(0,1,0,-1)$ then $\tau \in \{(0,1,0,-1),0\}$, if $L_4(\tau)$ is a composition factor of $Y_4(2,0,0,-1)$ then $\tau \in \{(2,0,0,-1),(0,1,0,-1),0\}$ and if $L_4(\tau)$ is a composition factor of $Y_4(1,1,0,-2)$ different from $L_4(1,1,0,-2)$ then, by (7.6.8), $\tau \in \{(0,0,1,-2),(1,0,0,-1)\}$. Thus if $\overline{\theta}$ is not a monomorphism then $L_4(\tau)$ occurs in the socle of Y/M for some

$$\tau \in \{(0,0,1,-2),(1,0,0,-1),0,(0,1,0,-1),(2,0,0,-1)\}$$

and so it suffices to show that $((Y/M) \otimes \tau)^B = 0$ for $\tau \in \mathcal{U} = \{(0,0,1,-1),(1,0,0,0),0,(0,1,0,-1),(2,0,0,-1)\}$. However $(Y \otimes \tau)^B = 0$ for all $\tau \in \mathcal{U}$ so (7.6.10) follows from:

(7.6.11) $\quad H(B, M \otimes \tau) = 0$ *for all* $\tau \in \mathcal{U}$ *and* $i > 0$.

We must now prove (7.6.11). For $i > 0$ and $\tau \in \mathcal{U}$, $H^i(B, Y(\lambda_1) \otimes (\tau - \lambda_4)) = 0$ by (1.5.1), (1.5.2), (2.1.6) and (2.3.3) and so, by (7.5.5), it suffices to show that $H^i(B, Y_4(1,0,0,-2) \otimes \tau) = 0$ for all $\tau \in \mathcal{U}$ and $i > 1$. By (2.1.3), $H^i(B, Y_4(1,0,0,-2) \otimes (0,0,1,-2)) = H^i(B, Y_4(1,0,0,-2) \otimes Y_4(0,0,1,-2))$. However, $Y_4(1,0,0,-2) \otimes Y_4(0,0,1,-2)$ has a good filtration with successive quotients $Y_4(0,1,0,-3)$, $Y_4(1,0,1,-4)$ and $H^i(B, Y_4(0,1,0,-3)) = H^i(B, Y_4(1,0,1,-4)) = 0$ for all $i \geq 0$, by (2.1.3), (2.1.6), (2.3.3) and (5.4). Similarly one can show that $H^i(B, Y_4(1,0,0,-2) \otimes \tau) = 0$ for all $i \geq 0$ and $\tau \in \mathcal{U}$, completing the proof of (7.6.11).

We let N/M be the image of $\bar{\theta}$. By (7.6.4) and (7.6.10), N/M has a good filtration with successive quotients $Y_4(1,1,0,-2)$, $Y_4(0,1,0,-1)$, $Y_4(2,0,0,-1)$. Now by characters $N \subseteq F_\Omega^0 Y$ and by (7.2.5), the character of $F_\Omega^0 Y/N$ is $\chi_4(1,0,1,-2)$. To show

(7.6.12) $F_\Omega^0 Y/N$ *is isomorphic to* $Y_4(1,0,1,-2)$

it suffices to show, by (1.5.4), that $((F_\Omega^0 Y/N \otimes \tau)^B = 0$ for all $\tau \in X_\Omega$ such that $-\tau$ is a weight of $Y_4(1,0,1,-2)$ and $-\tau$ is not equal to the bottom weight of $Y_4(1,0,1,-2)$. Thus it suffices to show that $((Y/N \otimes \tau)^B = 0$ for $\tau \in \{0,(0,1,0,-1),(2,0,0,-1)\}$. However, for such τ, $H^i(B,Y \otimes \tau) = 0$ for all $i \geq 0$ so it suffices to show that $H^1(B,N \otimes \tau) = 0$. Moreover, we have already **shown** $H^1(B,M \otimes \tau)$ to be zero for $\tau \in \mathcal{U}$, (7.6.11), so that it suffices to show that $H^1(B,F(Q \otimes (\lambda_1 - \lambda_4)) \otimes \tau) = 0$ for $\tau \in \mathcal{U} = \{0,(0,1,0,-1),(2,0,0,-1)\}$. By (2.1.3), $H^1(B,F(Q \otimes (\lambda_1 - \lambda_4)) \otimes \tau) = H^1(B,F(Q \otimes (\lambda_1 - \lambda_4)) \otimes Y_4(\tau))$ for $\tau \in \mathcal{U}$. Thus it suffices to show that the latter is zero. In fact we shall show:

(7.6.13) $H^i(B,F(Q \otimes (\lambda_1 - \lambda_4)) \otimes Y_4(\tau)) = 0$ *for all* $i \geq 0$ *and* $\tau \in \mathcal{U}$.

By (2.1.3)
$H^i(B,F(Q \otimes (\lambda_1 - \lambda_4)) \otimes Y_4(\tau)) \cong H^i(P_4,F(Q \otimes (\lambda_1 - \lambda_4) \otimes Y_4(\tau)))$. By (1.1.6), (1.1.7) and (7.6.4) the latter is isomorphic to $H^i(B,Q \otimes (\lambda_1 - \lambda_4) \otimes Y_4(\tau))$. By (1.1.6), (1.1.5) it suffices to show that $H^i(P_1,R^j G((\lambda_1 - \lambda_4) \otimes Y_4(\tau)) \otimes Q)$ for all $i,j \geq 0$, $\tau \in \mathcal{U}$. For $\tau = 0$ this is true by the tensor identity and (2.1.5). By (7.6.6), (7.6.7) it thus suffices to prove that $H^i(P_1,Q) = 0$ for all $i \geq 0$. Now Q has a filtration with quotients $Y_1(1,0,0,0)$, $Y_1(-1,1,0.0)$, $H^i(P_1,Y_1(-1,1,0,0)) = 0$ for $i \geq 0$ by (2.1.6) and $H^i(P_1,Y_1(1,0,0,0)) = 0$ for $i \geq 0$ by (2.1.3) and (1.5.2). Thus $H^i(P_1,Q) = 0$ for all $i \geq 0$, completing the proof of (7.6.13) and

(7.6.12).

We now have that $F^{O}_{\Omega}Y/N$ is isomorphic to $Y_4(1,0,1,-2)$. Thus $F^{O}_{\Omega}Y$ has a good filtration and so, by (7.2.6) and the argument of Proposition 4.7.1:

(7.6.14) $Y(\lambda_2)|_{P_\Omega}$ *has a good filtration.*

By Proposition 3.2.4, $Y/F^{O}_{\Omega}Y$ has a good filtration. By (7.5.1) the character of $Y/F^{O}_{\Omega}Y$ is $\chi_4(1,0,0,0) + \chi_4(1,1,0,-1) + \chi_4(0,1,0,0)$ and so, by Proposition 3.2.6, there is a P_Ω submodule S of Y containing $F^{O}_{\Omega}Y$ such that $S/F^{O}_{\Omega}Y$ is isomorphic to $Y_4(1,0,0,0)$. We conclude this section with a proof that

(7.6.15) *the extension* S/N *of* $Y_4(1,0,1,-2)$ *by* $Y_4(1,0,0,)$ *is non-split.*

If the extension is split then Y/N contains a copy of $Y_4(1,0,0,0)$ and so $((Y/N) \otimes (1,0,0,-1))^B$ is non-zero ($(-1,0,0,1)$ is the lowest weight of $Y_4(1,0,0,0)$). Now $H^i(B,Y \otimes (\lambda_1 - \lambda_4)) = 0$ for all $i \geq 0$ so to establish (7.6.16) it suffices to show that $H^1(B,N \otimes (\lambda_1 - \lambda_4)) = 0$. We have $H^i(B,Y(\lambda_1) \otimes (\lambda_1 - 2\lambda_4)) = 0$ for all $i \geq 0$ by (2.1.6) and (2.3.3). Hence by (7.5.5), $H^i(B,M \otimes (\lambda_1 - \lambda_4)) \cong H^{i+1}(B,Y_4(1,0,0,-2) \otimes (\lambda_1 - \lambda_4))$ for all $i \geq 0$. Now, for $j \geq 0$,
$H^j(B,Y_4(1,0,0,-2) \otimes (\lambda_1 - \lambda_4)) \cong H^j(B,Y_4(1,0,0,-2) \otimes Y_4(1,0,0,-1))$, by (2.1.3), and $Y_4(1,0,0,-2) \otimes Y_4(1,0,0,-1)$ has a good filtration with quotients $Y_4(2,0,0,-3)$, $Y_4(0,1,0,-3)$, $Y_4(0,0,0,-2)$. Using (2.1.6) and (2.3.3) one can now see that $H^j(B,Y_4(\tau)) = 0$ for all $j \geq 0$ and each successive quotient $Y_4(\tau)$. Hence $H^i(B,M \otimes (\lambda_1 - \lambda_4)) = 0$ for all $i \geq 0$, in particular $H^1(B,M \otimes (\lambda_1 - \lambda_4)) = 0$. Thus to prove (7.6.15) it suffices to show that $H^1(B,(N/M) \otimes (\lambda_1 - \lambda_4)) = 0$. By (7.5.10), N/M is isomorphic to $F(Q \otimes (\lambda_1 - \lambda_4))$ so it suffices to show that $H^1(B,F(Q \otimes (\lambda_1 - \lambda_4)) \otimes (\lambda_1 - \lambda_4)) = 0$. By (2.1.3), for $i \geq 0$,

$H^i(B, F(Q \otimes (\lambda_1 - \lambda_4)) \otimes (\lambda_1 - \lambda_4)) = H^i(P_4, F(Q \otimes (\lambda_1 - \lambda_4)) \otimes Y_4(\lambda_1 - \lambda_4))$ and by (7.6.2) and (1.1.6), (1.1.7) this is isomorphic to $H^i(B, Q \otimes (\lambda_1 - \lambda_4) \otimes Y_4(\lambda_1 - \lambda_4))$. Hence by (1.1.6), (1.1.7) it suffices to show that $H^i(P_1, Q \otimes R^j G(Y_4(\lambda_1 - \lambda_4) \otimes (\lambda_1 - \lambda_4)) = 0$ for all $i, j \geq 0$. Now, as a P_Γ module, $Y_4(\lambda_1 - \lambda_4)$ has a filtration with quotients $Y_\Gamma(1,0,0,-1)$, $Y_\Gamma(-1,1,0,-1)$, $Y_\Gamma(-1,0,0,0)$ (here $\Gamma = \{\alpha_2, \alpha_3\}$) . Hence $Y_4(\lambda_1 - \lambda_4) \otimes (\lambda_1 - \lambda_4)$ has a filtration with quotients $Y_\Gamma(2,0,0,-2)$, $Y_\Gamma(0,1,0,-2)$, $Y_\Gamma(0,0,0,-1)$. We now obtain $R^j G(Y_4(\lambda_1 - \lambda_4) \otimes (\lambda_1 - \lambda_4)) = 0$ for all $j \geq 0$ by (2.1.1), (2.1.5), (2.3.1). This completes the proof of (7.6.15).

7.7 M , N/E and Y/S

By (7.6.4) and (7.6.10), N/M has a good filtration with quotients $Y_4(1,1,0,-2)$, $Y_4(0,1,0,-1)$, $Y_4(2,0,0,-1)$. Hence by Proposition 3.2.6, there is a P_Ω submodule E/M with an ascending filtration with quotients $Y_4(1,1,0,-2)$, $Y_4(0,1,0,-1)$ and such that N/E is isomorphic to $Y_4(2,0,0,-1)$. Let $\Sigma = \Delta \backslash \{\alpha_j\}$ for some $1 \leq j < 4$ and $\mathcal{D} = Ind_B^{P_\Sigma}$. We wish to show that $Y|_{P_\Sigma} \cong \mathcal{D}(Y|_B)$ has a good filtration. By the long exact sequence of derived functors it will be enough to show that $\mathcal{D}(M)$, $\mathcal{D}(E/M)$, $\mathcal{D}(N/E)$, $\mathcal{D}(S/N)$, $\mathcal{D}(Y/S)$ each has a good filtration and that $R\mathcal{D}(M) = R\mathcal{D}(E/M) = R\mathcal{D}(N/E) = R\mathcal{D}(S/N) = 0$. In this section we deal with the simpler cases.

(7.7.1) *$R^i Ind_B^{P_\Sigma}(M)$ has a good filtration for $i = 0$ and is 0 for $i > 0$, where $\Sigma = \Delta \backslash \{\alpha_j\}$ and $1 \leq j < 4$.*

We have $R^i\mathcal{D}(Y(\lambda_1) \otimes - \lambda_4) = 0$ for all $i \geq 0$ by (2.1.5) so that, by (7.5.5), $R^i\mathcal{D}(M) \cong R^{i+1}\mathcal{D}(Y_4(1,0,0,-2))$. Now if τ is any weight of $Y_4(1,0,0,-2)$ then $(\tau, \alpha_4^\vee) = -1$ or -2 . Hence $\mathcal{D}(\tau) = 0$ for any weight τ . Moreover, using (2.1.5), (2.3.1), (2.1.3) and (1.7.4) we have $R^i\mathcal{D}(\tau) = 0$ for any weight τ of $Y_4(1,0,0,-2)$ and $i \geq 2$.

Hence $H^i\mathcal{D}(Y_4(1,0,0,-2)) = 0$ for $i \neq 1$ and $R\mathcal{D}(Y_4(1,0,0,-2))$ has a filtration with successive quotients $R\mathcal{D}(\tau)$ for τ a weight of $Y_4(1,0,0,-2)$. By (2.3.1), (2.1.5) this is a good filtration. Hence $R^i\mathcal{D}(M) = 0$ for $i > 0$ and $\mathcal{D}(M)$ has a good filtration.

We also have, from (5.2):

(7.7.2) *$R^i Ind_B^P$ (N/E) and $R^i Ind_B^P$ (Y/S) each has a good filtration for $i = 0$ and are both zero for $i > 0$, where $\Sigma = \Delta\backslash\{\alpha_j\}$ and $1 \leq j < 4$.*

7.8 $\Sigma = \Delta\backslash\{\alpha_3\}$

We now suppose that $\Sigma = \Delta\backslash\{\alpha_3\}$ and let $\Gamma = \Sigma \cap \Omega = \{\alpha_1,\alpha_2\}$ and put $\mathcal{D} = Ind_B^{P_\Sigma}$.

(7.8.1) *For $\Sigma = \Delta\backslash\{\alpha_3\}$, $R^i Ind_B^{P_\Sigma}(E/M)$ has a good filtration for $i = 0$ and is 0 for $i > 0$.*

We identify $Y_4(1,1,0,-2)$ and hence $M_\Gamma(1,1,0,-2)$ (see (5.8)) with a submodule of $\bar{E} = E/M$. Using (7.6.8), (7.6.9) and (2.2.3), (2.2.5) it is not difficult to show that when $\chi_4(1,1,0,-2)$ is expressed as a sum of $\chi_\Gamma(\tau)$'s for various $\tau \in X_\Gamma$ any occurring τ different from $(1,1,0,-2)$ satisfies $(\tau,\alpha_4^\vee) \geq -1$. By (5.8) and (5.2) $R^i\mathcal{D}(M_\Gamma(1,1,0,-2))$ has a good filtration for $i = 0$ and is 0 for $i > 0$. Thus to prove (7.8.1) it suffices to show that $R^i\mathcal{D}(F)$ has a good filtration if $i = 0$ and is 0 for $i > 0$, where $F = \bar{E}/M_\Gamma(1,1,0,-2)$. We know, by (5.1.2), (7.6.5) and (7.6.10), that F is a non-split P_Γ-extension of $Y_\Gamma(1,1,0,-2)$ by $Y_4(0,1,0,-1)$. Now $Y_4(0,1,0,-1)$ has an ascending filtration with quotients $Y_\Gamma(1,0,-1,1)$, $Y_\Gamma(1,1,-1,0)$, $Y_\Gamma(0,1,0,-1)$. We have $Ext^1_{P_\Gamma}(Y_\Gamma(1,0,-1,1), Y_\Gamma(1,1,0,-2)) = 0$ by Lemma 3.2.1. Also $Ext^1_{P_\Gamma}(Y_\Gamma(0,1,0,-1), Y_\Gamma(1,1,0,-2)) \cong H^1(P_\Gamma, Y_\Gamma(1,0,-1,1) \otimes Y_\Gamma(1,1,0,-2))$ and $Y_\Gamma(1,0,-1,1) \otimes Y_\Gamma(1,1,0,-2)$ has a filtration with quotients

$Y_\Gamma(1,0,0,-1)$, $Y_\Gamma(0,2,-1,-1)$, $Y_\Gamma(2,1,-1,-1)$ so that $Ext^1_{P_\Gamma}(Y_\Gamma(0,1,0,-1)$, $Y_\Gamma(1,1,0,-2)) = 0$ by (2.1.3) and (2.1.6). Thus, as a P_Γ module, F has an ascending filtration $0 = F_0, F_1, F_2, F_3 = F$ where $F_1/F_0 \cong Y_\Gamma(1,0,-1,1)$, F_2/F_1 is a non-split extension of $Y_\Gamma(1,1,0,-2)$ by $Y_\Gamma(1,1,-1,0)$ and $F_3/F_2 \cong Y_\Gamma(0,1,0,-1)$. Thus, to prove $R^i\mathcal{D}(F)$ has a good filtration for $i = 0$ and is 0 for $i > 0$ it suffices, by (2.1.3) and (2.1.5), to show that $R^i\mathcal{D}(\overline{F})$ has a good filtration for $i = 0$ and is 0 for $i > 0$, where $\overline{F} = F_2/F_1$. Now we have, by (5.10), $Ext^1_{P_\Gamma}(Y_\Gamma(1,1,-1,0), Y_\Gamma(1,1,0,-2)) = Ext^1_B((1,1,-1,0),(1,1,0,-2)) \cong H^1(B,-\alpha_4)$ which is isomorphic to k by (1.4.7). By (5.11), $\overline{F}$ is the unique non-split extension of $Y_\Gamma(1,1,0,-2)$ by $Y_\Gamma(1,1,-1,0)$. However, $Y_3(1,1,-1,1) \otimes -\lambda_4$ is a non-split extension of $Y_\Gamma(1,1,0,-2)$ by $Y_\Gamma(1,1,-1,0)$ and so $\overline{F} \cong Y_3(1,1,-1,1) \otimes -\lambda_4$. Hence $R^i\mathcal{D}(\overline{F}) = 0$ for all i by (2.1.5) and the tensor identity. This completes the proof of (7.8.2).

We now show:

(7.8.2) $R^i Ind_B^{P_\Sigma}(S/N)$ *has a good filtration for* $i = 0$ *and is* 0 *for* $i > 0$, *where* $\Sigma = \Delta\backslash\{\alpha_3\}$.

We identify $Y_4(1,0,1,-2)$ and hence $M_\Gamma(1,0,1,-2)$ (see (5.8)) with a submodule of $\overline{S} = S/N$. It is easy to check that, when $\chi_4(1,0,1,-2)$ is expressed as a sum of $\chi_\Gamma(\tau)$'s for $\tau \in X_\Gamma$, any occurring τ other than $(1,0,1,-2)$ satisfies $(\tau,\alpha_4^\vee) \geq -1$. By (5.8) and 5.2), $R^i\mathcal{D}(M_\Gamma(1,0,1,-2))$ has a good filtration for $i = 0$ and is 0 for $i > 0$. Thus it suffices to prove that $R^i\mathcal{D}(L)$ has a good filtration for $i = 0$ and is 0 for $i > 0$, where $L = \overline{S}/M_\Gamma(1,0,1,-2)$ - a non-split extension of $Y_\Gamma(1,0,1,-2)$ by $Y_4(1,0,0,0)$, by (5.12) and (7.1.15). Now $Y_4(1,0,0,0)$ as a P_Γ-module has an ascending filtration with quotients $Y_\Gamma(0,1,-1,1)$, $Y_\Gamma(1,0,0,0)$. Moreover, $Ext^1_{P_\Gamma}(Y_\Gamma(0,1,-1,1), Y_\Gamma(1,0,1,-2)) = 0$ by Lemma 3.2.1 so that L contains a P_Γ-submodule L_1 isomorphic to $Y_\Gamma(0,1,-1,1)$ and

$\overline{L} = L/L_1$ is a non-split extension of $Y_\Gamma(1,0,1,-2)$ by $Y_\Gamma(1,0,0,0)$. By (2.1.3), (1.7.4), $R^i\mathcal{D}(Y_\Gamma(0,1,-1,1))$ is isomorphic to $Y_\Omega(0,1,-1,1)$ if $i = 0$ and is 0 if $i > 0$. Thus it suffices to show that $R^i\mathcal{D}(\overline{L})$ has a good filtration if $i = 0$ and is 0 if $i > 0$. By (5.10),

$Ext^1_{P_\Gamma}(Y_\Gamma(1,0,0,0),Y_\Gamma(1,0,1,-2)) \cong Ext^1_B((1,0,0,0),(1,0,1,-2)) \cong H^1(B,-\alpha_4)$,

which is k by (1.4.7). By (5.11), $\overline{L}$ is the unique non-split extension of $Y_\Gamma(1,0,1,-2)$ by $Y_\Gamma(1,0,0,0)$. However, $Y_3(1,0,0,1) \otimes -\lambda_4$ is such an extension and so $\overline{L}$ is isomorphic to $Y_3(1,0,0,1) \otimes -\lambda_4$. Thus $R^i\mathcal{D}(\overline{L}) = 0$ for all i , by (2.1.5) and the tensor identity. The pooof is complete.

7.9 $\Sigma = \Delta\backslash\{\alpha_2\}$

We now take $\Sigma = \Delta\backslash\{\alpha_2\}$, $\Gamma = \Omega \cap \Sigma = \{\alpha_1,\alpha_3\}$ and $\mathcal{D} = Ind_B^{P_\Sigma}$. We shall show:

(7.9.1) $R^i Ind_B^{P_\Sigma}(E/M)$ *has a good filtration if* $i = 0$ *and is* 0 *if* $i > 0$, *where* $\Sigma = \Delta\backslash\{\alpha_2\}$.

We leave it to the reader to check that

$$\chi_4(1,1,0,-2) = \chi_\Gamma(1,1,0,-2) + \chi_\Gamma(2,-1,1,-2) + \chi_\Gamma(1,-2,2,-2) + \chi_\Gamma(0,0,1,-2) + \chi$$

where χ is a sum of $\chi_\Gamma(\tau)$'s for various $\tau \in X_\Gamma$ and any occurring τ satisfies $\tau \not\leq (0,0,1,-2)$ and $(\tau,\alpha_4^\vee) \geq -1$. By Proposition 3.2.4 there is a P_Γ submodule C of $\overline{E} = E/M$ with a good filtration such that, identifying $Y_4(1,1,0,-2)$ with a submodule of E, $Y_4(1,1,0,-2)/C$ has a good filtration with successive quotients $Y_\Gamma(0,0,1,-2)$, $Y_\Gamma(1,-2,2,-2)$, $Y_\Gamma(2,-1,1,-2)$, $Y_\Gamma(1,1,0,-2)$. By (5.2), $R^i\mathcal{D}(C)$ has a good filtration for $i = 0$ and is 0 for $i > 0$. Thus, to prove (7.9.1), it suffices to show that $R^i\mathcal{D}(F)$ has a good filtration for $i = 0$ and is 0 for $i > 0$, where $F = \overline{E}/C$. Now $Y_4(0,1,0,-1)$, as a P_Γ-module, has an ascending filtration in which the successive quotients are $Y_\Gamma(1,-2,1,0)$, $Y_\Gamma(0)$, $Y_\Gamma(2,-1,0,0)$, $Y_\Gamma(1,-1,1,-1)$, $Y_\Gamma(0,1,0,-1)$. Thus F has a descending filtration $\{F_i : i = 0,1,\dots,9\}$ such that $F_0 = F$, $F_0/F_1 \cong Y_\Gamma(0,1,0,-1)$, $F_1/F_2 \cong Y_\Gamma(1,-1,1,-1)$, $F_2/F_3 \cong Y_\Gamma(2,-1,0,0)$, $F_3/F_4 \cong Y_\Gamma(0)$, $F_4/F_5 \cong Y_\Gamma(1,-1,1,0)$, $F_5/F_6 \cong Y_\Gamma(1,1,0,-2)$, $F_6/F_7 \cong Y_\Gamma(2,-1,1,-2)$, $F_7/F_8 = Y_\Gamma(1,-2,2,-2)$, $F_8/F_9 = Y_\Gamma(0,0,1,-2)$, $F_9 = 0$. Moreover $F_6 = M_\Gamma(1,1,0,-2)/C$, so that the extension F/F_6 of F_5/F_6 by F/F_5 is non-split, by (5.12). However, $Ext^1_{P_\Gamma}(Y_\Gamma(\tau),Y_\Gamma(1,1,0,-2)) = 0$ for $\tau = (1,-2,1,0)$, 0 , $(2,-1,0,0)$ by Lemma 3.2.1. Also $Ext^1_{P_\Gamma}(Y_\Gamma(0,1,0,-1),Y_\Gamma(1,1,0,-2)) = H^1(P_\Gamma,Y_\Gamma(1,0,0,-1))$ which is zero by (2.1.3) and (2.1.6). Thus we may rechoose the descending filtration F_i such that $F_0/F_1 \cong Y_\Gamma(0,1,0,-1)$, F_1/F_2 is a non-split extension of $Y_\Gamma(1,1,0,-2)$ by $Y_\Gamma(1,-1,1,-1)$, $F_2/F_3 \cong Y_\Gamma(2,-1,0,0)$, $F_3/F_4 \cong Y_\Gamma(0)$, $F_4/F_5 \cong Y_\Gamma(1,-2,1,0)$, $F_5/F_6 \cong Y_\Gamma(2,-1,1,-2)$, $F_6/F_7 \cong Y_\Gamma(1,-2,2,-2)$, $F_7/F_8 \cong Y_\Gamma(0,0,1,-2)$ and $F_8 = 0$. Now $Ext^1_{P_\Gamma}(Y_\Gamma(1,-1,1,-1),Y_\Gamma(1,1,0,-2)) \cong H^1(P_\Gamma,Y_\Gamma(1,-2,1,0) \otimes Y_\Gamma(1,1,0,-2))$ and $Y_\Gamma(1,-2,1,0) \otimes Y_\Gamma(1,1,0,-2)$ has a good filtration with quotients $Y_\Gamma(2,-1,1,-2)$ and $Y_\Gamma(0,0,1,-2)$. Since $H^i(P_\Gamma,Y_\Gamma(2,-1,1,-2)) = 0$ for all $i \ge 0$, by (2.1.3) and (2.1.6), we have $Ext^1_{P_\Gamma}(Y_\Gamma(1,-1,1,-1),Y_\Gamma(1,1,0,-2)) = H^1(P_\Gamma,Y_\Gamma(0,0,1,-2))$ which, by (2.1.3) and (1.4.7) is k . Thus, by (5.11), F_1/F_2 is the unique non-split extension of $Y_\Gamma(1,1,0,-2)$ by $Y_\Gamma(1,-1,1,-1)$ and hence isomorphic to $Y_2(1,-1,1,0) \otimes -\lambda_4$.

We have $Ext^1_P(Y_\Gamma(\tau),Y_\Gamma(2,-1,1,-2)) = 0$ for $\tau = (1,-2,1,0)$, 0 by Lemma 3.2.1, $Ext^1_{P_\Gamma}(Y_2(1,-1,1,0) \otimes -\lambda_4,Y_\Gamma(2,-1,1,-2))$ $= H^1(P_\Gamma,Y_2(1,-1,1,0)^* \otimes Y_\Gamma(2,-1,1,-1))$ which is zero by (2.1.3) and (2.1.6) and $Ext^1_{P_\Gamma}(Y_\Gamma(0,1,0,-1)Y_\Gamma(2,-1,1,-2)) \cong H^1(P_\Gamma,Y_\Gamma(2,-2,1,-1))$ which is zero by (2.1.3) and (2.1.6). Moreover,

$Hom_{P_\Gamma}(\bar{E},Y_\Gamma(2,-2,1,-1)) \cong Hom_{P_\Omega}(\bar{E},Ind_{P_\Gamma}^{P_\Omega}(Y_\Gamma(2,-2,1,-1)) = 0$ by reciprocity and transitivity of induction and (1.7.1) so that F does not have a P_Γ quotient isomorphic to $Y_\Gamma(2,-2,1,-1)$. Thus we may rechoose the filtration such that $F_o/F_1 \cong Y_\Gamma(0,1,0,-1)$, $F_1/F_2 \cong Y_2(1,-1,1,0) \otimes -\lambda_4$, F_2/F_3 is a non-split extension of $Y_\Gamma(2,-1,1,-2)$ by $Y_\Gamma(2,-1,0,0)$, $F_3/F_4 \cong Y_\Gamma(0)$, $F_4/F_5 \cong Y_\Gamma(1,-2.1,0$, $F_5/F_6 \cong Y_\Gamma(1,-2,2,-2)$, $F_6/F_7 \cong Y_\Gamma(0,0,1,-2)$ and $F_7 = 0$. By (5.10) we have

$Ext^1_{P_\Gamma}(Y_\Gamma(2,-1,0,0),Y_\Gamma(2,-1,1,-2)) \cong Ext^1_B((2,-1,0,0),(2,-1,1,-2))$

$\cong H^1(B,-\alpha_4) = k$. Thus F_2/F_3 is the unique non-split extension of $Y_\Gamma(2,-1,1,-2)$ by $Y_\Gamma(2,-1,0,0)$ and hence isomorphic to $Y_2(2,-1,0,1) \otimes -\lambda_4$.

We have $Ext^1_{P_\Gamma}(Y_\Gamma()),Y_\Gamma(0,0,1,-2)) = 0$ for $\tau = (1,-2,2,-2)$, $(1,-2,1,0)$ by Lemma 3.2.1 and

$Ext^1_{P_\Gamma}(Y_2(\tau) \otimes -\lambda_4,Y_\Gamma(0,0,1,-2)) = H^1(P_\Gamma,Y_2(\tau)^* \otimes Y_\Gamma(0,0,1,-1)) = 0$

for $\tau = (2,-1,0,1)$, $(1,-1,1,0)$, $(0,1,0,0)$ by (2.1.3) and (2.1.6). Moreover $Hom_{P_\Gamma}(\bar{E},Y\ (0,0,1,-2)) = 0$ so that we may rechoose the filtration such that F_i/F_{i+1} is unchanged for $i = 0,1,2,F_3$ F_3/F_4 is a non-split extension of $Y_\Gamma(0,0,1,-2)$ by $Y_\Gamma(0)$, $F_4/F_5 \cong Y_\Gamma(1,-2,1,0)$, $F_5/F_6 \cong Y_\Gamma(1,-2,2,-2)$ and $F_6 = 0$. Now $Ext^1_{P_\Gamma}(Y_\Gamma(0),Y_\Gamma(0,0,1,-2)) \cong H^1(B,-\alpha_4) = k$, by (2.1.3) and (1.4.7). Hence, by (5.11), F_3/F_4 is the unique non-split extension of $Y_\Gamma(0,0,1,-2)$ by $Y_\Gamma(0)$ and so isomorphic to $Y_2(0,0,0,1) \otimes -\lambda_4$.

We have

$Ext^1_{P_\Gamma}(Y_2(\tau) \otimes -\lambda_4,Y_\Gamma(1,-2,2,-2)) \cong H^1(P_\Gamma,Y_2(\tau)^* \otimes Y_\Gamma(1,-2,2,-1))$,

which is zero, for $\tau = (0,1,0,0),(1,-1,1,0),(2,-1,0,o),(0,0,0,1)$ by (2.1.3) and (2.1.6). Moreover $Hom_{P_\Gamma}(\bar{E},Y_\Gamma(1,-2,2,-2)) = 0$ so that the extension F_4 of $Y_\Gamma(1,-2,2,-2)$ by $Y_\Gamma(1,-2,1,0)$ is non-split. Now $R^i\mathcal{D}(Y_2(\tau) \otimes -\lambda_4) = 0$ for all $i \geq 0$ and

$\tau = (0,1,0,0),(1,-1,1,0),(2,-1,0,1),(0,0,0,1)$ by the tensor identity and (2.1.6). Thus, $R^i\mathcal{D}(F_j/F_{j+1}) = 0$ for $0 \le j \le 3$. Hence, to show that $R^i\mathcal{D}(F)$ has a good filtration for $i = 0$ and is zero for $i > 0$, and hence complete the proof of (7.9.1), it suffices to show that $R^i\mathcal{D}(F_4)$ has a good filtration for $i = 0$ and is 0 for $i > 0$.

Now $Ext^1_{P_\Gamma}(Y_\Gamma(1,-2,2,-2),Y_\Gamma(1,-2,1,0)) = Ext^1_B((1,-2,2,-2),(1,-2,1,0))$ $= H^1(B,-\alpha_4) = k$, by (5.10). Hence, (5.11), F_4 is the unique non-split extension of $Y_\Gamma(1,-2,2,-2)$ by $Y_\Gamma(1,-2,1,0)$. Now $Y_2(1,-2,1,1)$, as a P_Γ module, has a good filtration with quotients $Y_\Gamma(1,0,1,-2)$, $Y_\Gamma(1,0,0,0)$, $Y_\Gamma(1,-2,2,-1)$, $Y_\Gamma(1,-2,1,1)$. Hence, by Proposition 3.2.6 there is a P_Γ submodule, say V, of $Y_2(1,-2,1,1)$ such that V is an extension of $Y_\Gamma(1,0,1,-2)$ by $Y_\Gamma(1,0,0,0)$ and $Y_2(1,-2,1,1)/V$ is an extension of $Y_\Gamma(1,-2,2,-1)$ by $Y_\Gamma(1,-2,1,1)$. Since the B socle of $Y_2(1,-2,1,1)$, and hence V, is $(-1,3,-1,-1)$, V is a non-split extension. Since $Hom_{P_\Gamma}(Y_2(1,-2,1,1),Y_\Gamma(1,-2,2,-1)) = 0$, $Y_2(1,-2,1,1)/V$ is also a non-split extension. Thus $(Y_2(1,-2,1,1)/V) \otimes -\lambda_4$ is a non-split extension of $Y_\Gamma(1,-2,2,-2)$ by $Y_\Gamma(1,-2,1,0)$ and hence isomorphic to F_4. Thus we have a short exact sequence of P_Γ modules

$$0 \to V \otimes -\lambda_4 \to Y_2(1,-2,1,1) \otimes -\lambda_4 \to F_4 \to 0 .$$

This gives $R^i\mathcal{D}(F_4) = R^{i+1}\mathcal{D}(V \otimes -\lambda_4)$ for all $i \ge 0$ so to complete the proof of (7.9.1) it suffices to show that $R^i\mathcal{D}(V \otimes -\lambda_4)$ has a good filtration for $j = 1$ and is zero for $j > 1$. Now $Ext^1_{P_\Gamma}(Y_\Gamma(1,0,0,-1),Y_\Gamma(1,0,1,-3)) \cong Ext^1_B((1,0,0,-1),(1,0,1,-3)) \cong$ $H^1(B,-\alpha_4) = k$, by (5.10) and so, by (5.11), $V \otimes -\lambda_4$ is the unique non-split extension of $Y_\Gamma(1,0,1,-3)$ by $Y_\Gamma(1,0,0,-1)$ and so isomorphic to $Y_2(1,0,0,1) \otimes - 2\lambda_4$. We now obtain $R^i\mathcal{D}(V \otimes - \lambda_4) = 0$ for all j by the tensor identity, (1.7.1) and (2.1.5).

We now show:

(7.9.2) $R^i Ind_B^P (S/N)$ *has a good filtration for* $i = 0$ *and is zero for* $i > 0$, *where* $\Sigma = \Delta\backslash\{\alpha_2\}$.

It is not difficult to check that

$$\chi_4(1,0,1,-2) = \chi_\Gamma(1,0,1,-2) + \chi_\Gamma(0,-1,2,-2) + \chi$$

where χ is a sum of $\chi_\Gamma(\tau)$'s for various $\tau \in X_\Gamma$ satisfying $(\tau,\alpha_4^\vee) \geq -1$ and $\tau \not\geq (0,-1,2,-2)$. Thus by Proposition 3.2.6, $Y_4(1,0,1,-2)$ has a P_Γ submodule, say J , with a good filtration such that $Y_4(1,0,1,-2)/J$ has a good filtration with successive quotients $Y_\Gamma(0,-1,2,-2)$, $Y_\Gamma(1,0,1,-2)$. Moreover $R^i\mathcal{D}(J)$ has a good filtration for $i = 0$ and is 0 for $i > 0$ by (5.2). Thus, identifying $Y_4(1,0,1,-2)$ and hence J with a submodule of $\bar{S} = S/N$ we see that to prove (7.9.2), it is enough to show that $R^i\mathcal{D}(L)$ has a good filtration for $i = 0$ and is 0 for $i > 0$, where $L = \bar{S}/J$.

Now $Y_4(1,0,0,0)$, as a P_Γ module, has a good filtration with quotients $Y_\Gamma(1,-1,0,1)$, $Y_\Gamma(0,-1,1,0)$, $Y_\Gamma(1,0,0,0)$. Thus L has a descending filtration $L = L_0,L_1,\ldots,L_5 = 0$ with quotients (in order) $Y_\Gamma(1,0,0,0)$, $Y_\Gamma(0,-1,1,0)$, $Y_\Gamma(1,-1,0,1)$, $Y_\Gamma(1,0,1,-2)$, $Y_\Gamma(0,-1,2,-2)$. Moreover, by (5.8), $L_4 = M(1,0,1,-2)/J$ so that, by (5.12), the extension L/L_4 of L_3/L_4 of L/L_3 is non-split. Now $Ext^1_{P_\Gamma}(Y_\Gamma(\tau),Y_\Gamma(1,0,1,-2)) = 0$ for $\tau = (1,-1,0,1),(0,-1,1,0)$ so that, by Lemma 3.2.1, we may rechoose the filtration such that L_0/L_1 is a non-split extension of $Y_\Gamma(1,0,1,-2)$ by $Y_\Gamma(1,0,0,0)$, $L_1/L_2 \cong Y_\Gamma(0,-1,1,0)$,, $L_2/L_3 \cong Y_\Gamma(1,-1,0,1)$, $L_3/L_4 \cong Y_\Gamma(0,-1,2,-2)$ and $L_4 = 0$. It follows from (5.10) and (5.11) that L_0/L_1 is isomorphic to $Y_2(1,0,0,1) \otimes -\lambda_4$. Also, we have $Ext^1_{P_\Gamma}(Y_\Gamma(1,-1,0,1),Y_\Gamma(0,-1,2,-2)) = 0$ by Lemma 3.2.1 and

$Ext^1_{P_\Gamma}(Y_2(1,0,0,1) \otimes -\lambda_4,Y_\Gamma(0,-1,2,-2)) = H^1(P_\Gamma,Y_2(1,0,0,1)^* \otimes Y_\Gamma(0,-1,2,-1))$ - which is 0 by (2.1.3) and (2.1.6). It is thus possible to rechoose the filtration such that

$L_0/L_1 \cong Y_2(1,0,0,1) \otimes -\lambda_4$, L_1/L_2 is an extension of $Y_\Gamma(0,-1,2,-2)$ by $Y_\Gamma(0,-1,1,0)$ and $L_2 \cong Y_\Gamma(1,-1,0,1)$. Moreover $Hom_{P_\Gamma}(\bar{S},Y_\Gamma(0,-1,2,-2)) = 0$ so that L_1/L_2 is a non-split extension. Now $R^i\mathcal{D}(L_2)$ is isomorphic to $Y_2(1,-1,0,1)$ for $i = 0$ and is zero for $i > 0$, by (2.1.2), (1.7.4) and $R^i\mathcal{D}(Y_2(1,0,0,1) \otimes -\lambda_4) = 0$ for all $i \geq 0$ by (2.1.5) and the tensor identity so that, to complete the proof of (7.9.2) it suffices to show that $R^i\mathcal{D}(\bar{L})$ has a good filtration for $i = 0$ and is zero for $i > 0$, where $\bar{L} = L_1/L_2$. By (5.10), $\bar{L}$ is the unique non-split extension of $Y_\Gamma(0,-1,2,-2)$ by $Y_\Gamma(0,-1,1,0)$ and therefore isomorphic to $(Y_2(0,-1,1,1) \otimes -\lambda_4)/Z$, where Z is a non-split extension of $Y_\Gamma(0,1,1,-3)$ by $Y_\Gamma(0,1,0,-1)$. Thus, by the long exact sequence of derived functors $R^i\mathcal{D}(\bar{L}) = R^{i+1}(Z)$ for all $i \geq 0$ and it suffices to prove that $R^j\mathcal{D}(Z)$ has a good filtration for $j = 1$ and is 0 for $j > 1$. By (5.10) and (5.11), Z is the unique non-split extension of $Y_\Gamma(0,1,1,-3)$ by $Y_\Gamma(0,1,0,-1)$ and hence isomorphic to $Y_2(0,1,0,1) \otimes -2\lambda_4$. It now follows from the tensor identity, (2.3.1) and (2.1.5) that $R^j\mathcal{D}(Z) = 0$ for all j .

<u>7.10 $\Sigma = \Delta\backslash\{\alpha_1\}$</u>

We now examine the remaining possibility $\Sigma = \Delta\backslash\{\alpha_1\}$, $\Gamma = \Omega \cap \Sigma = \{\alpha_2,\alpha_3\}$, $\mathcal{D} = Ind_B^{P_\Sigma}$. We shall prove:

(7.10.1) $R^i Ind_B^{P_\Sigma}(E/M)$ *has a good filtration for* $i = 0$ *and is* 0 *for* $i > 0$, *where* $\Sigma = \Delta\backslash\{\alpha_1\}$.

We leave it to the reader to check that

$$\chi_4(1,1,0,-2) = \chi_\Gamma(1,1,0,-2) + \chi_\Gamma(-1,2,0,-2) + \chi_\Gamma(-2,1,1,-2) + \chi$$

where χ is a sum of $\chi_\Gamma(\tau)$'s for various elements τ of X_Γ and any occurring τ satisfies $(\tau,\alpha_4^\vee) \geq -1$ and $\tau \not\geq (-2,1,1,-2)$. Thus, by Proposition 3.2.6, there is a P_Γ submodule C of $\bar{E} = E/M$ with a good filtration such that, identifying $Y_4(1,1,0,-2)$ with a submodule of $\bar{E}$, $Y_4(1,1,0,-2)/C$ has a good filtration with successive quotients $Y_\Gamma(-2,1,1,-2)$, $Y_\Gamma(0,0,1,-2)$, $Y_\Gamma(-1,2,0,-2)$, $Y_\Gamma(1,1,0,-2)$. By (5.2), $R^i\mathcal{D}(F)$ has a good filtration for $i = 0$ and is 0 for $i > 0$. Thus, to prove (7.10.1), it suffices to show that $R^i\mathcal{D}(F)$ has a good filtration for $i = 0$ and is 0 for $i > 0$, where $F = \bar{E}/C$. Now $Y_4(0,1,0,-1)$, as a P_Γ module, has an ascending filtration in which the successive quotients are $Y_\Gamma(-2,1,0,0)$, $Y_\Gamma(0)$, $Y_\Gamma(-1,0,1,-1)$, $Y_\Gamma(0,1,0,-1)$. Thus F has a descending filtration $\{F_i : i = 0,1,\ldots,8\}$ with successive quotients $Y_\Gamma(0,1,0,-1)$, $Y_\Gamma(-1,0,1,-1)$, $Y_\Gamma(0)$, $Y_\Gamma(-2,1,0,0)$, $Y_\Gamma(1,1,0,-2)$, $Y_\Gamma(-1,2,0,-2)$, $Y_\Gamma(0,0,1,-2)$, $Y_\Gamma(-2,1,1,-2)$. Moreover $F_5 = M_\Gamma(1,1,0,-2)/C$, by (5.8), so that the extension F/F_5 of F_4/F_5 by F/F_4 is non-split, by (5.12). However, $\mathcal{E}xt^1_{P_\Gamma}(Y_\Gamma(\tau),Y_\Gamma(1,1,0,-2)) = 0$ for $\tau = (-2,1,0,0)$, $(-1,0,1,-1)$ by Lemma 3.2.1. Hence we may rechoose the filtration such that F_0/F_1 is a non-split extension of $Y_\Gamma(1,1,0,-2)$ by $Y_\Gamma(0,1,0,-1)$ and the successive quotients F_i/F_{i+1} for $1 \le i \le 6$ are $Y_\Gamma(-1,0,1,-1)$, $Y_\Gamma(0)$, $Y_\Gamma(-2,1,0,0)$, $Y_\Gamma(-1,2,0,-2)$, $Y_\Gamma(0,0,1,-2)$, $Y_\Gamma(-2,1,1,-2)$. Now $\mathcal{E}xt^1_{P_\Gamma}(Y_\Gamma(0,1,0,-1),Y_\Gamma(1,1,0,-2)) \cong H^1(P_\Gamma,Y_\Gamma(-2,1,0,0) \otimes Y_\Gamma(1,1,0,-2))$ and $Y_\Gamma(-2,1,0,0) \otimes Y_\Gamma(1,1,0,-2)$ has a good filtration with quotients $Y_\Gamma(-1,2,0,-2)$, $Y_\Gamma(0,0,1,-2)$, $Y_\Gamma(1,0,0,-1)$. By (2.1.3) and (2.1.6), $H^i(P_\Gamma,Y_\Gamma(\tau)) = 0$ for all $i \ge 0$ and $\tau = (-1,2,0,-2),(1,0,0,-1)$. Hence $\mathcal{E}xt^1_{P_\Gamma}(Y_\Gamma(0,1,0,-1),Y_\Gamma(1,1,0,-2)) \cong H^1(P_\Gamma,Y_\Gamma(0,0,1,-2))$, which is k, by (2.1.3) and (1.4.7). Thus F_0/F_1 is, by (5.11), the unique non-split extension of $Y_\Gamma(1,1,0,-2)$ by $Y_\Gamma(0,1,0,-1)$ and hence isomorphic to $Y_1(0,1,0,0) \otimes -\lambda_4$.

Now $\mathcal{E}xt^1_{P_\Gamma}(Y_\Gamma(\tau),Y_\Gamma(-1,2,0,-2)) = 0$ for $\tau = (-2,1,0,0),0$ by Lemma 3.2.1 and $\mathcal{E}xt^1_{P_\Gamma}(Y_1(0,1,0,0) \otimes -\lambda_4,Y_\Gamma(-1,2,0,-2)) \cong H^1(P_\Gamma,Y_1(0,1,0,0)^* \otimes Y_\Gamma(-1,2,0,-1))$ which is 0 by (2.1.3) and (2.1.6).

Moreover $Hom_{P_\Gamma}(\bar{E}, Y_\Gamma(-1,2,0,-1)) = 0$ so that F does not have an epimorphic image isomorphic to $Y_\Gamma(-1,2,0,-1)$ so we may rechoose the filtration so that F_o/F_1 is unchanged, F_1/F_2 is a non-split extension of $Y_\Gamma(-1,2,0,-2)$ by $Y_\Gamma(-1,0,1,-1)$ and the quotients F_i/F_{i+1} for $i = 2,3,3,4$ are $Y_\Gamma(0)$, $Y_\Gamma(-2,1,0,0)$, $Y_\Gamma(0,0,1,-2)$, $Y_\Gamma(-2,1,1,-2)$. Let $\pi = \{\alpha_3\}$.

We have $Ext^1_{P_\Gamma}(Y_\Gamma(-1,0,1,-1), Y_\Gamma(-1,2,0,-2)) \cong$ $Ext^1_{P_\pi}(Y_\pi(-1,0,1,-1), Y_\pi(-1,2,0,-2)) \cong H^1(P_\pi, Y_\pi(1,-2,1,0) \otimes Y_\pi(-1,2,0,-1))$ $\cong H^1(P_\pi, Y\ (0,0,1,-2)) \cong H^1(B, -\alpha_4) = k$. Hence F_1/F_2 is the unique non-split extension of $Y_\Gamma(-1,2,0,-2)$ by $Y_\Gamma(-1,0,1,-1)$. Now $Y_1(-1,0,1,0)$, as a P_Γ module has a good filtration with quotients $Y_\Gamma(1,0,1,-2)$, $Y_\Gamma(1,0,0,0)$, $Y_\Gamma(-1,2,0,-1)$, $Y_\Gamma(-1,0,1,0)$ so that, by Proposition 3.2.6, there is a P_Γ submodule, A' of $Y_1(-1,0,1,0)$ which is an extension of $Y_\Gamma(1,0,1,-2)$ by $Y_\Gamma(1,0,0,0)$. Of course $Y_1(-1,0,1,0)/A'$ is a non-split extension since $Hom_{P_\Gamma}(Y_1(-1,0,1,0), Y_\Gamma(-1,2,0,-1)) = 0$. Hence F_1/F_2 is isomorphic to $(Y_1(-1,0,1,0)/A') \otimes -\lambda_4$ and so

(7.10.2) *there is a short exact sequence of* P_Γ *modules*

$$0 \to A \to Y_1(-1,0,1,0) \otimes -\lambda_4 \to F_1/F_2 \to 0$$

where A *is an extension of* $Y_\Gamma(1,0,1,-3)$ *by* $Y_\Gamma(1,0,0,-1)$.

We have $Ext^1_{P_\Gamma}(Y_\Gamma(-2,1,0,0), Y_\Gamma(0,0,1,-2)) = 0$ by Lemma 3.2.1. It follows from (7.10.2) that $Ext^1_{P_\Gamma}(F_1/F_2, Y_\Gamma(0,0,1,-2)) \cong Ext^2_{P_\Gamma}(A, Y_\Gamma(0,0,1,-2))$ and the latter is zero, by Lemma 3.2.1 and (7.10.2). Also $Ext^1_{P_\Gamma}(F_o/F_1, Y_\Gamma(0,0,1,-2)) \cong H^1(P_\Gamma, Y_1(0,1,0,0)^* \otimes Y_\Gamma(0,0,1,-1))$, which is zero by (2.1.3) and (2.1.6). Moreover $Hom_{P_\Gamma}(F, Y_\Gamma(0,0,1,-2)) = 0$ and so we may rechoose the filtration leaving F_i/F_{i+1} unchanged for

$i = 0,1$ and such that F_2/F_3 is a non-split extension of $Y_\Gamma(0,0,1,-2)$ by $Y_\Gamma(0)$, F_3/F_4 is isomorphic to $Y_\Gamma(-2,1,0,0)$, F_4/F_5 is isomorphic to $Y_\Gamma(-2,1,1,-2)$ and $F_5 = 0$. Now $\mathit{Ext}^1_{P_\Gamma}(Y_\Gamma(0),Y_\Gamma(0,0,1,-2)) = k$ so that F_2/F_3 is the unique non-split extension of $Y_\Gamma(0,0,1,-2)$ by $Y_\Gamma(0)$ and it follows:

(7.10.3) *there is a short exact sequence of* P_Γ *modules*

$$0 \to Y_\Gamma(2,0,0,-2) \to Y_2(0,0,0,1) \otimes -\lambda_4 \to F_2/F_3 \to 0 .$$

Now $R^i\mathcal{D}(F_0/F_1) \cong R^i\mathcal{D}(Y_1(0,1,0,0) \otimes -\lambda_4$, which is zero, for all $i \geq 0$, by the tensor identity and (2.1.5). Also, for $i \geq 0$, $R^i\mathcal{D}(F_1/F_2 \cong R^{i+1}\mathcal{D}(A)$, by (7.10.2). But $R^j\mathcal{D}(Y_\Gamma(1,0,0,-1)) = 0$ for all $j \geq 0$ by (2.1.5) and $R^j\mathcal{D}(Y_\Gamma(1,0,1,-3)) = 0$ for all $j \geq 0$ by (2.1.3) and (2.1.5) and (5.5). Thus $R^j\mathcal{D}(A) = 0$ for all $j \geq 0$ and so $R^i\mathcal{D}(F_1/F_2) = 0$ for all $i \geq 0$. Thus to prove that $R^i\mathcal{D}(F)$ has a good filtration if $i = 0$ and is 0 if $i > 0$, and so to complete the proof of (7.10.3) it suffices to prove that $R^i\mathcal{D}(F_3)$ has a good filtration for $i = 0$ and is 0 for $i > 0$.

We claim that F_3 is a non-split extension of $Y_\Gamma(-2,1,1,-2)$ by $Y_\Gamma(-2,1,0,0)$. We have, by (7.10.3),
$\mathit{Ext}^1_{P_\Gamma}(F_2/F_3,Y_\Gamma(-2,1,1,-2)) \cong \mathit{Ext}^2_{P_\Gamma}(Y_\Gamma(2,0,0,-2),Y_\Gamma(-2.1,1,-2))$, which is zero by Lemma 3.2.1. Also, by (7.10.2),
$\mathit{Ext}^1_{P_\Gamma}(F_1/F_2,Y_\Gamma(-2,1,1,-2)) \cong \mathit{Ext}^2_{P_\Gamma}(A,Y_\Gamma(-2,1,1,-2))$, which is zero by Lemma 3.2.1 and (7.10.2). We have
$\mathit{Ext}^1_{P_\Gamma}(F_0/F_1,Y_\Gamma(-2,1,1,-2)) \cong H^1(P_\Gamma,Y_1(0,1,0,0)^* \otimes Y_\Gamma(-2,1,1,-1))$,
which is zero by (2.1.3) and (2.1.6). Thus, if F_3 is.a split extension F has a homomorphic image isomorphic to $Y_\Gamma(-2,1,1,-2)$. But
$\mathit{Hom}_{P_\Gamma}(\bar{E},Y_\Gamma(-2,1,1,-2)) \cong \mathit{Hom}_{P_\Sigma}(\bar{E},\mathit{Ind}_{P_\Gamma}^{P_\Sigma}(-2,1,1,-2))$, which is 0 by (1.7.1), so F_3 is a non-split extension of $Y_\Gamma(-2,1,1,-2)$ by $Y_\Gamma(-2,1,0,0)$. Let $\pi = \{\alpha_2\}$. It is easy to check that

(7.10.4) $\chi_\Gamma(-2,1,1,-2) = \chi_\pi(-2,1,1,-2) + \chi_\pi(-2,3,-1,-1)$

$+ \chi_\pi(-1,3,-2,0) + \chi_\pi(-1,1,0,-1) + \chi_\pi(0,1,-1,0)$

$+ \chi_\pi(1,1,-2,1)$.

Hence $M_\pi(-2,1,1,-2)$ has a good filtration with quotients $Y_\pi(1,1,-2,1)$, $Y_\pi(0,1,-1,0)$, $Y_\pi(-1,1,0,-1)$, $Y_\pi(-1,3,-2,0)$, $Y_\pi(-2,3,-1,-1)$. Now applications of (1.7.1), (2.1.5) and (2.3.1) yield $R^i\mathcal{D}(M_\pi(-2,1,1,-2)) = 0$ for all $i \geq 0$. Hence, to complete the proof of (7.10.1), it suffices to show that $R^i\mathcal{D}(\overline{F})$ has a good filtration if $i = 0$ and is 0 if $i > 0$, where $\overline{F} = F_3/M_\pi(-2,1,1,-2)$, $Y_\pi(-2,1,1,-2)$ being identified with a submodule of F_3 . By (5.12), F is a non-split extension of $Y_\pi(-2,1,1,-2)$ by $Y_\Gamma(-2,1,0,0)$. Now $Y_\Gamma(-2,1,0,0)$, as a P_π module, has a good filtration with quotients $Y_\pi(-1,1,-1,1)$, $Y_\pi(-2,1,0,0)$ and $Ext^1_{P_\pi}(Y_\pi(-1,1,-1,1),Y_\pi(-2,1,1,-2)) = 0$ by Lemma 3.2.1 so that $\overline{F}$ has a submodule, say Z , isomorphic to $Y_\pi(-1,1,-1,1)$ and $\overline{F}/Z$ is a non-split extension of $Y_\pi(-2,1,1,-2)$ by $Y_\pi(-2,1,0,0)$. By (2.1.5), $R^i\mathcal{D}(Z) = 0$ for all $i \geq 0$ so to complete the proof of (7.10.1) we need only show that $R^i\mathcal{D}(\overline{F}/Z)$ has a good filtration for $i = 0$ and is 0 for $i > 0$. Now $Ext^1_{P_\pi}(Y_\pi(-2,1,0,0),Y_\pi(-2,1,1,-2)) = k$ by (5.10) so that $\overline{F}/Z$ is the unique non-split extension of $Y_\pi(-2,1,1,-2)$ by $Y_\pi(-2,1,0,0)$, and hence isomorphic to $Y_\Lambda(-2,1,0,1) \otimes -\lambda_4$, where $\Lambda = \{\alpha_2,\alpha_4\}$. Thus $R^iInd_B^{P_\Lambda}(\overline{F}/Z) = 0$ for all $i \geq 0$, by (2.1.5) and the tensor identity so that $R^i\mathcal{D}(\overline{F}/Z) = 0$ for all $i \geq 0$ by (1.1.3)'. The proof of (7.10.1) is complete.

(7.10.5) $R^iInd_B^P(S/N)$ *has a good filtration for* $i = 0$ *and is* 0 *for* $i > 0$, *where* $\Sigma = \Delta\backslash\{\alpha_1\}$.

We leave it to the reader to check that

$$\chi_4(1,0,1,-2) = \chi_\Gamma(1,0,1,-2) + \chi_\Gamma(-1,1,1,-2) + \chi$$

where χ is a sum of $\chi_\Gamma(\tau)$'s for various $\tau \in X_\Gamma$ and any occurring τ satisfies $(\tau,\alpha_4^\vee) \geq -1$ and $\tau \not\geq (-1,1,1,-2)$. Thus by Proposition 3.2.6, $Y_4(1,0,1,-2)$, has a P_Γ submodule, say J, with a good filtration such that $Y_4(1,0,1,-2)/J$ has a good filtration with successive quotients $Y_\Gamma(-1,1,1,-2)$, $Y_\Gamma(1,0,1,-2)$. Moreover $R^i\mathcal{D}(J)$ has a good filtration for $i = 0$ and is 0 for $i > 0$, by (5.2). Thus identifying $Y_4(1,0,1,-2)$, and hence J, with a submodule of $\bar{S} = S/N$ we see that, to prove (7.10.5), it is enough to show that $R^i\mathcal{D}(L)$ has a good filtration for $i = 0$ and is 0 for $i > 0$, where $L = \bar{S}/J$.

Now $Y_4(1,0,0,0)$, as a P_Γ module, has a good filtration with quotients $Y_\Gamma(-1,0,0,1)$, $Y_\Gamma(-1,1,0,0)$, $Y_\Gamma(1,0,0,0)$. Hence L has a descending filtration $\{L_i : i = 0,1,\ldots,5\}$ with quotients $Y_\Gamma(1,0,0,0)$, $Y_\Gamma(-1,1,0,0)$, $Y_\Gamma(-1,0,0,1)$, $Y_\Gamma(1,0,1,-2)$, $Y_\Gamma(-1,1,1,-2)$. Moreover, by (5.8), $L_4 = M(1,0,1,-2)/J$ so that, by (5.12), the extension L/L_4, of L_3/L_4 by L/L_3, is non-split. However, $Ext^1_{P_\Gamma}(Y_\Gamma(\tau),Y_\Gamma(1,0,1,-2)) = 0$ for $\tau = (-1,0,0,1),(-1,1,0,0)$ by Lemma 3.2.1 so that we may rechoose the filtration such that L_o/L_1 is a non-split extension of $Y_\Gamma(1,0,1,-2)$ by $Y_\Gamma(1,0,0,0)$, $L_1/L_2 = Y_\Gamma(-1,0,0,1)$ $L_1/L_2 = Y_\Gamma(-1,0,0,1)$, $L_3/L_4 = Y_\Gamma(-1,1,1,-2)$ and $L_4 = 0$. Now $Ext^1_{P_\Gamma}(Y_\Gamma(1,0,0,0),Y_\Gamma(1,0,1,-2)) \cong k$ by (5.10) and it follows from (5.11) that L_o/L_1 is the unique non-split P_Γ extension of $Y_\Gamma(1,0,1,-2)$ by $Y_\Gamma(1,0,0,0)$. Now $Y_1(1,0,0,1)$, as a P_Γ-module, has a filtration with quotients $Y_\Gamma(3,0,0,-1)$, $Y_\Gamma(1,0,1,-1)$, $Y_\Gamma(1,0,0,1)$ and it follows:

(7.10.6) *there is a short exact sequence of* P_Γ *modules*

$$0 \to Y_\Gamma(3,0,0,-2) \to Y_1(1,0,0,1) \otimes -\lambda_4 \to L_1/L_1 \to 0 .$$

We have $Ext^1_{P_\Gamma}(Y_\Gamma(-1,0,0,1),Y_\Gamma(-1,1,1,-2)) = 0$ by Lemma 3.2.1. Also, by (7.10.6),

$Ext^1_{P_\Gamma}(F_o/F_1,Y_\Gamma(-1,1,1,-2)) = Ext^2_{P_\Gamma}(Y_\Gamma(3,0,0,-2),Y_\Gamma(-1,1,1,-2))$, which is isomorphic to $H^2(P_\Gamma, Y_\Gamma(-4,1,1,0))$. By (2.1.3), $H^2(P_\Gamma, Y_\Gamma(-4,1,1,0)) \cong H^2(B,(-4,1,1,0))$ which, as one may easily check with the aid of (5.4), is 0 . But $Hom_{P_\Gamma}(\bar{S}, Y_\Gamma(-1,1,1,-2)) \cong Hom_{P_\Gamma}(\bar{S}, Ind_{P_\Gamma}^{P_\Sigma}(Y_\Gamma(-1,1,1,-2)) = 0$, so L does not have an epimorphic image isomorphic to $Y_\Gamma(-1,1,1,-2)$ and we may rechoose the filtration such that L_o/L_1 is unchanged, L_1/L_2 is a non-split extension of $Y_\Gamma(-1,1,1,-2)$ by $Y_\Gamma(-1,1,0,0)$, $L_2/L_3 \cong Y_\Gamma(-1,0,0,1)$ and $L_3 = 0$.

Now to complete the proof of (7.10.5) we show that $R^i\mathcal{D}(L_j/L_{j+1})$ has a good filtration for $i = 0$ and is zero for $i > 0$, for any $0 \le j \le 2$. We have $R^i\mathcal{D}(L_o/L_1 \cong R^{i+1}\mathcal{D}(Y_\Gamma(3,0,0,-2))$ for any $i \ge 0$, by (7.10.6). Moreover $R^j\mathcal{D}(Y_\Gamma(3,0,0,-2)) = 0$ for all $j \ge 0$ by (2.3.1) and (2.1.5). Also $R^i\mathcal{D}(Y_\Gamma(-1,0,0,1)) = Y_\Sigma(-1,0,0,1)$ for $i = 0$ and is 0 for $i > 0$, by (2.1.2) and (1.7.4). Thus it remains to show that $R^i\mathcal{D}(\bar{L})$ has a good filtration for $i = 0$ and is 0 for $i > 0$, where $\bar{L} = L_1/L_2$. Let $\pi = \{\alpha_2\}$. From (7.10.4) it follows that $M_\pi(-1,1,1,-2)$ has a good filtration with quotients $Y_\pi(2,1,-2,1)$, $Y_\pi(1,1,-1,0)$, $Y_\pi(0,1,0,-1)$, $Y_\pi(0,3,-2,0)$, $Y_\pi(-1,3,-1,-1)$. Now applications of (1.7.1), (2.1.5) and (2.3.1) yield $R^i\mathcal{D}(M_\pi(-1,1,1,-2)) = 0$ for all $i \ge 0$. Thus, to complete the proof of (7.10.5), it suffices to show that $R^i\mathcal{D}(\bar{L}/M_\pi(-1,1,1,-2))$ has a good filtration for $i = 0$ and is zero for $i > 0$, where $Y_\Gamma(-1,1,1,-2)$ is identified with a submodule of $\bar{L}$. By (5.12), $Q = \bar{L}/M\ (-1,1,1,-2)$ is a non-split extension of $Y_\pi(-1,1,1,-2)$ by $Y_\Gamma(-1,1,0,0)$. Now $Y_\Gamma(-1,1,0,0)$, as a P_π module, has a good filtration with quotients $Y_\pi(0,1,-1,1)$, $Y_\pi(-1,1,0,0)$. We have $Ext^1_{P_\pi}(Y_\pi(0,1,-1,1), Y_\pi(-1,1,1,-2)) = 0$ by Lemma 3.2.1 so that Q has a submodule Q' isomorphic to $Y_\pi(0,1,-1,1)$ and Q/Q' is a non-split extension of $Y_\pi(-1,1,1,-2)$ by $Y_\pi(-1,1,0,0)$. It follows from (5.10) and (5.11) that Q/Q' is isomorphic to $Y_\Lambda(-1,1,0,1) \otimes -\lambda_4$, where $\Lambda = \{\alpha_2,\alpha_4\}$. Now $R^i\mathcal{D}(Q') = 0$ for all $i \ge 0$ by (2.1.5).

Also, by (2.1.5) and the tensor identity, $R^i Ind_B^{P_\Lambda}(Q/Q') = 0$ for all $i \geq 0$ and so, by (1.1.3)' of $R^i \mathcal{D}(Q/Q') = 0$ for all i. Hence $R^i \mathcal{D}(Q) = 0$ for all i, completing the proof of (7.10.5).

7.11 Σ arbitrary

We are now able to show:

(7.11.1) *For any* $\Sigma \subseteq \Delta$, $Y(\lambda_2)|_{P_\Sigma}$ *has a good filtration.*

Since we know $\underline{H}(\Phi_\Sigma)$ to be true for every proper subset Σ of Δ we may assume, by induction on $|\Sigma|$, that $\Sigma = \Delta\backslash\{\alpha_j\}$ for some $1 \leq j \leq 4$. We know that (7.11.1) holds for $j = 4$, by (7.6.14) so we may assume that $1 \leq j \leq 3$. Let $Y_0 = 0$, $Y_1 = M$, $Y_2 = E$, $Y_3 = N$, $Y_4 = S$, $Y_5 = Y$. We have shown, in sections 7.7, 7.8, 7.9 and 7.10 that

$Ind_B^{P_\Sigma}(Y_i/Y_{i-1})$ has a good filtration for $1 \leq i \leq 5$ and

$RInd_B^{P_\Sigma}(Y_i/Y_{i-1}) = 0$ for $1 \leq i \leq 4$. Thus, by the long exact sequence of derived functors, $Y|_{P_\Sigma} \cong Ind_B^{P_\Sigma}(Y|_B)$ has a filtration $\{Ind_B^{P_\Sigma}(Y_i) : i = 0,1,\ldots,5\}$, which may be refined to a good filtration.

7.12 $Y(\lambda_2) \otimes Y(\lambda_2)$

In this section we shall show that $Y(\lambda_2) \otimes Y(\lambda_2)$ has a good filtration. This turns out to be crucial in establishing $\underline{H}(G)$. We let $\mathcal{E} = Ind_B^G$.

(7.12.1) $R^i\mathcal{E}(M \otimes \lambda_2)$ *has a good filtration for* $i = 0$ *and is* 0 *for* $i > 0$.

By (7.5.5), the tensor identity and (2.1.5), $R^i\mathcal{E}(M \otimes \lambda_2) \cong R^{i+1}\mathcal{E}(Y_4(1,0,0,-2) \otimes \lambda_2)$ which, by the tensor identity and (2.1.2) is isomorphic to $R^{i+1}\mathcal{E}(Y_4(1,0,0,-2) \otimes Y_4(0,1,0,0))$. Now

$Y_4(1,0,0,-2) \otimes Y_4(0,1,0,0)$ has a good filtration with quotients $Y_4(1,0,0,-2)$, $Y_4(0,0,1,-2)$, $Y_4(1,1,0,-2)$. Also $R^j E(Y_4(\tau)) = 0$ for all $j \geq 0$ and $\tau = (1,0,0,-1),(1,1,0,-2)$ by (2.1.5), (2.1.2) and (2.3.1). Hence $R^{i+1} E(Y_4(1,0,0,-2) \otimes Y_4(0,1,0,0)) \cong R^{i+1}E(Y_4(0,0,1,-2))$ which, by (2.1.2), (2.3.1) and the vanishing theorem is k if $i = 0$ and 0 if $i > 0$.

(7.12.2) $R^i E((N/M) \otimes \lambda_2)$ *has a good filtration if* $i = 0$ *and is* 0 *if* $i > 0$.

We have, by (7.6.4), $N/M \cong F(Q \otimes (\lambda_1 - \lambda_4))$, where $F = Ind_B^{P_4}$. Now $E = Ind_{P_4}^G \circ F$ and $R^j F(F(Q \otimes (\lambda_1 - \lambda_4)) \otimes \lambda_2) = 0$ for all $j > 0$, by the tensor identity and (2.1.2), hence by (1.1.6), $R^i E(F(Q \otimes (\lambda_1 - \lambda_4)) \otimes \lambda_2) \cong R^i Ind_{P_4}^G (F(Q \otimes (\lambda_1 - \lambda_4)) \otimes Y_4(\lambda_2))$ for all $i \geq 0$. Also $R^j F(Q \otimes (\lambda_1 - \lambda_4)) = 0$ for all $j > 0$ by (7.6.2) so that, by the tensor identity and (1.1.6), $R^i E(F(Q \otimes (\lambda_1 - \lambda_2)) \otimes \lambda_2) \cong R^i E(Q \otimes (\lambda_1 - \lambda_4) \otimes Y_4(\lambda_2))$. Now let $G = Ind_B^{P_1}$ and $\Gamma = \{\alpha_2, \alpha_3\}$. Now $Y_4(\lambda_2) \otimes (\lambda_1 - \lambda_4)$, as a P_Γ module, has a filtration with quotients $Y_\Gamma(-1,1,0,0)$, $Y_\Gamma(1,0,0,0)$, $Y_\Gamma(0,0,1,-1)$, $Y_\Gamma(1,1,0,-1)$. There is thus a P_Γ submodule V of $Y_4(\lambda_2) \otimes (\lambda_1 - \lambda_4)$ such that V is an extension of $Y_\Gamma(-1,1,0,0)$ by $Y_\Gamma(1,0,0,0)$ and $(Y_4(\lambda_2) \otimes (\lambda_1 - \lambda_4))/V$ has a good filtration with quotients $Y_\Gamma(0,0,1,-1)$, $Y_\Gamma(1,1,0,-1)$. Moreover, since the B socle of $Y_4(\lambda_2)$ is simple, V is a non-split extension. It follows, from (2.1.2) and (2.1.5), that $R^i G(Y_4(0,1,0,0) \otimes (\lambda_1 - \lambda_4))$ is zero if $i > 0$ and isomorphic to $G(V)$ for $i = 0$. By (5.14), $G(V)$ is a non-split extension of $Y_1(-1,1,0,0)$ by $Y_1(1,0,0,0)$. It follows, from (5.14), that $G(Y_4(0,1,0,0)$ is isomorphic to $G(Z)$ for a non-split B-module extension Z of $(-1,1,0,0)$ by $(1,0,0,0)$. Thus $G(Y_4(0,1,0,0)) \cong G(Y_\Lambda(\lambda_1))$, where $\Lambda = \{\alpha_1\}$.

We have, by (1.1.6) and the fact that $R^i G(Y_4(0,1,0,0) \otimes (\lambda_1 - \lambda_4)) = 0$ for $i > 0$,

$R^iE((N/M) \otimes \lambda_2) \cong R^iE(Q \otimes (\lambda_1 - \lambda_4) \otimes Y_4(\lambda_2)) \cong R^i Ind_{P_1}^G (Q \otimes G(Y_\Lambda(\lambda_1)))$.

Now the weights of $Y_\Lambda(\lambda_1)$ are $(1,0,0,0),(-1,1,0,0)$ so that $R^jG(Y_\Lambda(\lambda_1)) = 0$ for $j > 0$ by the vanishing theorem, (2.1.2) and (2.1.5). Thus by the tensor identity and (1.1.6), $R^i Ind_{P_1}^G (Q \otimes G(Y_\Lambda(\lambda_1)))$, and hence $R^iE((N/M) \otimes \lambda_2)$ is isomorphic to $R^iE(Q \otimes Y_\Lambda(\lambda_1))$. Now let $H = Ind_B^{P_\Lambda}$. The weights of Q are $(1,0,0,0),(-1,1,0,0),(0,-1,1,0),(0,1,-1,1),(1,-1,0,1),(0,1,0,-1),$ $(1,-1,1,-1),(1,1,-1,0),(2,-1,0.0)$. It follows from (2.1.2) and (2.1.5) that $H(Q)$ has a good filtration with quotients $Y_\Lambda(2,-1,0,0)$, $Y_\Lambda(1,1,-1,0)$, $Y_\Lambda(1,-1,1,-1)$, $Y_\Lambda(0,1,0,-1)$, $Y_\Lambda(1,-1,0,1)$, $Y_\Lambda(0,1,-1,1)$, $Y_\Lambda(0,-1,1,0)$, $Y_\Lambda(1,0,0,0)$ and that $R^jH(Q) = 0$ for $j > 0$. We obtain, from the tensor identity and (1.1.6), that $R^iE(Q \otimes Y_\Lambda(\lambda_1))$, and hence $R^iE((N/M) \otimes \lambda_2)$, is isomorphic to $R^i Ind_{P_1}^G (H(Q) \otimes Y_\Lambda(\lambda_1))$. Now $H(Q) \otimes Y_\Lambda(\lambda_1)$ has a filtration with quotients $Y_\Lambda(1,0,0,0)$, $Y_\Lambda(3,-1,0,0)$, $Y_\Lambda(0,2,-1,0)$, $Y_\Lambda(2,1,-1,0)$, $Y_\Lambda(0,0,1,-1)$,, $Y_\Lambda(2,-1,1,-1)$, $Y_\Lambda(1,1,0,-1)$, $Y_\Lambda(0,0,0,1)$, $Y_\Lambda(2,-1,0,0)$, $Y_\Lambda(1,1,-1,1)$, $Y_\Lambda(1,-1,1,0)$, $Y_\Lambda(0,1,0,0)$, $Y_\Lambda(2,0,0,0)$. It follows from (2.1.2) and (2.1.5) that $R^i Ind_{P_1}^G (H(Q) \otimes Y_\Lambda(\lambda_1))$, and hence $R^iE((N/M) \otimes \lambda_2)$ has a good filtration with quotients $Y(\lambda_1)$, $Y(\lambda_4)$, $Y(\lambda_2)$, $Y(2\lambda_1)$ if $i = 0$ and is 0 if $i > 0$. This completes the proof of (7.12.1).

(7.12.3) $R^iE((S/N) \otimes \lambda_2)$ *has a good filtration if* $i = 0$ *and is* 0 *if* $i > 0$.

We have, by the tensor identity, (1.1.6) and (2.1.3) that $R^iE((S/N) \otimes \lambda_2) \cong R^i Ind_{P_4}^G ((S/N) \otimes Y_4(\lambda_2))$. Now, by (7.6.15) and (7.3.11), $S/N \cong F(Y_\pi(1,0,0,1) \otimes -\lambda_4)$, where $\pi = \{\alpha_4\}$, and so $R^i Ind_{P_4}^G ((S/N) \otimes Y_4(\lambda_2))$, and therefore $R^iE((S/N) \otimes \lambda_2)$ is isomorphic to $R^i Ind_{P_4}^G (F(Y_\pi(1,0,0,1) \otimes -\lambda_4 \otimes Y_4(\lambda_2))$. The weights of

$Y_\pi(1,0,0,1) \otimes -\lambda_4$ are $(1,0,0,0),(1,0,1,-2)$ so that $R^i F(Y_\pi(1,0,0,1) \otimes -\lambda_4) = 0$ for $i > 0$, by the vanishing theorem and (2.1.2). Hence by (1.1.6), $R^i Ind_{P_4}^G (F(Y_\pi(1,0,0,1) \otimes -\lambda_4 \otimes Y_4(\lambda_2))$, and hence $R^i E((S/N) \otimes \lambda_2)$, is isomorphic to $R^i E(Y_4(\lambda_2) \otimes Y_\pi(1,0,0,1) \otimes -\lambda_4)$. Let $\Gamma = \{\alpha_1,\alpha_2\}$. As a P_Γ module, $Y_4(\lambda_2)$ has a good filtration with quotients $Y_\Gamma(0,1,0,0)$, $Y_\Gamma(1,1,-1,1)$, $Y_\Gamma(1,0,-1,2)$ and so, by (7.3.15), $Y_4(\lambda_2) \otimes Y_\pi(1,0,0,1) \otimes -\lambda_4$ has a filtration with quotients $Y_3(0,1,0,1) \otimes (\lambda_1 - \lambda_4)$, $Y_3(1,1,-1,1) \otimes \lambda_1$, $Y_\Gamma(1,0,-1,1) \otimes Y_\Gamma(1,0,0,1)$. Now one can see that $R^i E(Y_3(0,1,0,0) \otimes (\lambda_1 - \lambda_4)) = 0$ for all $i \geq 0$ from (2.1.5) and (1.1.6). Also by the tensor identity, (2.1.2) and (1.1.6), $R^i E(Y_3(1,1,-1,1) \otimes \lambda_1) \cong R^i Ind_{P_3}^G (Y_3(1,1,-1,1) \otimes Y_3(\lambda_1))$. Moreover $Y_3(1,1,-1,1) \otimes Y_3(\lambda_1)$ has a good filtration with quotients $Y_3(2,1,-1,1)$, $Y_3(0,2,-1,1)$, $Y_3(1,0,0,1)$. It follows, from (2.1.2), (2.1.5) that $R^i E(Y_3(1,1,-1,1) \otimes \lambda_1)$ is isomorphic to $Y(1,0,0,1)$, if $i = 0$ and is 0 if $i > 0$. Thus, to complete the proof of (7.12.3) it suffices to show that $R^i E(Y_\Gamma(1,0,-1,1) \otimes Y_\pi(1,0,0,1))$ has a good filtration if $i = 0$ and is 0 if $i > 0$.

As a B module $Y_\Gamma(1,0,-1,1) \otimes Y_\pi(1,0,0,1)$ has a filtration with quotients $Y_\Gamma(1,0,-1,2) \otimes \lambda_1$, $Y_\Gamma(1,0,0,0) \otimes \lambda_1$. Thus it suffices to show that $R^i E(Z)$ has a good filtration for $i = 0$ and is 0 for $i > 0$, for $Z = Y_\Gamma(1,0,-1,2) \otimes \lambda_1$, $Y_\Gamma(1,0,0,0) \otimes \lambda_1$. It follows, from (1.1.6) and (2.1.2) that $R^i E(Y_\Gamma(1,0,-1,2) \otimes \lambda_1) \cong R^i Ind_{P_\Gamma}^G (Y_\Gamma(1,0,-1,2) \otimes Y_\Gamma(\lambda_1))$. Now $Y_\Gamma(1,0,-1,2) \otimes Y_\Gamma(\lambda_1)$ has a filtration with quotients $Y_\Gamma(2,0,-1,2)$, $Y_\Gamma(0,1,-1,2)$. It follows, from (2.1.5) that $R^i E(Y_\Gamma(1,0,-1,2) \otimes \lambda_1) = 0$ for all $i \geq 0$. Similar considerations show that $R^i E(Y_\Gamma(\lambda_1) \otimes \lambda_1)$ has a good filtration with quotients $Y(\lambda_2)$, $Y(2\lambda_1)$ if $i = 0$ and is 0 if $i > 0$. The proof of (7.12.3) is now complete.

(7.12.4) $R^i E((Y/S) \otimes \lambda_2)$ *has a good filtration if* $i = 0$ *and is* 0 *if* $i > 0$.

The P_4 module Y/S has a filtration with quotients $Y_4(1,1,0,-1)$, $Y_4(0,1,0,0)$. Thus it suffices to show that $R^i E(Y_4(\tau) \otimes \lambda_2)$ has a good filtration for $i = 0$ and is 0 for $i > 0$, for $\tau = (1,1,0,-1),(0,1,0,0)$. We have, by (1.1.6) and (2.1.2), $R^i E(Y_4(\tau) \otimes \lambda_2) \cong R^i Ind_{P_4}^G (Y_4(\tau) \otimes Y_4(\lambda_2))$. Now $Y_4(1,1,0,-1) \otimes Y_4(0,1,0,0)$ has a good filtration with successive quotients $Y_4(1,0,0,1)$, $Y_4(0,0,1,0)$, $Y_4(1,1,0,0)$ (twice), $Y_4(3,0,0,0)$, $Y_4(0,1,1,-1)$, $Y_4(2,0,1,-1)$, $Y_4(1,2,0,-1)$. Thus, by (2.1.2) and (2.1.5), $R^i E(Y_4(1,1,0,-1) \otimes \lambda_2)$ has a good filtration if $i = 0$ and is 0 if $i > 0$. Also $Y_4(0,1,0,0) \otimes Y_4(0,1,0,0)$ has a good filtration with quotients $Y_4(0,0,0,2)$, $Y_4(0,1,0,1)$, $Y_4(2,0,0,1)$, $Y_4(1,0,1,0)$, $Y_4(0,2,0,0)$ so that, by the vanishing theorem, (2.1.2) and (2.1.5), $R^i E(Y_4(\lambda_2) \otimes Y_4(\lambda_2))$ has a good filtration if $i = 0$ and is 0 if $i > 0$.

Now putting together (7.12.1), (7.12.2), (7.12.3), (7.12.4) we have, by the long exact sequence of derived functors, and the fact that $Y(\lambda_2) \otimes Y(\lambda_2) \cong E(Y(\lambda_2) \otimes \lambda_2)$:

(7.12.5) $Y(\lambda_2) \otimes Y(\lambda_2)$ *has a good filtration.*

7.13 $\underline{H}_1(G)$ and $\underline{H}_2(G)$ are true

We shall now show that G satisfies $\underline{H}_1(G)$ and $\underline{H}_2(G)$. To do this we need:

(7.13.1) $\chi(\lambda_1)^2 = \chi(2\lambda_1) + \chi(\lambda_2) + \chi(\lambda_4) + \chi(\lambda_1) + \chi(0)$;

(7.13.2) $\chi(\lambda_1)\chi(\lambda_4) = \chi(\lambda_1 + \lambda_4) + \chi(\lambda_2) + \chi(\lambda_1)$;

(7.13.3) $\chi(\lambda_1)\chi(\lambda_2) = \chi(\lambda_1 + \lambda_2) + \chi(\lambda_3) + \chi(\lambda_1 + \lambda_4)$

$$+ \chi(2\lambda_1) + \chi(\lambda_2) + \chi(\lambda_4) + \chi(\lambda_1) .$$

We leave it to the reader to check the above using (7.2.3), (2.2.3) and (2.2.5).

We now assume that the characteristic p of k is not 0 and let

$$C = \{\lambda \in X : 0 \leq (\lambda, \alpha^{\vee}) \leq p \text{ for all } \alpha \in \Phi^{+}\} .$$

Of course $\lambda \in X$ belongs to C if and only if $\lambda \in X^{+}$ and $(\lambda, \beta_o^{\vee}) \leq p$ where β_o is the highest short root, in this case $2\alpha_1 + 3\alpha_2 + 2\alpha_3 + \alpha_4$. Any element of X is conjugate under the affine Weyl group W_p (generated by W and all elements of the form $p\alpha, \alpha \in \Phi$) to precisely one element of C. We say a G-module V belongs to $\lambda \in C$ if, for any composition factor $L(\tau)$ of V, $\tau + \rho$ is conjugate to λ under W_p. It follows from the strong linkage principle, [4], that any indecomposable rational G-module belongs to λ for some $\lambda \in C$ and, if M is any G-module then

(7.13.4) $\quad M = \bigoplus_{\lambda \in C} M_{\lambda}$

where M_{λ} is the largest submodule of M belonging to λ. (This theory applies generally to semisimple, simply connected groups - a fuller discussion is to be found in section one of [44]).

We now assume that $p = 2$. We have $(\lambda_1, \beta_o^{\vee}) = (\lambda_4, \beta_o^{\vee}) = 2$ so that λ_1 and λ_4 belong to C. Since λ_1 and λ_4 are roots, $\lambda_3 + \rho$ is conjugate under W_p to $(-1,1,2,-1)$, which is conjugate under W to $(1,0,1,1)$. This is conjugate, under W_p, to $(-1,0,1,-1)$, which is conjugate under W to λ_1. Similar observations reveal that, for $\tau = \lambda_1 + \lambda_4, 2\lambda_1, \lambda_2, \lambda_4, \lambda_1$ the weight $\tau + \rho$ is W_p conjugate to λ_1 and $\lambda_1 + \lambda_4 + \rho$ is conjugate to λ_4. Now $Y(\lambda_1) \otimes Y(\lambda_2)$ has a good filtration with quotients $Y(\lambda_1)$, $Y(\lambda_4)$, $Y(\lambda_2)$, $Y(2\lambda_1)$, $Y(\lambda_1 + \lambda_4)$, $Y(\lambda_3)$, $Y(\lambda_1 + \lambda_2)$ (by (7.13.3), (7.2.15) and Proposition 3.2.6). From Corollary 3.2.5 and (7.13.4)

(7.13.5) $\quad Y(\lambda_1) \otimes Y(\lambda_2) = Y(\lambda_1 + \lambda_2) \oplus N$

where N has a good filtration with successive quotients $Y(\lambda_4)$, $Y(\lambda_2)$, $Y(2\lambda_1)$, $Y(\lambda_1 + \lambda_4)$, $Y(\lambda_3)$.

For $\xi \in X^+$ we let $S(\xi) = \{\tau \in X^+ : Y(\xi) \otimes Y(\tau)$ has a good filtration$\}$. Suppose that $\tau \in S(\lambda_2)$. Then, by (7.2.15), $Y(\lambda_1) \otimes Y(\lambda_1) \otimes Y(\lambda_2)$ has a good filtration. However, by (7.13.1) and Proposition 3.2.4, $Y(\lambda_1) \otimes Y(\lambda_1)$ has a submodule, say D , such that $(Y(\lambda_1) \otimes Y(\lambda_1))/D \cong Y(2\lambda_1)$ and D has a good filtration with quotients $Y(0)$, $Y(\lambda_1)$, $Y(\lambda_4)$, $Y(\lambda_2)$. Now $D \otimes Y(\tau)$ has a good filtration by (7.2.15), (7.3.13) and the assumption $\tau \in S(\lambda_2)$. Hence $Y(2\lambda_1) \otimes Y(\tau)$, which is isomorphic to $(Y(\lambda_1) \otimes Y(\lambda_1) \otimes Y(\tau))/(D \otimes Y(\tau))$, has a good filtration by Proposition 3.2.4, i.e. $\tau \in S(2\lambda_1)$. We have shown:

(7.13.6) $S(\lambda_2) \subseteq S(2\lambda_1)$.

Similarly, using (7.13.2), we have:

(7.13.7) $S(\lambda_2) \subseteq S(\lambda_1 + \lambda_4)$

and, using (7.13.3), (7.13.4), (7.13.6), (7.13.7)

(7.13.8) $S(\lambda_2) \subseteq S(\lambda_3)$.

We shall now show:

(7.13.9) $\underline{H}_1(G)$ *is true.*

If not then we must have, by Proposition 3.5.4, that $Y(\lambda_i) \otimes Y(\tau)$ does not have a good filtration for some $1 \leq i \leq 4$ and some $\tau \in X^+$. We choose τ to be minimal such that this occurs for some $1 \leq i \leq 4$, that is minimal with respect to the partial order $\prec$ such that, for $\lambda , \mu \in X$, $\lambda \prec \mu$ if $\mu - \lambda$ is a linear combination of dominant weights with non-negative coefficients and some coefficient positive. By (7.2.15), (7.3.13) we must have $i = 2$ or 3 and so, by (7.13.8), $Y(\lambda_2) \otimes Y(\tau)$ does not have a good filtration. Suppose that $(\tau,\alpha_1^\vee) > 0$ and let $\tau' = \tau - \lambda_1$. Then $Y(\tau') \otimes Y(\lambda_1)$ has a good filtration and so, by Proposition 3.2.6, has a submodule, N , which has a good filtration and such that $(Y(\tau') \otimes Y(\lambda_1))/N$ is isomorphic to $Y(\tau)$. Now $Y(\lambda_2) \otimes N$ has a good filtration, by the minimality of τ and

$Y(\lambda_2) \otimes Y(\tau') \otimes Y(\lambda_1)$ has a good filtration, by (7.2.15) and the minimality of τ. Thus $Y(\lambda_2) \otimes Y(\tau) \cong (Y(\lambda_2) \otimes Y(\tau') \otimes Y(\lambda_1))/(Y(\lambda_2) \otimes N)$ has a good filtration by Proposition 3.2.4. Hence we have $(\tau,\alpha_1^\vee) = 0$ and similarly $(\tau,\alpha_4^\vee) = 0$. Now suppose that $(\tau,\alpha_3^\vee) > 0$ and let $\tau' = \tau - \lambda_3$. By the above argument $Y(\lambda_2) \otimes Y(\tau') \otimes Y(\lambda_3)$ cannot have a good filtration. By the minimality of τ, $Y(\lambda_2) \otimes Y(\tau')$ has a good filtration and, since $\lambda_2 < \lambda_3$, for any successive quotient $Y(\xi)$ in a good filtration of $Y(\lambda_2) \otimes Y(\tau')$, $Y(\xi) \otimes Y(\lambda_2)$ has a good filtration. Thus it follows from (7.13.8) that $Y(\lambda_2) \otimes Y(\tau') \otimes Y(\lambda_3)$ does have a good filtration. Hence $(\tau,\alpha_3^\vee) = 0$, that is $\tau = m\lambda_2$, for some $m > 0$. By (7.12.5) $m \geq 2$. It is an easy matter to check that, for any weight ξ of $Y(\lambda_2)$, $(\xi,\alpha_2^\vee) \geq -3$. It follows that if $Y_2(\eta)$ is a successive quotient in a good filtration of $Y(\lambda_2)|_{P_2} \otimes \tau$, then $(\eta,\alpha_2^\vee) \geq -1$ and so, by (5.2), $Y(\lambda_2) \otimes Y(\tau) = Ind_{P_2}^G (Y(\lambda_2)|_{P_2} \otimes \tau)$ has a good filtration. This completes the proof of (7.13.9).

We now show:

(7.13.10) $\underline{H}(G)$ *is true.*

Thanks to (7.13.9) and Proposition 3.5.4 (iii) we need only show that $Y(\lambda_i)|_{P_\Sigma}$ has a good filtration for all $1 \leq i \leq 4$, $\Sigma \subseteq \Delta$, $\Sigma \neq \Delta$. Moreover, by (5.3), (7.3.12) and (7.11.1) it only remains to show that $Y(\lambda_3)|_{P_\Sigma}$ has a good filtration. However $(Y(\lambda_1) \otimes Y(\lambda_1))|_{P_\Sigma}$ has a good filtration by $\underline{H}(P_\Sigma)$ since $Y(\lambda_1)|_{P_\Sigma}$ has a good filtration. Moreover $D|_{P_\Sigma}$ has a good filtration by (7.2.13), (7.3.12) and (7.11.1). Hence, by Proposition 3.2.4, $Y(2\lambda_1)|_{P_\Sigma}$ has a good filtration. Similarly $Y(\lambda_1 + \lambda_4)|_{P_\Sigma}$ has a good filtration and now using (7.13.5) together with Proposition 3.2.4 and Corollary 3.2.5 we obtain that $Y(\lambda_3)|_{P_\Sigma}$ has a good filtration.

There has been a standing assumption throughout most of this section that the characteristic of k is 2. However we already have,

(7.4.8), that $\underline{H}(G)$ is true if the characteristic of p is not 2. Thus we have:

(7.13.11) *For* G *a semisimple, simply connected affine algebraic group of type* F_4, *over an algebraically closed field of arbitrary characteristic,* H(G) *is true.*

8. $\underline{E}_6$

For a group G of type E_n $(n = 6, 7 \text{ or } 8)$ we label the elements α_i $(1 \le i \le n)$ of Δ such that $(\alpha_i, \alpha_{i+1}^{\vee}) = -1$ for $1 \le i \le n-2$. Let G be semisimple, simply connected of type E_6, over an algebraically closed field k. We use the notation $\langle a_1\ a_2\ \overset{a_6}{a_3}\ a_4\ a_5 \rangle$ for $a_i \in \mathbb{Z}$, $1 \le i \le 6$ to denote $\Sigma a_i \alpha_i$. Also, we denote by $a_1\ a_2\ \overset{a_6}{a_3}\ a_4\ a_5$ the weight $\Sigma a_i \lambda_i$ (λ_i is the fundamental weight corresponding to α_i) and a zero occurring in such an expression may be replaced by a dot. Similar conventions apply to the groups of type E_7 and E_8 considered in chapters 9 and 10. We have, from p.69 of [34] (with suitable relabelling),

(8.1) $\lambda_1 = \frac{1}{3}\langle 4\ 5\ \overset{3}{6}\ 4\ 2 \rangle$, $\qquad \lambda_2 = \frac{1}{3}\langle 5\ 10\ \overset{6}{12}\ 8\ 4 \rangle$,

$\lambda_3 = \langle 2\ 4\ \overset{3}{6}\ 4\ 2 \rangle$, $\qquad \lambda_4 = \frac{1}{3}\langle 4\ 8\ \overset{6}{12}\ 10\ 5 \rangle$,

$\lambda_5 = \frac{1}{3}\langle 2\ 4\ \overset{3}{6}\ 5\ 4 \rangle$, $\qquad \lambda_6 = \langle 1\ 2\ \overset{2}{3}\ 2\ 1 \rangle$.

We note

(8.2) the W-orbit of λ_1 is $1\ \cdot\ \overset{\cdot}{\cdot}\ \cdot\ \cdot$, $-1\ 1\ \overset{\cdot}{\cdot}\ \cdot\ \cdot$, $\cdot\ -1\ \overset{\cdot}{1}\ \cdot\ \cdot$,

$\cdot\ \cdot\ \overset{1}{-1}\ 1\ \cdot$, $\cdot\ \cdot\ \overset{-1}{\cdot}\ 1\ \cdot$, $\cdot\ \cdot\ \overset{1}{\cdot}\ -1\ 1$, $\cdot\ \cdot\ \overset{-1}{1}\ -1\ 1$, $\cdot\ \cdot\ \overset{1}{\cdot}\ \cdot\ -1$,

$\cdot\ 1\ \overset{\cdot}{-1}\ \cdot\ 1$, $\cdot\ \cdot\ \overset{-1}{1}\ \cdot\ -1$, $1\ -1\ \overset{\cdot}{\cdot}\ \cdot\ 1$, $\cdot\ 1\ \overset{\cdot}{-1}\ 1\ -1$, $-1\ \cdot\ \overset{\cdot}{\cdot}\ \cdot\ 1$,

$1\ -1\ \overset{\cdot}{\cdot}\ 1\ -1$, $\cdot\ 1\ \overset{\cdot}{\cdot}\ -1\ \cdot$, $-1\ \cdot\ \overset{\cdot}{\cdot}\ 1\ -1$, $1\ -1\ \overset{\cdot}{1}\ -1\ \cdot$, $-1\ \cdot\ \overset{\cdot}{1}\ -1\ \cdot$,

$1\ \cdot\ \overset{1}{-1}\ \cdot\ \cdot$, $-1\ 1\ \overset{1}{-1}\ \cdot\ \cdot$, $1\ \cdot\ \overset{-1}{\cdot}\ \cdot\ \cdot$, $\cdot\ -1\ \overset{1}{\cdot}\ \cdot\ \cdot$, $-1\ 1\ \overset{-1}{\cdot}\ \cdot\ \cdot$,

$\cdot\ -1\ \overset{-1}{1}\ \cdot\ \cdot$, $\cdot\ \cdot\ \overset{\cdot}{-1}\ 1\ \cdot$, $\cdot\ \cdot\ \overset{\cdot}{\cdot}\ -1\ 1$, $\cdot\ \cdot\ \overset{\cdot}{\cdot}\ \cdot\ -1$.

Throughout this section Ω denotes $\Delta \setminus \{\alpha_1\}$. It follows from (2.2.3) and (2.2.5) that

(8.3) $\chi(\lambda_1) = \chi_\Omega(\lambda_1) + \chi_\Omega(\lambda_2 - \lambda_1) + \chi_\Omega(\lambda_5 - \lambda_1)$.

Since λ_1 and λ_5 are miniscule we have, by Proposition 4.2.1, the following result.

(8.4) For $i = 1$ or 5 , every $\Sigma \subseteq \Delta$ and every $\tau \in X^+$, $Y(\lambda_i)|_{P_\Sigma}$ has a good filtration and $Y(\lambda_i) \otimes Y(\tau)$ has a good filtration.

In particular $Y(\lambda_1)|_{P_\Omega}$ has a good filtration. We put $M = F_\Omega^{-2/3} Y(\lambda_1) \cong Y_\Omega(\lambda_5 - \lambda_1)$. By (8.3), $Y(\lambda_1)/M$ is a non-split extension of $Y_\Omega(\lambda_2 - \lambda_1)$ by $Y_\Omega(\lambda_1)$. Moreover $Hom_B(Y(\lambda_1), Y_\Omega(\lambda_2 - \lambda_1)) = Hom_{P_\Omega}(Y(\lambda_1), Y_\Omega(\lambda_2 - \lambda_1)) \cong Hom_G(Y(\lambda_1), Y(\lambda_2 - \lambda_1)) = 0$, by (2.1.3), (1.5.1) and the transitivity of induction. We have shown:

(8.5) $Y(\lambda_1)/M$ *is a non-split* B-*module extension of* $Y_\Omega(\lambda_2 - \lambda_1)$ *by* $Y_\Omega(\lambda_1)$.

We now analyse $Y(\lambda_6)$. The character $\chi(\lambda_6)$ is the character of the adjoint representation (see p.45 of [51]) and it is not difficult to check that

(8.6) $\chi(\lambda_6) = \chi_\Omega(\lambda_6) + \chi_\Omega(\lambda_4 - \lambda_1) + \chi_\Omega(0) + \chi_\Omega(\lambda_2 - 2\lambda_1)$.

Let $Y = Y(\lambda_6)$. Since $F_\Omega^{-1} Y$ has simple B-socle $-\lambda_6$ and its character is $\chi_\Omega(\lambda_2 - 2\lambda_1)$ we have $F_\Omega^{-1} Y \cong Y_\Omega(\lambda_2 - 2\lambda_1)$. Now $H^i(B,Y) = 0$ for all $i \geq 0$, by (1.5.2) and (2.1.4), so that $(Y/F_\Omega^{-1}Y)^B \cong H^1(B, Y_\Omega(\lambda_2 - 2\lambda\ ))$ by the long exact sequence of cohomology. Moreover $H^1(B, Y_\Omega(\lambda_2 - 2\lambda_1)) \cong k$ by (2.1.3) and (1.4.7) so that, by (2.1.3), $(Y/F_\Omega^{-1}Y)^{P_\Omega} = k$. We let D be the P_Ω-submodule of Y with $D/F_\Omega^{-1}Y = k$, the trivial P_Ω-module. Since $Y^B = 0$, $Y \otimes \lambda_1$ is a non-split extension of $Y_\Omega(\lambda_2 - \lambda_1)$ by $\ _1$ and thus by (8.5), (5.10), (5.11) and (1.4.7), D is isomorphic to $(Y(\lambda_1)/M \otimes -\lambda_1$. We have now shown:

(8.7) *there is a short exact sequence of* B-*modules*

$$0 \to Y_\Omega(\lambda_5 - 2\lambda_1) \to Y(\lambda_1) \otimes -\lambda_1 \to D \to 0 .$$

Let $E = F_\Omega^0 Y$. We have $ch\, E/D = \chi_\Omega(\lambda_4 - \lambda_1)$ so that, by (1.5.4), to prove that E/D is isomorphic to $Y_\Omega(\lambda_4 - \lambda_1)$ it suffices to show that E/D has a simple B-socle $\lambda_1 - \lambda_4$. Since $\lambda_1 - \lambda_4$ occurs with multiplicity one as a weight of E/D, it is enough to prove that $((E/D) \otimes \tau)^B = 0$ for any τ in X such that $-\tau$ is a weight of E/D. The only such τ is 0 so it suffices to show that $(E/D)^B = 0$. By the long exact sequence of cohomology $(Y/D)^B = H^1(B,D)$. We have $H^1(B,D) = H^2(B,Y_\Omega(\lambda_5 - 2\lambda_1))$ by (8.7) and (2.1.6). The latter is 0 by (2.3.3) and (2.1.4). Thus $(Y/D)^B$ and hence $(E/D)^B = 0$ and E/D is isomorphic to $Y_\Omega(\lambda_4 - \lambda_1)$. Now $ch\, Y_\Omega(\lambda_6) = s_\Omega(\lambda_6)$ so that $Y_\Omega(\lambda_6)$ is a simple module and Y/E is isomorphic to $Y_\Omega(\lambda_6)$. We have shown:

(8.8) $Y(\lambda_6)|_{P_\Omega}$ has a good filtration.

Now let $\Sigma = \Delta \backslash \{\alpha_j\}$, $j = 1$ and $F = Ind_{P_\Gamma}^{P_\Sigma}$, where $\Gamma = \Sigma \cap \Omega$. Thus $Y(\lambda_6)|_{P_\Sigma}$ is isomorphic to $F(Y(\lambda_6)|_{P_\Gamma})$ and, to show that $Y(\lambda_6)|_{P_\Sigma}$ has a good filtration, it suffices to prove that $F(Y(\lambda_6)|_{P_\Gamma})$ has a good filtration. Thus it suffices to prove that $F(D)$, $F(Y(\lambda_6)/D)$ each has a good filtration and that $RF(D) = 0$. From (8.7), (2.1.5) and the tensor identity we obtain, via the long exact sequence of derived functors,

$$R^i F(D) = R^{i+1} F(Y_\Omega(\lambda_5 - 2\lambda_1)) \qquad \text{for all } i \geq 0$$

so we need to show that $R^i F(Y_\Omega(\lambda_5 - 2\lambda_1))$ has a good filtration for $i = 1$ and is 0 for $i = 2$. However, if τ is a weight of $Y_\Omega(\lambda_5 - 2\lambda_1)$ then $\tau = \lambda_5 - 2\lambda_1 - \sum_{i=2}^{6} m_i \alpha_i$ for some $m_i \geq 0$, by (1.5.2) (applied to G_Ω). Thus $(\tau, \alpha_1^\vee) \geq -2$. Also $-\lambda_5$ is the lowest weight of $Y(\lambda_1)$ and hence of $Y_\Omega(\lambda_5 - \lambda_1)$. Thus the lowest weight of $Y_\Omega(\lambda_5 - 2\lambda_1)$ is $-\lambda_5 - \lambda_1$ and we obtain, by (1.5.2), $\tau = -\lambda_5 - \lambda_1 + \sum_{i=2}^{6} n_i\alpha_i$ with $n_i \geq 0$. Hence $(\tau, \alpha_1^\vee) \leq -1$ and so $RF(Y_\Omega(\lambda_5 - 2\lambda_1))$ has a good filtration and $R^2F(Y_\Omega(\lambda_5 - 2\lambda_1)) = 0$ by (5.7). Also, $Y(\lambda_6)/D$ has a filtration with successive quotients $Y_\Omega(\lambda_4 - \lambda_1), Y_\Omega(\lambda_6)$ so $F(Y(\lambda_6)/D)$ has a good filtration be Lemma 3.7.1. We have now shown:

(8.9) $Y(\lambda_6)|_{P_\Sigma}$ *has a good filtration for any* $\Sigma \subseteq \Delta$.

Let τ be a weight of $Y(\lambda_6)$. Then τ is either 0 or $\tau = w\lambda_6$ for some $w \in W$. For $\alpha \in \Phi$, $(w\lambda_6, \alpha^\vee) = (\lambda_6, w^{-1}\alpha^\vee) \geq (\lambda_6, -\beta_0^\vee) = -2$, where β_0 is the highest root (i.e. λ_6). Thus $(\tau, \alpha^\vee) \geq -2$ and we now have, by Proposition 3.5.2:

(8.10) $Y(\lambda_6) \otimes Y(\tau)$ *has a good filtration for any* $\tau \in X^+$.

We are now able to deduce that $H(\underline{G})$ is true but, as with F_4 , the deduction is not uniform in p . We need to consider separately the cases $p \neq 2$ and $p = 2$. The highest weight of $\Lambda^2 Y(\lambda_1)$ is λ_2 and $\dim \Lambda^2 Y(\lambda_1) = \dim Y(\lambda_2)$ (see p.45 of [51]) so that

(8.11) $\Lambda^2 \chi(\lambda_1) = \chi(\lambda_2)$.

Similar dimension arguments (using p.45 of [51]) imply that

(8.12) $\Lambda^2 \chi(\lambda_6) = \chi(\lambda_3) + \chi(\lambda_6)$.

Now assume that $p \neq 2$. Then $\Lambda^2 Y(\lambda_1)$ is a component of $Y(\lambda_1) \otimes Y(\lambda_1)$. Thus, by (8.4) the truth of $\underline{H}(\Phi_\Sigma)$ for proper subsets Σ of Δ and Corollary 3.2.5, $\Lambda^2 Y(\lambda_1)|_{P_\Sigma}$ has a good filtration for any $\Sigma \subseteq \Delta$. In particular the G-module $\Lambda^2 Y(\lambda_1)$ has a good filtration and so, by (8.11), is isomorphic to $Y(\lambda_2)$. Thus $Y(\lambda_2)|_{P_\Sigma}$ has a good filtration for any $\Sigma \subseteq \Delta$. By symmetry $Y(\lambda_4)|_{P_\Sigma}$ has a good filtration for any $\Sigma \subseteq \Delta$. Also, for any $\tau \in X^+$, $Y(\lambda_1) \otimes Y(\lambda_1) \otimes Y(\tau)$

has a good filtration by (8.4) and so, by Corollary 3.2.5, $Y(\lambda_2) \otimes Y(\tau)$ has a good filtration. Similarly $Y(\lambda_4) \otimes Y(\tau)$ has a good filtration for any $\tau \in X^+$. Let $V = \Lambda^2 Y(\lambda_6)$. By (8.10) above and Corollary 3.2.5, V has a good filtration. Thus, by (8.12) and Proposition 3.2.6, V has a submodule U isomorphic to $Y(\lambda_6)$ with V/U isomorphic to $Y(\lambda_3)$. Now by (8.9), the truth of $\underline{H}(\Phi_\Sigma)$ for proper subsets Σ of Δ and Corollary 3.2.5, $V|_{P_\Sigma}$ has a good filtration for any $\Sigma \subseteq \Delta$. Hence, by (8.9) above and Proposition 3.2.4, $Y(\lambda_3)|_{P_\Sigma}$ has a good filtration for any $\Sigma \subseteq \Delta$. Also, for $\tau \in X^+$, $Y(\lambda_6) \otimes Y(\lambda_6) \otimes Y(\tau)$ has a good filtration by (8.10). Thus $V \otimes Y(\tau)$ has a good filtration by Corollary 3.2.5, and $Y(\lambda_3)$ has a good filtration by (8.10) and Proposition 3.2.4. We have now shown that for every $1 \le i \le 6$, $\Sigma \subseteq \Delta$ and $\tau \in X^+$ $Y(\lambda_i) \otimes Y(\tau)$ has a good filtration and $Y(\lambda_i)|_{P_\Sigma}$ has a good filtration. By Proposition 3.5.4 we thus have the following.

(8.13) *If the characteristic of* k *is not* 2 *then* $\underline{H}(G)$ *is true.*

To prove $\underline{H}(G)$ in characteristic 2 we use the method of section 7.13. Using (2.2.3), (2.2.5) and (8.2) it is easy to check that

(8.14) $\chi(\lambda_1)\chi(\lambda_6) = \chi(\lambda_1 + \lambda_6) + \chi(\lambda_4) + \chi(\lambda_1)$.

Suppose now the characteristic p of k is 2. We have $\lambda_1 + \lambda_2 + \rho = 2\;1\;\overset{2}{1}\;1\;2$ which is conjugate under W_p (translating by $-p(\lambda_3 + \lambda_6)$) to $2\;1\;\overset{\cdot}{-1}\;1\;1$. The latter weight is conjugate under W to $2\;\cdot\;\overset{1}{\cdot}\;\cdot\;1$ which is conjugate (subtracting $2\lambda_6$) to $2\;\cdot\;\overset{-1}{\cdot}\;\cdot\;1$. This weight is conjugate under W to $1\;\cdot\;\overset{1}{\cdot}\;\cdot\;\cdot$ which is W_p conjugate to $1\;\cdot\;\overset{-1}{\cdot}\;\cdot\;\cdot$ and hence to $1\;\cdot\;\overset{\cdot}{\cdot}\;\cdot\;\cdot$ which belongs to C since $(\lambda_1, \beta_o^\vee) = 1$ where $\beta_o = \langle 12\overset{2}{3}21 \rangle$ is the highest root. Similar calculations reveal that $\lambda_4 + \rho$ is W_p conjugate to λ_4 and that $\lambda_1 + \rho$ is also conjugate to λ_4. Now $M = Y(\lambda_1) \otimes Y(\lambda_6)$ has a good filtratration by (8.4) and so, by Corollary 3.2.5, M_λ (see section 7.13) has a good filtration for each $\lambda \in C$. Thus (see (7.13.4)) $M = M_{\lambda_1} \oplus M_{\lambda_4}$, M_{λ_1} is isomorphic to $Y(\lambda_1 + \lambda_6)$ and M_{λ_4} has a good filtration with quotients $Y(\lambda_1)$ and $Y(\lambda_4)$. Let $N = M_{\lambda_4}$. By Proposition 3.2.6, N has a submodule, say K, isomorphic to $Y(\lambda_1)$ such that N/K is isomorphic to $Y(\lambda_4)$.

Let S be the set of G-modules V such that $V \otimes V'$ has a good filtration for every G-module V' with a good filtration and $V|_{P_\Sigma}$ has a good filtration for every $\Sigma \subseteq \Delta$. We have $Y(\lambda_1), Y(\lambda_5), Y(\lambda_6) \in S$ by (8.4), (8.9) and (8.10). Hence, by (1.12) Proposition, $M \in S$ and so, also by (5.15), $N \in S$. However, $K \cong Y(\lambda_1)$ so by a further application of (5.15), $Y(\lambda_4) \cong N/K \in S$. By symmetry we also have $Y(\lambda_2) \in S$. It follows from (8.2) and (2.2.3), (2.2.5) that

$$(8.15) \qquad \chi(\lambda_1)\chi(\lambda_5) = \chi(\lambda_1 + \lambda_5) + \chi(\lambda_6) + \chi(0) .$$

Thus, by (8.4) and Proposition 3.2.6, $Y(\lambda_1) \otimes Y(\lambda_5)$ has a submodule S with a good filtration with quotients $Y(0)$, $Y(\lambda_6)$ and $(Y(\lambda_1) \otimes Y(\lambda_5))/S$ isomorphic to $Y(\lambda_1 + \lambda_5)$. Since $S \in S$ we have, by (5.15), $Y(\lambda_1 + \lambda_5) \in S$.

We shall now show that $Y(\lambda_3) \in S$. We have, by (8.2) and (2.2.3), (2.2.5),

$$(8.16) \qquad \chi(\lambda_1)\chi(\lambda_2) = \chi(\lambda_1 + \lambda_2) + \chi(\lambda_3) + \chi(\lambda_1 + \lambda_5) + \chi(\lambda_6) .$$

We leave it to the reader to check that $\lambda_1 + \lambda_2 + \rho$ is W_p conjugate to $\lambda_1 + \lambda_5$ and $\lambda_3 + \rho$ is W_p conjugate to λ_6. Let $Z = Y(\lambda_1) \otimes Y(\lambda_2)$. Then, by (8.16) and (7.13.4), $Z' = Z_{\lambda_6}$ is a G-component of Z with a good filtration not involving $Y(\lambda_1 + \lambda_2)$ (Z' has a good filtration by (8.4) and Corollary 3.2.5). Thus, by Proposition 3.2.6, Z' has a G-submodule Z'' with a good filtration such that Z'/Z'' is isomorphic to $Y(\lambda_3)$. By (8.16) the possible quotients in a good filtration of Z'' are $Y(\lambda_6)$ and $Y(\lambda_1 + \lambda_5)$.

Hence Z" belongs to S . Also $z \in S$ since $Y(\lambda_1), Y(\lambda_2) \in S$. Thus Z' belongs to S by (5.15) and, also by (5.15), $Y(\lambda_3) \cong Z'/Z''$ belongs to S . We have now shown that $Y(\lambda_i) \in S$ for $1 \le i \le 6$ and by Proposition 3.5.4, the proof of $\underline{H}(G)$, when the characteristic of k is 2, is complete. Combining this with (8.13) we have:

(8.17) *Let* G *be a semisimple, simply connected algebraic group of type* E_6 *over an algebraically closed field (of arbitrary characteristic). Then* $\underline{H}(G)$ *is true.*

Combining the main results on filtrations obtained so far, Theorem 4.3.1, (6.7), (7.13.10) and (8.17) we obtain, thanks to Corollary 3.4.5 and Corollary 3.4.7, what we consider to be the most attractive result of this monograph.

(8.18) Theorem *For every root system* Φ *which has no indecomposable component of type* E_7 *or* E_8 , $\underline{H}(\Phi)$ *is true. That is, for every connected algebraic group* G , *over an algebraically closed field, such that the root system of* G *has no component of type* E_7 *or* E_8 *the hypotheses* $\underline{H}_1(G)$ *and* $\underline{H}_2(G)$ *are true.*

For groups of type E_7 and E_8 , with which we are concerned in the next two chapters, we are only able to prove the hypotheses when the characteristic of the field involved is not 2 .

9. $\underline{E}_7$

9.1 $Y(\lambda_1)$

Let G be a semisimple, simply connected group of type E_7 over an algebraically closed field k. In this chapter we show that, provided that the characteristic of k is not 2, $\underline{H}(G)$ is true. The proof is quite straightforward in characteristic different from 2 and 7. The extra analysis needed to deal with characteristic 7 is quite unpleasant (as with F_4 in characteristic 2). So far we have no complete proof in characteristic 2. By p.69 of [34] we have (with suitable relabelling to conform with the labelling of the Dynkin-Diagram explained at the beginning of Chapter 8):

(9.1.1) $\lambda_1 = \frac{1}{2}\langle 3\,4\,5\,\overset{3}{6}\,4\,2\rangle$, $\lambda_2 = \langle 2\,4\,5\,\overset{3}{6}\,4\,2\rangle$

$\lambda_3 = \frac{1}{2}\langle 5\ 10\ 15\ \overset{9}{18}\ 12\ 6\rangle$, $\lambda_4 = \langle 3\ 6\ 9\ \overset{6}{12}\ 8\ 4\rangle$,

$\lambda_5 = \langle 2\,4\,6\,\overset{4}{8}\,6\,3\rangle$, $\lambda_6 = \langle 1\,2\,3\,\overset{2}{4}\,3\,2\rangle$,

$\lambda_7 = \frac{1}{2}\langle 3\ 6\ 9\ \overset{7}{12}\ 8\ 4\rangle$.

It is convenient to note:

(9.1.2) *The* W-*orbit of* λ_1 *is* $1\ \cdot\ \cdot\ \overset{\cdot}{\cdot}\ \cdot\ \cdot$, $-1\ 1\ \cdot\ \overset{\cdot}{\cdot}\ \cdot\ \cdot$,

$\cdot\ -1\ 1\ \overset{\cdot}{\cdot}\ \cdot\ \cdot$, $\cdot\ \cdot\ -1\ \overset{\cdot}{1}\ \cdot\ \cdot$, $\cdot\ \cdot\ \cdot\ \overset{1}{-1}\ 1\ \cdot$, $\cdot\ \cdot\ \cdot\ \overset{-1}{\cdot}\ 1\ \cdot$,

$\cdot\ \cdot\ \cdot\ \overset{1}{\cdot}\ -1\ 1$, $\cdot\ \cdot\ \cdot\ \overset{-1}{1}\ -1\ 1$, $\cdot\ \cdot\ \cdot\ \overset{1}{\cdot}\ \cdot\ -1$, $\cdot\ \cdot\ 1\ \overset{\cdot}{-1}\ \cdot\ 1$,

$\cdot\ \cdot\ \cdot\ \overset{-1}{1}\ \cdot\ -1$, $\cdot\ 1\ -1\ \overset{\cdot}{\cdot}\ \cdot\ 1$, $\cdot\ \cdot\ 1\ \overset{\cdot}{-1}\ 1\ -1$, $1\ -1\ \cdot\ \overset{\cdot}{\cdot}\ \cdot\ 1$,

$\cdot\ 1\ -1\ \overset{\cdot}{\cdot}\ 1\ -1$, $\cdot\ \cdot\ 1\ \overset{\cdot}{\cdot}\ -1\ \cdot$, $-1\ \cdot\ \cdot\ \overset{\cdot}{\cdot}\ \cdot\ 1$, $1\ -1\ \cdot\ \overset{\cdot}{\cdot}\ 1\ -1$,

$\cdot\ 1\ -1\ \overset{\cdot}{1}\ -1\ \cdot$, $-1\ \cdot\ \cdot\ \overset{\cdot}{\cdot}\ 1\ -1$, $1\ -1\ \cdot\ \overset{\cdot}{1}\ -1\ \cdot$, $\cdot\ 1\ \cdot\ \overset{1}{-1}\ \cdot\ \cdot$,

$-1\ \cdot\ \cdot\ \overset{\cdot}{1}\ -1\ \cdot$, $1\ -1\ 1\ \overset{1}{-1}\ \cdot\ \cdot$, $\cdot\ 1\ \cdot\ \overset{-1}{\cdot}\ \cdot\ \cdot$, $-1\ \cdot\ 1\ \overset{1}{-1}\ \cdot\ \cdot$,

$1\ \cdot\ -1\ \overset{1}{\cdot}\ \cdot\ \cdot$, $1\ -1\ 1\ \overset{-1}{\cdot}\ \cdot\ \cdot$, $-1\ 1\ -1\ \overset{1}{\cdot}\ \cdot\ \cdot$, $-1\ \cdot\ 1\ \overset{-1}{\cdot}\ \cdot\ \cdot$,

$\begin{smallmatrix} &&&-1&& \\ 1&\cdot&-1&1&\cdot&\cdot \end{smallmatrix}$, $\begin{smallmatrix} &&&1&& \\ \cdot&-1&\cdot&\cdot&\cdot&\cdot \end{smallmatrix}$, $\begin{smallmatrix} &&&-1&& \\ -1&1&-1&1&\cdot&\cdot \end{smallmatrix}$, $\begin{smallmatrix} &&&\cdot&& \\ 1&\cdot&\cdot&-1&1&\cdot \end{smallmatrix}$,

$\begin{smallmatrix} &&&-1&& \\ \cdot&-1&\cdot&1&\cdot&\cdot \end{smallmatrix}$, $\begin{smallmatrix} &&&\cdot&& \\ -1&1&\cdot&-1&1&\cdot \end{smallmatrix}$, $\begin{smallmatrix} &&&\cdot&& \\ 1&\cdot&\cdot&\cdot&-1&1 \end{smallmatrix}$, $\begin{smallmatrix} &&&\cdot&& \\ \cdot&-1&1&-1&1&\cdot \end{smallmatrix}$,

$\begin{smallmatrix} &&&\cdot&& \\ -1&1&\cdot&\cdot&-1&1 \end{smallmatrix}$, $\begin{smallmatrix} &&&\cdot&& \\ 1&\cdot&\cdot&\cdot&\cdot&-1 \end{smallmatrix}$, $\begin{smallmatrix} &&&\cdot&& \\ \cdot&\cdot&-1&\cdot&1&\cdot \end{smallmatrix}$, $\begin{smallmatrix} &&&1&& \\ \cdot&-1&1&\cdot&-1&1 \end{smallmatrix}$,

$\begin{smallmatrix} &&&\cdot&& \\ -1&1&\cdot&\cdot&\cdot&-1 \end{smallmatrix}$, $\begin{smallmatrix} &&&\cdot&& \\ \cdot&\cdot&-1&1&-1&1 \end{smallmatrix}$, $\begin{smallmatrix} &&&\cdot&& \\ \cdot&-1&1&\cdot&\cdot&-1 \end{smallmatrix}$, $\begin{smallmatrix} &&&1&& \\ \cdot&\cdot&\cdot&-1&\cdot&1 \end{smallmatrix}$,

$\begin{smallmatrix} &&&\cdot&& \\ \cdot&\cdot&-1&1&\cdot&-1 \end{smallmatrix}$, $\begin{smallmatrix} &&&-1&& \\ \cdot&\cdot&\cdot&\cdot&\cdot&1 \end{smallmatrix}$, $\begin{smallmatrix} &&&1&& \\ \cdot&\cdot&\cdot&-1&1&-1 \end{smallmatrix}$, $\begin{smallmatrix} &&&-1&& \\ \cdot&\cdot&\cdot&\cdot&1&-1 \end{smallmatrix}$,

$\begin{smallmatrix} &&&1&& \\ \cdot&\cdot&\cdot&\cdot&-1&\cdot \end{smallmatrix}$, $\begin{smallmatrix} &&&-1&& \\ \cdot&\cdot&\cdot&1&-1&\cdot \end{smallmatrix}$, $\begin{smallmatrix} &&&\cdot&& \\ \cdot&\cdot&1&-1&\cdot&\cdot \end{smallmatrix}$, $\begin{smallmatrix} &&&\cdot&& \\ \cdot&1&-1&\cdot&\cdot&\cdot \end{smallmatrix}$,

$\begin{smallmatrix} &&&\cdot&& \\ 1&-1&\cdot&\cdot&\cdot&\cdot \end{smallmatrix}$, $\begin{smallmatrix} &&&\cdot&& \\ -1&\cdot&\cdot&\cdot&\cdot&\cdot \end{smallmatrix}$.

Let $\Omega = \Delta\backslash\{\alpha_1\}$. It follows from (2.2.3) and (2.2.5) that

(9.1.3) $\chi(\lambda_1) = \chi_\Omega(\lambda_1) + \chi_\Omega(\lambda_2 - \lambda_1) + \chi_\Omega(\lambda_6 - \lambda_1) + \chi_\Omega(-\lambda_1)$.

Since λ_1 is miniscule we have, by Proposition 4.2.1, the following result.

(9.1.4) *For every* $\Sigma \subseteq \Delta$ *and* $\tau \in X^+$, $Y(\lambda_1)|_{P_\Sigma}$ *has a good filtration and* $Y(\lambda_1) \otimes Y(\tau)$ *has a good filtration.*

Let $Y = Y(\lambda_1)$. By Proposition 3.2.6 and (9.1.3), there is a P_Ω submodule M of Y such that M has a good filtration with quotients (in order) $Y_\Omega(-\lambda_1)$, $Y_\Omega(\lambda_6 - \lambda_1)$ and Y/M has a good filtration with quotients $Y_\Omega(\lambda_2 - \lambda_1)$, $Y_\Omega(\lambda_1)$. We shall need:

(9.1.5) Y/M *has a simple* B *socle* $\lambda_1 - \lambda_6$.

Since the B socle of $Y_\Omega(\lambda_2 - \lambda_1)$ is $\lambda_1 - \lambda_6$ and the B socle of $Y_\Omega(\lambda_1)$ is λ_1 it suffices to show that $((Y/M) \otimes -\lambda_1)^B = 0$. Now $H^i(B, Y \otimes -\lambda_1) = 0$ for all i by (2.1.6) so it suffices to prove that $H^1(B, M \otimes -\lambda_1) = 0$. However $H^1(B, -2\lambda_1) = H^1(B, Y_\Omega(\lambda_6 - 2\lambda_1)) = 0$ by (2.1.3) and Corollary 2.3.5 so that $H^1(B, M \otimes -\lambda_1)$ is indeed 0 and the proof is complete.

9.2 $Y(\lambda_6)$

We now analyse $Y(\lambda_6)$. We leave it to the reader to check that

(9.2.1) $\chi(\lambda_6) = \chi_\Omega(\lambda_6) + \chi_\Omega(\lambda_7 - \lambda_1) + \chi_\Omega(0) + \chi_\Omega(\lambda_2 - 2\lambda_1)$

and

(9.2.2) $\chi(\lambda_1)\chi(\lambda_6) = \chi(\lambda_1 + \lambda_6) + \chi(\lambda_7) + \chi(\lambda_1)$.

By (9.1.4), (9.2.2) above and Proposition 3.2.6, $Y(\lambda_1) \otimes Y(\lambda_6)$ has a good filtration with quotients (in order) $Y(\lambda_1)$, $Y(\lambda_7)$, $Y(\lambda_1 + \lambda_6)$. Hence by (1.5.3), $\mathrm{Hom}_G(Y(\lambda_1), Y(\lambda_1) \otimes Y(\lambda_6)) \cong k$ and so, using reciprocity and the tensor identity, $\mathrm{Hom}_{P_\Omega}(Y(\lambda_1) \otimes -\lambda_1, Y(\lambda_6)) \cong k$. Let $\theta : Y(\lambda_1) \otimes -\lambda_1 \to Y(\lambda_6)$ be a non-zero P_Ω map and let N denote the image of θ . Since the B socle of $Y(\lambda_6)$ is $-\lambda_6$ we have $\theta(v) \neq 0$ for some element v of $Y(\lambda_1) \otimes -\lambda_1$ of weight $-\lambda_6$. Now $\lambda_1 - \lambda_6$ is conjugate to λ_1 and so occurs exactly once as a weight of $Y(\lambda_1)$. Thus, by (9.1.5), $v + M \otimes -\lambda_1$ spans the B socle of $(Y(\lambda_1) \otimes -\lambda_1)/(M \otimes -\lambda_1)$. Moreover $-\lambda_6$ is not a weight of $M \otimes -\lambda_1$ so that $\theta(M \otimes -\lambda_1) = 0$ and θ induces an epimorphism $\bar\theta : (Y(\lambda_1) \otimes -\lambda_1)/(M \otimes -\lambda_1) \to N$. Since $\bar\theta$ is injective on the B socle $kv + M \otimes -\lambda_1$ of $(Y(\lambda_1) \otimes -\lambda_1)/(M \otimes -\lambda_1)$ $\bar\theta$ is injective and hence an isomorphism. Thus:

(9.2.3) *there is a short exact sequence of* P_Ω *modules*

$$0 \to M \otimes -\lambda_1 \to Y(\lambda_1) \otimes -\lambda_1 \to N \to 0 .$$

It follows that the character of N is $\chi_\Omega(\lambda_2 - 2\lambda_1) + \chi_\Omega(0)$ and, by (9.2.1), $F_\Omega^0 Y(\lambda_6)/N$ has character $\chi_\Omega(\lambda_7 - \lambda_1)$. By (1.5.4) to show that V/N is isomorphic to $Y_\Omega(\lambda_7 - \lambda_1)$, where $V = F_\Omega^0 Y(\lambda_6)$, it suffices to show that the B socle of $\bar V = V/N$ is $\lambda_1 - \lambda_7$. Since the multiplicity of $\lambda_1 - \lambda_7$ in $\bar V$ is one it suffices to show that $(\bar V \otimes \tau)^B = 0$ for any $\tau \in X_\Omega$ such that $-\tau$ is a weight of

$Y_\Omega(\lambda_7 - \lambda_1)$ and $\tau \neq \lambda_7 - \lambda_1$. However $\tau = 0$ is the only weight to satisfy these conditions so that it is enough to prove that $\overline{V}^B = 0$. Now $\overline{V}$ embeds in $Y(\lambda_6)/N$ so that it suffices to show that $(Y(\lambda_6)/N)^B = 0$. Moreover $Y(\lambda_6)^B = 0$ so that it suffices to prove that $H^1(B,N) = 0$. By (9.2.3) above and (2.1.6) we have $H^1(B,N) \cong H^2(B,M \otimes -\lambda_1)$. However $M \otimes -\lambda_1$ has a filtration with quotients $-2\lambda_1$, $Y_\Omega(\lambda_6 - 2\lambda_1)$ so the result follows from (2.1.3) and, for example Corollary 2.3.5. Thus V/N is isomorphic to $Y_\Omega(\lambda_7 - \lambda_1)$. Since $\chi_\Omega(\lambda_6) = s_\Omega(\lambda_6)$, $Y_\Omega(\lambda_6)$ is simple and hence isomorphic to $Y(\lambda_6)/V$. Thus we have shown:

(9.2.4) $Y(\lambda_6)|_{P_\Sigma}$ *has a good filtration.*

Let $\Sigma = \Delta\backslash\{\alpha_i\}$ for $2 \le i \le 7$, $\Gamma = \Omega \cap \Sigma$ and $F = Ind_{P_\Gamma}^{P_\Sigma}$. We have $Y(\lambda_6)|_{P_\Sigma} = F(Y(\lambda_6)|_{P_\Gamma})$. By (5.1), to prove that $Y(\lambda_6)|_{P_\Sigma}$ has a good filtration, it is enough to show that $F(N)$, $F(Y(\lambda_6)/N)$ each has a good filtration and that $RF(N) = 0$. By (5.2), $F(Y(\lambda_6)/N)$ has a good filtration. Thus, by (9.2.3) and the long exact sequence of derived functors it is enough to prove that $RF(M \otimes -\lambda_1)$ has a good filtration and that $R^2F(M \otimes -\lambda_1) = 0$. However these facts follow from (5.7). Thus we obtain, using the fact that $\underline{H}(\Psi)$ is true for proper subroot systems Ψ of E_7:

(9.2.5) $Y(\lambda_6)|_{P_\Sigma}$ *has a good filtration for all* $\Sigma \subseteq \Delta$

We obtain from Proposition 3.5.2:

(9.2.6) $Y(\lambda_6) \otimes Y(\tau)$ *has a good filtration for every* $\tau \in X^+$.

9.3 $p \neq 2$ or 7

In the later sections we shall show that, for every G-module V with a good filtration, $Y(\lambda_7) \otimes V$ has a good filtration and also that $Y(\lambda_7)|_{P_\Sigma}$ has a good filtration for every subset Σ of Δ. From this it may be quickly deduced that G satisfies the hypotheses provided that the characteristic of k is not 2 or 3. In this section we use a little block theory to show that $Y(\lambda_7)$ satisfies the above conditions when the characteristic p of k is not 2 or 7 and deduce $\underline{H}(G)$ when $p \neq 2$ or 7 (using a little extra block theory in the case $p = 3$). Some of this section is therefore redundant, but may be of value to the reader wishing to postpone the unpleasantly technical later sections.

It follows from (9.1.2), (2.2.3) and (2.2.5) that

$$\chi(\lambda_1)\chi(\lambda_6) = \chi(\lambda_1 + \lambda_6) + \chi(\lambda_7) + \chi(\lambda_1) \tag{9.3.1}$$

As in section 7.13 we put

$$C = \{\lambda \in X: 0 \leq (\lambda,\alpha^\vee) \leq p \text{ for all } \alpha \in \Phi^+\}$$

where we assume that the characteristic p of k is not zero. (In characteristic zero every finite dimensional rational module for each parabolic subgroup has a good filtration and so $\underline{H}(G)$ is certainly true). Let $M = Y(\lambda_1) \otimes Y(\lambda_6)$ and, as in section 7.13, for $\lambda \in C$, M_λ denotes the largest submodule of M belonging to λ. We wish to show that, when p is not very small, $Y(\lambda_7)$ occurs at the top of a good filtration of some G component of M which does not involve $Y(\lambda_1 + \lambda_6)$. To do so, as in section 7.13, we use the strong linkage principle, [4].

For $\lambda \in X$ we denote by $\tilde{\lambda}$ the element of C to which $\lambda + \rho$ is conjugate under the affine Weyl group W_p. It is not difficult to calculate $\tilde{\lambda}$ for $\lambda = \lambda_1 + \lambda_6$ and $\lambda = \lambda_7$, as in section 7.13. The results are as follows: for $p > 19$, $\widetilde{\lambda_1 + \lambda_6} = \lambda_1 + \lambda_6 + \rho$, $\tilde{\lambda}_7 = \lambda_7 + \rho$; for $p = 19$, $\widetilde{\lambda_1 + \lambda_6} = 21\overset{1}{1}11$, $\tilde{\lambda}_7 = \lambda_7 + \rho$; for $p = 17$, $\widetilde{\lambda_1 + \lambda_6} = 211\overset{1}{1}\cdot 1$, $\tilde{\lambda}_7 = 111\overset{2}{1}\cdot 1$; for $p = 13$, $\widetilde{\lambda_1 + \lambda_6} = 11\cdot\overset{1}{1}\cdot 1$, $\tilde{\lambda}_7 = \cdot 11\overset{1}{\cdot}11$; for $p = 11$, $\widetilde{\lambda_1 + \lambda_6} = 1\cdot 1\overset{\cdot}{1}\cdot 1$, $\tilde{\lambda}_7 = 2\cdot 1\overset{1}{\cdot}1\cdot$; for $p = 7$, $\widetilde{\lambda_1 + \lambda_6} = 1\cdot\overset{\cdot}{1}\cdot\cdot$, $\tilde{\lambda}_7 = \cdot 1\cdot\overset{\cdot}{1}\cdot\cdot$; for $p = 5$, $\widetilde{\lambda_1 + \lambda_6} = 1\cdot\cdot\overset{1}{\cdot}\cdot 1$, $\tilde{\lambda}_7 = \cdot\cdot\cdot\overset{\cdot}{1}\cdot\cdot$; for $p = 3$,

$\widetilde{\lambda_1 + \lambda_6} = 1\cdot\overset{1}{\cdot}\cdot\cdot\cdot\cdot$, $\tilde{\lambda}_7 = \cdot\cdot\overset{\cdot}{\cdot}\cdot 1$; for $p = 2$, $\widetilde{\lambda_1 + \lambda_6} = \cdot\cdot\overset{\cdot}{\cdot}\cdot\cdot 1$, $\tilde{\lambda}_7 = \cdot\cdot\overset{\cdot}{\cdot}\cdot\cdot 1$.

The interesting point to notice is:

(9.3.2) *if* p *is not equal to* 2 *or* 7 *then* $\widetilde{\lambda_1 + \lambda_6}$ *is not equal to* $\tilde{\lambda}_7$.

So we now assume that p is neither 2 nor 7. Then we obtain, from (9.3.2) that $M_{\tilde{\lambda}_7}$ has no composition factor of the form $L(\lambda_1 + \lambda_6)$. We let S be the set of rational G-module V such that $V \otimes V'$ has a good filtration for every rational G-module V' with a good filtration and $V|_{P_\Sigma}$ has a good filtration for every subset Σ of Δ . Thus by (9.1.4), (9.2.5) and (9.2.6), $Y(\lambda_1)$ and $Y(\lambda_6)$ belong to S . Hence by (5.15), M and hence $N = M_{\tilde{\lambda}_7}$ belongs to S . In particular N has a good filtration, this cannot involve $L(\lambda_1 + \lambda_6)$ so by (9.3.1) there is a submodule K of N such that N/K is isomorphic to $Y(\lambda_7)$ and K is either zero or isomorphic to $Y(\lambda_1)$. Since K belongs to S , by (5.15), $Y(\lambda_7)$ belongs to S .

Now assume that p is not equal to 2, 3 or 7. We shall prove that $\underline{H}(G)$ is true. Suppose, for a contradiction that $\underline{H}(G)$ is false. Then by Proposition 3.1.1, $Y(\tau) \notin S$ for some $\tau \in X^+$. Consider the order $\prec$ on X: we say $\mu \prec \lambda$ if, when $\lambda - \mu$ is expressed as a linear combination of simple roots in $R \otimes_{\mathbb{Z}} X$, each coefficient is non-negative and some coefficient is positive. If $\tau \in X^+$ is minimal with respect to this ordering such that $Y(\tau) \notin S$ then it follows, from the argument of Proposition 3.5.4 (ii) and the truth of $\underline{H}(P_\Sigma)$ for proper subsets Σ of Δ ((8.18) Theorem) that $\tau = \lambda_i$ for some $1 \le i \le 7$. Certainly $i \ne 1$, by (9.1.4). Suppose that $2 \le i \le 4$. Now $\Lambda^i Y(\lambda_1)$ is a direct summand of the ith tensor power $\otimes^i Y(\lambda_1)$. Thus $\Lambda^i Y(\lambda_1)$ belongs to S by (9.1.4) and (5.15). Moreover, the highest weight of $\Lambda^i Y(\lambda_1)$ is λ_i so that, by Proposition 3.2.6, there is a submodule S of $\Lambda^i Y(\lambda_1)$ such that S has a good filtration and $(\Lambda^i Y(\lambda_1))/S$ is isomorphic to

$Y(\lambda_i)$. Further, by the minimality of λ_i , S belongs to S and so, by (5.15), $Y(\lambda_i)$ belongs to S - a contradiction. Now $i \neq 6$ or 7 by (9.2.5), (9.2.6) and the above discussion (for $Y(\lambda_7)$) . Thus the only remaining possibility is $i = 5$. However by applying the above argument (for $\otimes^i Y(\lambda_1)$, $2 \le i \le 4$) to $\Lambda^2 Y(\lambda_6)$ we see that this is also impossible. We have now shown:

(9.3.3) *if the characteristic of* k *is not equal to* 2, 3 *or* 7 *then* $\underline{H}(G)$ *is true.*

Now assume $p = 3$. We have, as above, $Y(\lambda_1), Y(\lambda_6), Y(\lambda_7) \in S$. Hence $Y(\lambda_1) \otimes Y(\lambda_1)$ belongs to S and $\Lambda^2 Y(\lambda_1)$ (being a direct summand of $Y(\lambda_1) \otimes Y(\lambda_1)$) also belongs to S . Now λ_2 is the unique highest weight of $\Lambda^2 Y(\lambda_1)$ and so, by Proposition 3.2.6, there is a submodule S such that S has a good filtration and $(\Lambda^2 Y(\lambda_1))/S \cong Y(\lambda_2)$. The possibilities for the successive quotients in a good filtration of S are $Y(0)$ and $Y(\lambda_6)$ and so $S \in S$. Hence, by (5.15), $Y(\lambda_2) \in S$. Similarly $\Lambda^2 Y(\lambda_6) \in S$, $\Lambda^2 Y(\lambda_6)$ has a submodule, say J , with a good filtration and $(\Lambda^2 Y(\lambda_6))/J$ is isomorphic to $Y(\lambda_5)$. The possibilities for the sections in a good filtration for K are $Y(0)$, $Y(\lambda_6)$ and $Y(\lambda_2)$ and so $K \in S$. Hence, by (5.15), $Y(\lambda_5) \in S$.

Since $Y(\lambda_1), Y(\lambda_6) \in S$, $Y(\lambda_1) \otimes Y(\lambda_6) \in S$. In particular $Y(\lambda_1) \otimes Y(\lambda_6)$ has a submodule, say K , with $(Y(\lambda_1) \otimes Y(\lambda_1))/K$ isomorphic to $Y(\lambda_1 + \lambda_6)$. The only possibilities for successive quotients in a good filtration of K is $Y(\lambda_7)$ and so $K \in S$. Hence, by (5.15), $Y(\lambda_1 + \lambda_6) \in S$. We now have that $Y(\lambda_1)$, $Y(\lambda_6)$, $Y(\lambda_7)$, $Y(\lambda_2)$ and $Y(\lambda_1 + \lambda_6)$ belongs to S .

Using (9.1.2), (2.2.3) and (2.2.5) it is easy to check that

$$\chi(\lambda_1)\chi(\lambda_5) = \chi(\lambda_1 + \lambda_5) + \chi(\lambda_6 + \lambda_7) + \chi(\lambda_3) + \chi(\lambda_1 + \lambda_6) + \chi(\lambda_7) .$$

It is also easy to check that $\widetilde{\lambda_1 + \lambda_5} = \widetilde{\lambda_6 + \lambda_7} = \lambda_2$ and $\tilde{\lambda}_3 = \lambda_5$. It follows that $Y(\lambda_1) \otimes Y(\lambda_5)$ has a component, say L , with highest weight λ_3 . By (5.15), L belongs to S . By Proposition 3.2.6, L has a submodule, say L' , with a good filtration and L/L' is isomorphic to $Y(\lambda_3)$. The possibilities for the successive quotients in a good filtration of L' are $Y(\lambda_7)$ and $Y(\lambda_1 + \lambda_6)$. Thus $L' \in S$ and, by (5.15), $Y(\lambda_3) \cong L/L' \in S$.

Suppose, for a contradiction, that $\underline{H}(G)$ is false and choose $\tau \in X^+$ minimal such that $Y(\tau) \notin S$. Then $\tau = \lambda_i$ for some i and the only possibility remaining is $\tau = \lambda_4$. However a, by now familiar, argument applied to $\Lambda^2 Y(\lambda_7)$ shows that this is impossible. We have now shown $\underline{H}(G)$ to be true in characteristic 3. Combining this with (9.3.3) we obtain the following.

(9.3.4) *If the characteristic of* k *is not* 2 *or* 7 *then* $\underline{H}(G)$ *is true.*

9.4 The restriction of $Y(\lambda_7)$ to P_Ω

We begin our analysis of $Y(\lambda_7)$. There is no restriction on the characteristic of k . We first show:

(9.4.1) $Hom_{P_\Omega}(Y(\lambda_6) \otimes -\lambda_1, Y(\lambda_7)) \cong k$.

By reciprocity and the tensor identity we have $Hom_{P_\Omega}(Y(\lambda_6) \otimes -\lambda_1, Y(\lambda_7)) = Hom_G(Y(\lambda_6), Y(\lambda_1) \otimes Y(\lambda_7))$. Now by (2.2.3), (2.2.5) and (9.1.2) $\chi(\lambda_1)\chi(\lambda_7) = \chi(\lambda_1 + \lambda_7) + \chi(\lambda_5) + \chi(\lambda_2) + \chi(\lambda_6)$. Thus by (9.1.4) and Proposition 3.2.6, $Y(\lambda_1) \otimes Y(\lambda_7)$ has an ascending filtration with quotients $Y(\lambda_6)$, $Y(\lambda_2)$, $Y(\lambda_5)$, $Y(\lambda_1 + \lambda_7)$ and it follows from (1.5.3) that $Hom_G(Y(\lambda_6), Y(\lambda_1) \otimes Y(\lambda_7)) \cong k$.

Let θ be a non-zero element of $Hom_{P_\Omega}(Y(\lambda_6) \otimes -\lambda_1, Y(\lambda_7))$ and let D be the image of θ . The B socle of $Y(\lambda_7)$ is $-\lambda_7$ and $-\lambda_7$ is not a weight of $N \otimes -\lambda_1$ (see section 9.2 for the definition of N)

hence θ induces a non-zero map $\overline{\theta}: (Y(\lambda_6) \otimes -\lambda_1)/(N \otimes -\lambda_1) \to D$. We claim:

(9.4.2) $Y(\lambda_6)/N$ *has simple* B *socle* $\lambda_1 - \lambda_7$.

From section 9.2, $Y(\lambda_6)/N$ has a good filtration with quotients $Y_\Omega(\lambda_7 - \lambda_1)$, $Y_\Omega(\lambda_6)$. The B socle of $Y_\Omega(\lambda_7 - \lambda_1)$ is $\lambda_1 - \lambda_7$ and the B socle of $Y_\Omega(\lambda_6)$ is $-\lambda_2 + 2\lambda_1$ so it suffices to show that $((Y(\lambda_6)/N) \otimes (\lambda_2 - 2\lambda_1))^B = 0$. We have $(Y(\lambda_6) \otimes (\lambda_2 - 2\lambda_1))^B = 0$ by (1.5.2) so in fact it suffices to show that $H^1(B, N \otimes (\lambda_2 - 2\lambda_1)) = 0$. Now $R^i Ind_B^G(\lambda_2 - 3\lambda_1) = 0$ for all $i \geq 0$ by (5.6) and (2.1.5) so that, by (1.1.8), $H^i(B, Y(\lambda_1) \otimes (\lambda_2 - 3\lambda_1)) = 0$ for all $i \geq 0$. We thus obtain, from (9.2.3) and the long exact sequence of cohomology, $H^1(B, N \otimes (\lambda_2 - 2\lambda_1)) \cong H^2(B, M \otimes (\lambda_2 - 3\lambda_1))$. It therefore suffices to show:

(9.4.3) $H^i(B, M \otimes (\lambda_2 - 3\lambda_1)) = 0$ *for all* $i \geq 0$.

Now M has a filtration with quotients $Y_\Omega(-\lambda_1)$, $Y_\Omega(\lambda_6 - \lambda_1)$ and $H^i(B, Y_\Omega(-\lambda_1) \otimes (\lambda_2 - 3\lambda_1)) = H^i(B, \lambda_2 - 4\lambda_1)$ is zero, for all $i \geq 0$, by (5.4). By (2.1.3) it thus suffices to show that $H^i(B, Y_\Omega(\lambda_6 - \lambda_1) \otimes Y_\Omega(\lambda_2 - 3\lambda_1)) = 0$ for all $i \geq 0$.

We record:

(9.4.4) *the* W_Ω*-orbit of* $\lambda_6 - \lambda_1$ *is* $-1 \cdot \cdot \overset{\cdot}{\cdot} \cdot 1$, $-1 \cdot \cdot \overset{\cdot}{\cdot} 1 -1$,

$-1 \cdot \cdot \overset{\cdot}{1} -1 \cdot$, $-1 \cdot 1 \overset{1}{-1} \cdot \cdot$, $-1\ 1 -1 \overset{1}{\cdot} \cdot \cdot$, $-1 \cdot 1 \overset{-1}{\cdot} \cdot \cdot$,

$\cdot -1 \cdot \overset{1}{\cdot} \cdot \cdot$, $-1\ 1 -1 \overset{-1}{1} \cdot \cdot$, $\cdot -1 \cdot \overset{-1}{1} \cdot \cdot$, $-1\ 1 \cdot \overset{\cdot}{-1}\ 1 \cdot$,

$\cdot -1\ 1 \overset{\cdot}{-1}\ 1 \cdot$, $-1\ 1 \cdot \overset{\cdot}{\cdot} -1\ 1$, $\cdot \cdot -1 \overset{\cdot}{\cdot}\ 1 \cdot$, $\cdot -1\ 1 \overset{\cdot}{\cdot} -1\ 1$,

$-1\ 1 \cdot \overset{\cdot}{\cdot} \cdot -1$, $\cdot \cdot -1 \overset{\cdot}{1} -1\ 1$, $\cdot -1\ 1 \overset{\cdot}{\cdot} \cdot -1$, $\cdot \cdot \cdot \overset{1}{-1} \cdot 1$,

$\cdot \cdot -1 \overset{\cdot}{1} \cdot -1$, $\cdot \cdot \cdot \overset{-1}{\cdot} \cdot 1$, $\cdot \cdot \cdot \overset{1}{-1}\ 1 -1$, $\cdot \cdot \cdot \overset{-1}{\cdot}\ 1 -1$,

$\cdot \cdot \cdot \overset{-1}{\cdot} -1 \cdot$, $\cdot \cdot \cdot \overset{-1}{1} -1 \cdot$, $\cdot \cdot 1 \overset{\cdot}{-1} \cdot \cdot$, $\cdot 1 -1 \overset{\cdot}{\cdot} \cdot \cdot$, $1 -1 \cdot \overset{\cdot}{\cdot} \cdot \cdot$

Thus, by (2.2.3), (2.2.5),

$$\chi_\Omega(\lambda_6 - \lambda_1)\chi_\Omega(\lambda_2 - 3\lambda_1) = \chi_\Omega(-4\lambda_1 + \lambda_2 + \lambda_6) + \chi_\Omega(-3\lambda_1 + \lambda_7) + \chi_\Omega(-2\lambda_1) .$$

Hence $Y_\Omega(\lambda_6 - \lambda_1) \otimes Y_\Omega(\lambda_2 - 3\lambda_1)$ has a good filtration with quotients $Y_\Omega(-2\lambda_1)$, $Y_\Omega(-3\lambda_1 + \lambda_7)$, $Y_\Omega(-4\lambda_1 + \lambda_2 + \lambda_6)$. Thus, by (2.1.3), to complete the proof of (7.4.3) (and hence of (7.4.2)) it suffices to show that $H^i(B,\tau) = 0$ for $\tau \in \{-2\lambda_1, -3\lambda_1 + \lambda_7, -4\lambda_1 + \lambda_2 + \lambda_6\}$. This is easily checked with the help of (5.4).

Now since $\overline{\theta}$ is non-zero and the B socle of $Y(\lambda_7)$ is $-\lambda_7$ we have that $\overline{\theta}(v) \neq 0$ for some element v of $((Y(\lambda_6) \otimes -\lambda_1)/(N \otimes -\lambda_1)$ of weight $-\lambda_7$. Now $\lambda_1 - \lambda_7$ is conjugate to λ_6 and so occurs with multiplicity one as a weight of $Y(\lambda_6)$. It follows that the k-span of v is, by (9.4.2), the B socle of $(Y(\lambda_6) \otimes -\lambda_1)/(N \otimes -\lambda_1)$. Thus $\overline{\theta}$ is injective on the B socle and therefore injective. Thus:

(9.4.5) *there is a short exact sequence of* P_Ω *modules*

$$0 \to N \otimes -\lambda_1 \to Y(\lambda_6) \otimes -\lambda_1 \to D \to 0 .$$

We next claim:

(9.4.6) $((Y(\lambda_7)/D) \otimes (-\lambda_1 + \lambda_5))^B = k$.

By (2.1.6) and the long exact sequence of cohomology this is equivalent to proving that $H^1(B,D \otimes (-\lambda_1 + \lambda_5)) \cong k$. Also $R^i Ind_B^G(-2\lambda_1 + \lambda_5) = 0$ for all $i \geq 0$ by (2.3.1) and (2.1.5) so that, by (1.1.7) and (1.1.8), $H^i(B,Y(\lambda_6) \otimes (-2\lambda_1 + \lambda_5)) = 0$ for all $i \geq 0$. Thus by (9.4.5) and the long exact sequence of cohomology, we have $H^1(B,D \otimes (-\lambda_1 + \lambda_5)) \cong H^2(B,N \otimes (-2\lambda_1 + \lambda_5))$. A similar argument, involving (5.6) and (9.2.3), shows that $H^2(B,N \otimes (-2\lambda_1 + \lambda_5))$ is isomorphic to $H^3(B,M \otimes (-3\lambda_1 + \lambda_5))$. Moreover M has a filtration with quotients $Y_\Omega(-\lambda_1)$, $Y_\Omega(\lambda_6 - \lambda_1)$ and $H^i(B,-4\lambda_1 + \lambda_5) = 0$ for all $i \geq 0$ by (5.4) so it suffices, by (2.1.3), to show that

$H^3(B,Y_\Omega(\lambda_6 - \lambda_1) \otimes Y_\Omega(-3\lambda_1 + \lambda_5)) \cong k$. From (2.2.3), (2.2.5) and (9.4.4)

$$\chi_\Omega(\lambda_6 - \lambda_1)\chi_\Omega(-3\lambda_1 + \lambda_5) = \chi_\Omega(-4\lambda_1 + \lambda_5 + \lambda_6) + \chi_\Omega(-4\lambda_1 + \lambda_4) + \chi_\Omega(-4\lambda_1 + \lambda_2 + \lambda_6) + \chi_\Omega(-3\lambda_1 + \lambda_7) .$$

so that $Y_\Omega(\lambda_6 - \lambda_1) \otimes Y_\Omega(-3\lambda_1 + \lambda_5)$ has a good filtration with quotients $Y_\Omega(-3\lambda_1 + \lambda_7)$, $Y_\Omega(-4\lambda_1 + \lambda_2 + \lambda_6)$, $Y_\Omega(-4\lambda_1 + \lambda_4)$, $Y_\Omega(-4\lambda_1 + \lambda_5 + \lambda_6)$. Moreover $H^i(B,Y_\Omega(\tau)) = 0$ for all $i \geq 0$ and $\tau = -3\lambda_1 + \lambda_7$, $-4\lambda_1 + \lambda_2 + \lambda_6$, $-4\lambda_1 + \lambda_5 + \lambda_6$ by (2.1.3) and (5.4) so that $H^3(B,Y_\Omega(\lambda_6 - \lambda_1) \otimes Y_\Omega(-3\lambda_1 + \lambda_5)) = H^3(B,Y_\Omega(-4\lambda_1 + \lambda_4))$. Hence by (2.1.3) it suffices to show that $H^3(B,-4\lambda_1 + \lambda_4) \cong k$. We obtain $H^3(B,-4\lambda_1 + \lambda_4) \cong H^2(B,RInd_B^{P_{\{\alpha_1\}}}(-4\lambda_1 + \lambda_4))$ from the proof of (5.4) and the weights of $RInd^{P_{\{\alpha_1\}}}(-4\lambda_1 + \lambda_4)$ are $-2\lambda_1 - \lambda_2 + \lambda_4$, $-2\lambda_2 + \lambda_4$, $2\lambda_1 - 3\lambda_2 + \lambda_4$. However $H^i(B,\tau) = 0$ for all $i \geq 0$ for $\tau = -2\lambda_1 - \lambda_2 + \lambda_4$, $-2\lambda_2 + \lambda_4$ by (2.1.6) and (2.3.3) so that $H^3(B,Y_\Omega(-4\lambda_1 + \lambda_4)) = H^2(B,2\lambda_1 - 3\lambda_2 + \lambda_4)$. However the latter is isomorphic to k by (5.5) and (2.3.3). This completes the proof of (9.4.6).

We now turn our attention briefly to the case in which k has characteristic zero. In this case $Y_\Omega(\tau)$ is a simple P_Ω module for $\tau \in X_\Omega$. It follows that the dual $Y_\Omega(-\lambda_1 + \lambda_5)^*$ of $Y_\Omega(-\lambda_1 + \lambda_5)$ is $Y_\Omega(-2\lambda_1 + \lambda_3)$ so that, by (9.4.6) and the tensor identity, $Y(\lambda_7)/D$ contains a copy of $Y_\Omega(-2\lambda_1 + \lambda_3)$. Thus $Y(\lambda_7)|_{P_\Omega}$ contains composition factors $Y_\Omega(\lambda_7 - 2\lambda_1)$, $Y_\Omega(\lambda_6 - \lambda_1)$, $Y_\Omega(\lambda_3 - 2\lambda_1)$ and, since $Y(\lambda_7)$ is self dual as a G module, $Y(\lambda_7)|_{P_\Omega}$ also has composition factors $Y_\Omega(\lambda_7)$, $Y_\Omega(\lambda_2 - \lambda_1)$, $Y_\Omega(\lambda_5 - \lambda_1)$. However, one may check, from the information given on p.45 and p.47 of [57] that $\dim Y(\lambda_7)$ is the sum of the $\dim Y_\Omega(\tau)$'s for $\tau = \lambda_7 - 2\lambda_1$, $\lambda_6 - \lambda_1$, $\lambda_3 - 2\lambda_1$,

$\lambda_2 - \lambda_1$, $\lambda_5 - \lambda_1$, λ_7 and so we have determined all composition factors of $Y(\lambda_7)|_{P_\Omega}$. Of course $\chi(\lambda_7)$ and $\chi_\Omega(\tau)$, $\tau \in X_\Omega$, are independent of characteristic, since they are given by Weyl's Character Formula so we have (in all characteristics):

$$(9.4.7) \quad \chi(\lambda_7) = \chi_\Omega(\lambda_7) + \chi_\Omega(\lambda_5 - \lambda_1) + \chi_\Omega(\lambda_2 - \lambda_1) + \chi_\Omega(\lambda_3 - 2\lambda_1) + \chi_\Omega(\lambda_6 - \lambda_1) + \chi_\Omega(\lambda_7 - 2\lambda_1) .$$

We now continue in arbitrary characteristic. We wish to show that $\overline{V} = V/D$ is isomorphic to $Y_\Omega(\lambda_3 - 2\lambda_1)$, where $V = F_\Omega^{-\frac{1}{2}}Y(\lambda_7)$. By (9.4.7) the character of $\overline{V}$ is $\chi_\Omega(\lambda_3 - 2\lambda_1)$ and so, by (1.5.4), it suffices to show that $(\overline{V} \otimes \tau)^B = 0$ for $\tau \in X_\Omega$ such that $-\tau$ is a weight of $\overline{V}$ and $\tau \neq \lambda_1 - \lambda_5$. The only possibility is $\tau = \lambda_2 - \lambda_1$ so it suffices to show that $(\overline{V} \otimes (\lambda_2 - \lambda_1))^B = 0$. Since $\overline{V}$ embeds in $Y(\lambda_7)/D$ it suffices to show that $((Y(\lambda_7)/D) \otimes (\lambda_2 - \lambda_1))^B = 0$. By the long exact sequence of cohomology it is thus enough to prove that $H^1(B,D \otimes (\lambda_2 - \lambda_1)) = 0$. By (2.3.3) and (1.5.2), $H^i(B,Y(\lambda_6) \otimes (\lambda_2 - 2\lambda_1)) = 0$ for all $i \geq 0$ so that, by (9.4.5), $H^1(B,D \otimes (\lambda_2 - \lambda_1)) \cong H^2(B,N \otimes (\lambda_2 - 2\lambda_1))$. Similarly, using (9.2.3), we obtain $H^2(B,N \otimes (\lambda_2 - 2\lambda_1)) \cong H^3(B,M \otimes (\lambda_2 - 3\lambda_1))$. This is 0 by (9.4.3) and thus $\overline{V}$ is isomorphic to $Y_\Omega(\lambda_3 - 2\lambda_1)$.

We next show:

(9.4.8) $H^i(B,D \otimes (\lambda_6 - \lambda_1)) = 0$ *for all* $i \geq 0$.

By (2.3.3) and (2.1.5), $H^i(B,Y(\lambda_6) \otimes (\lambda_6 - 2\lambda_1)) = 0$ for all $i \geq 0$ so, by (9.4.5), it is enough to show that $H^j(B,N \otimes (\lambda_6 - 2\lambda_1)) = 0$ for all $j > 0$. Similarly, using (9.2.3), we obtain that it suffices to show that $H^\ell(B,M \otimes (\lambda_6 - 3\lambda_1)) = 0$ for all $\ell > 1$. Moreover M has a good filtration with quotients $Y_\Omega(-\lambda_1)$, $Y_\Omega(\lambda_6 - \lambda_1)$ and $H^\ell(B,\lambda_6 - 4\lambda_1) = 0$ for all $\ell \geq 0$, by (5.4), so it suffices, by (2.1.3), to show that $H^\ell(B,Y_\Omega(\lambda_6 - \lambda_1) \otimes Y_\Omega(\lambda_6 - 3\lambda_1)) = 0$ for all $\ell > 1$.

By (9.4.4) and (2.2.3), (2.2.5),

$$\chi_\Omega(\lambda_6 - \lambda_1)\chi_\Omega(\lambda_6 - 3\lambda_1) = \chi_\Omega(2\lambda_6 - 4\lambda_1) + \chi_\Omega(-4\lambda_1 + \lambda_5)$$
$$+ \chi_\Omega(-4\lambda_1 + \lambda_2)$$

so that $Y_\Omega(\lambda_6 - \lambda_1) \otimes Y_\Omega(\lambda_6 - 3\lambda_1)$ has a filtration with quotients $Y_\Omega(-4\lambda_1 + \lambda_2)$, $Y_\Omega(-4\lambda_1 + \lambda_5)$, $Y_\Omega(-4\lambda_1 + 2\lambda_6)$. However, it is easy to check that $H^i(B,Y_\Omega(\tau)) = 0$ for all $i \geq 0$, $\tau = -4\lambda_1 + \lambda_2$, $-4\lambda_1 + \lambda_5$, $-4\lambda_1 + 2\lambda_6$, with the aid of (2.1.3) and (5.4). This completes the proof of (9.4.8).

We now show:

(9.4.9) $((Y(\lambda_7)/V) \otimes (\lambda_6 - \lambda_1))^B \cong k$.

We obtain, from (2.1.6),
$((Y(\lambda_7)/V) \otimes (\lambda_6 - \lambda_1))^B \cong H^1(B,V \otimes (\lambda_6 - \lambda_1))$ which, by (9.4.8), is isomorphic to $H^1(B,\bar{V} \otimes (\lambda_6 - \lambda_1))$ i.e.
$H^1(B,Y_\Omega(-2\lambda_1 + \lambda_3) \otimes (\lambda_6 - \lambda_1))$. By (2.1.3) we thus have
$((Y(\lambda_7)/V) \otimes (\lambda_6 - \lambda_1))^B \cong H^1(B,Y_\Omega(-2\lambda_1 + \lambda_3) \otimes Y_\Omega(-\lambda_1 + \lambda_6))$. By (9.4.4), (2.2.3) and (2.2.5),

$$\chi_\Omega(-2\lambda_1 + \lambda_3)\chi_\Omega(-\lambda_1 + \lambda_6) = \chi_\Omega(-3\lambda_1 + \lambda_3 + \lambda_6)$$
$$+ \chi_\Omega(-3\lambda_1 + \lambda_2 + \lambda_7) + \chi_\Omega(-2\lambda_1 + \lambda_5) + \chi_\Omega(-2\lambda_1 + \lambda_2) .$$

Thus $Y_\Omega(-2\lambda_1 + \lambda_3) \otimes Y_\Omega(-\lambda_1 + \lambda_6)$ has a filtration with quotients $Y_\Omega(-2\lambda_1 + \lambda_2)$, $Y_\Omega(-2\lambda_1 + \lambda_5)$, $Y_\Omega(-3\lambda_1 + \lambda_2 + \lambda_7)$, $Y_\Omega(-3\lambda_1 + \lambda_3 + \lambda_6)$. Using (2.1.3), (2.3.3), (2.3.5) and (5.6) one may easily check that $H^i(B,Y_\Omega(\tau)) = 0$ for all $i \geq 0$ and $\tau = -2\lambda_1 + \lambda_5$, $-3\lambda_1 + \lambda_2 + \lambda_7$, $-3\lambda_1 + \lambda_3 + \lambda_6$. Hence $H^1(B,Y_\Omega(-2\lambda_1 + \lambda_3) \otimes Y_\Omega(-\lambda_1 + \lambda_6) \cong H^1(B,-\alpha_1) \cong k$, as required. This completes the proof of (9.4.9).

By the reciprocity of induction and the tensor identity $((Y(\lambda_7)/V) \otimes (\lambda_6 - \lambda_1))^B = ((Y(\lambda_7)/V) \otimes Y_\Omega(\lambda_6 - \lambda_1))^{P_\Omega}$. We have $\chi_\Omega(\lambda_6 - \lambda_1) = s_\Omega(\lambda_6 - \lambda_1)$ so that $Y_\Omega(\lambda_6 - \lambda_1)$ is a simple module and the dual $Y_\Omega(\lambda_6 - \lambda_1)^*$ is isomorphic to $Y_\Omega(\lambda_2 - \lambda_1)$ and we have, by

(9.4.9), $Hom_{P_\Omega}(Y_\Omega(\lambda_2 - \lambda_1), Y(\lambda_7)/V) \cong k$. Thus there is a P_Ω sub-module, say E , of $Y(\lambda_7)$ containing V such that E/V is isomorphic to $Y_\Omega(\lambda_2 - \lambda_1)$.

We shall need in this and later sections

(9.4.10) *the* P_Ω *module extension* E/D *of* V/D *by* E/V *is non-split*

and it is convenient to observe this now. If not $Y(\lambda_7)/D$ contains a copy of $Y_\Omega(\lambda_2 - \lambda_1)$ and hence contains $-\lambda_6 + \lambda_1$ in the B socle. ɜut $((Y(\lambda_7)/D) \otimes (\lambda_6 - \lambda_1))^B \cong H^1(B, D \otimes (\lambda_6 - \lambda_1))$ which is zero by (9.4.8).

Let $Z/E = F_\Omega^{\frac{1}{2}} Y(\lambda_7)/E$. We wish to show that $\bar{Z} = Z/E$ is isomor-›hic to $Y_\Omega(\lambda_5 - \lambda_1)$. By (9.4.7) we have $ch\bar{Z} = \chi_\Omega(\lambda_5 - \lambda_1)$ so it ;uffices, by (1.5.4), to show that $(\bar{Z} \otimes \tau)^B = 0$ for every $\tau \in X_\Omega$;uch that $-\tau$ is a weight of $Y_\Omega(\lambda_5 - \lambda_1)$ different from $2\lambda_1 - \lambda_3$. :he only available weight is $\lambda_6 - \lambda_1$. To show that $\bar{Z} \otimes (\lambda_6 - \lambda_1))^B = 0$ it suffices to show that $H^1(B, E \otimes (\lambda_6 - \lambda_1)) = 0$. ›y the construction of E we have a short exact sequence

$$0 \to V \otimes (\lambda_6 - \lambda_1) \to E \otimes (\lambda_6 - \lambda_1) \to Y_\Omega(\lambda_2 - \lambda_1) \otimes (\lambda_6 - \lambda_1) \to 0$$

ɲnd hence an exact sequence

$$(E \otimes (\lambda_6 - \lambda_1))^B \to (Y_\Omega(\lambda_2 - \lambda_1) \otimes (\lambda_6 - \lambda_1))^B$$
$$\to H^1(B, V \otimes (\lambda_6 - \lambda_1)) \to H^1(B, E \otimes (\lambda_6 - \lambda_1))$$
$$\to H^1(B, Y_\Omega(\lambda_2 - \lambda_1) \otimes (\lambda_6 - \lambda_1)) .$$

ɦhe B socle of $Y(\lambda_7)$, and hence E , is $-\lambda_7$ so that ɛ $\otimes (\lambda_6 - \lambda_1))^B = 0$. The B socle of $Y_\Omega(\lambda_2 - \lambda_1)$ is $\lambda_1 - \lambda_6$ so ɲat $(Y_\Omega(\lambda_2 - \lambda_1) \otimes (\lambda_6 - \lambda_1))^B \cong k$, and by (1.5.2) and (1.4.6), $^1(B, Y_\Omega(\lambda_2 - \lambda_1) \otimes (\lambda_6 - \lambda_1)) = 0$ (this is also deducible from (2.1.3) ɲd Lemma 3.2.1). By the proof of (9.4.9), $H^1(B, Y \otimes (\lambda_6 - \lambda_1)) \cong k$ ɔ the above exact sequence becomes

$$0 \to k \to k \to H^1(B, E \otimes (\lambda_6 - \lambda_1)) \to 0$$

and so $H^1(B, E \otimes (\lambda_6 - \lambda_1)) = 0$. Thus $\tilde{Z}$ is isomorphic to $Y_\Omega(\lambda_5 - \lambda_1)$.

In order to show that $Y(\lambda_7)|_{P_\Omega}$ has a good filtration it only remains to show that $Y(\lambda_7)/Z$ is isomorphic to $Y_\Omega(\lambda_7)$. We have, by (9.4.7) and the construction of Z , that the character of $Y(\lambda_7)/Z$ is $\chi_\Omega(\lambda_7)$. Hence, by (1.5.4), we only need show that if $\tau \in X_\Omega$ is such that $-\tau$ is a weight of $Y_\Omega(\lambda_7)$ different from $-\lambda_7 + 2\lambda_1$ then $((Y(\lambda_7)/Z) \otimes \tau)^B = 0$. The only possible τ is $-\lambda_1$. The B socle of $Y(\lambda_7)$ is $-\lambda_7$ so that $(Y(\lambda_7) \otimes -\lambda_1)^B = 0$ and it suffices to prove that $H^1(B, D \otimes -\lambda_1) = H^1(B, E/D \otimes -\lambda_1) = H^1(B, Z/E \otimes -\lambda_1) = 0$. We have, by (2.1.3), $H^1(B, Y_\Omega(\tau) \otimes -\lambda_1) \cong H^1(B, \tau - \lambda_1)$ for $\tau \in X_\Omega$. From this we easily obtain, for example by (2.3.3), (2.3.5) and (5.6) that $H^1(B, Y_\Omega(\tau) \otimes -\lambda_1) = 0$ for $\tau = \lambda_7 - 2\lambda_1$, $\lambda_6 - \lambda_1$, $\lambda_5 - \lambda_1$. Thus it remains to prove that $H^1(B, K \otimes - 1_1) = 0$, where $K = E/D$.

By (5.10),

$Ext^1_{P_\Omega}(Y_\Omega(\lambda_2 - \lambda_1), Y_\Omega(\lambda_3 - 2\lambda_1)) \cong Ext^1_{P_\Lambda}(Y_\Lambda(\lambda_2 - \lambda_1), Y_\Lambda(\lambda_3 - 2\lambda_1))$ where $\Lambda = \{\alpha_2\}$. Now the latter is isomorphic to

$H^1(P_\Lambda, Y_\Lambda(\lambda_2 - \lambda_3) \otimes Y_\Lambda(\lambda_3 - 2\lambda_1)) \cong H^1(P_\Lambda, Y_\Lambda(-2\lambda_1 + \lambda_2)) \cong H^1(B, -\alpha_1) = k$, by (2.1.3) and (1.4.7). Thus by (5.11) and (9.4.10) K is the unique non-split extension of $Y_\Omega(\lambda_3 - 2\lambda_1)$ by $Y_\Omega(\lambda_2 - \lambda_1)$ and so, by (5.1.4), K is isomorphic to $F(Y_2(\lambda_2) \otimes -\lambda_1)$ where $\Sigma = \{\alpha_1, \alpha_2\}$ and $F = Ind_B^{P_\Sigma}$. By (2.1.3) and the tensor identity,

$H^1(B, K \otimes -\lambda_1) \cong H^1(P_\Omega, F(Y_\Sigma(\lambda_2) \otimes -2\lambda_1))$ so that, by (1.1.8) it suffices to show that $H^1(B, Y_\Sigma(\lambda_2) \otimes -2\lambda_1) = 0$. However $H^1(B, Y_\Sigma(\lambda_2) \otimes -2\lambda_1) \cong (Y_\Sigma(\lambda_2) \otimes -\lambda_2)^B$ by (2.3.3), which is zero by (2.1.6). This completes the proof that $Y(\lambda_7)/Z$ is isomorphic to $Y_\Omega(\lambda_7)$ and we now have:

(9.4.11) $Y(\lambda_7)|_{P_\Omega}$ *has a good filtration.*

9.5 $R^i Ind_B^{P_\Sigma}(D)$

Let $\Sigma = \Delta\backslash\{\alpha_j\}$ for $2 \le j \le 7$. In order to show that $Y(\lambda_7)|_{P_\Sigma} \cong Ind_B^{P_\Sigma}(Y(\lambda_7)|_B)$ has a good filtration we must show that $G(D)$, $G(E/D)$, $G(Y(\lambda_7)/D)$ each has a good filtration and $RG(D) = RG(E/D) = 0$, where $G = Ind_B^{P_\Sigma}$. We now show:

(9.5.1) *$R^i Ind_B^{P_\Sigma}(D)$ has a good filtration if $i = 0$ and is 0 if $i > 0$, where $\Sigma = \Delta\backslash\{\alpha_j\}$ for some $2 \le j \le 7$.*

We have $R^i G(Y(\lambda_6) \otimes -\lambda_1) = 0$ for all $i \ge 0$ by (2.1.5) and the tensor identity. Thus by (9.4.5):

(9.5.2) $R^i G(D) \cong R^{i+1} G(N \otimes -\lambda_1)$ *for all* $i \ge 0$.

If $j \ne 2$ then $R^i G(Y(\lambda_1) \otimes -2\lambda_1) = 0$ for all $i \ge 0$ by the tensor identity, (1.5.2), (2.3.1) and (2.1.5). Thus we have, from (9.2.3):

(9.5.3) *If* $j \ne 2$, $R^i G(N \otimes -\lambda_1) \cong R^{i+1} G(M \otimes -2\lambda_1)$ *for all* $i \ge 0$.

Now suppose that $j > 3$. Then $R^i G(-3\lambda_1) = 0$ for all $i \ge 0$ by (5.6) and, since M has a filtration with quotients $Y_\Omega(-\lambda_1)$, $Y_\Omega(-\lambda_1 + \lambda_6)$, $R^i G(M \otimes -2\lambda_1) \cong R^i G(Y_\Omega(-3\lambda_1 + \lambda_6))$ for all $i \ge 0$. Using (5.6), (9.4.4) and (2.3.1), (2.1.5) we have $R^i G(\tau) = 0$ for all $i \ge 0$ for weights τ of $Y_\Omega(-3\lambda_1 + \lambda_6)$ other than $-3\lambda_1 + \lambda_3 - \lambda_4 + \lambda_7$, $-3\lambda_1 + \lambda_3 - \lambda_7$. Moreover $R^i G(\tau)$ is $Y_\Omega(\tau + 2\lambda_1 + \lambda_2)$ if $i = 2$ and 0 otherwise for $\tau = -3\lambda_1 + \lambda_3 - \lambda_4 + \lambda_7$, $-3\lambda_1 + \lambda_3 - \lambda_7$. Hence $R^i G(M \otimes -2\lambda_1)$ has a good filtration if $i = 2$ and is 0 if $i \ne 2$. By (9.5.2), (9.5.3), this proves (9.5.1) in the case in which $j > 3$.

We now consider the case $j = 3$. Let $\Gamma = \Omega \cap \Sigma = \Delta\backslash\{\alpha_1, \alpha_3\}$. As a P_Γ module, $Y_\Omega(-\lambda_1 + \lambda_6)$ has an ascending filtration with quotients $Y_\Gamma(\lambda_2 - \lambda_3)$, $Y_\Gamma(-\lambda_3 + \lambda_5)$, $Y_\Gamma(-\lambda_1 + \lambda_2 - \lambda_3 + \lambda_7)$,

$Y_\Gamma(-\lambda_1 + \lambda_6)$. Hence $F = M \otimes -2\lambda_1$, as a P_Γ module, has a good filtration $0 = F_0, F_1, \ldots, F_5 = F$ with successive quotients $Y_\Gamma(-3\lambda_1)$, $Y_\Gamma(-2\lambda_1 + \lambda_2 - \lambda_3)$, $Y_\Gamma(-2\lambda_1 - \lambda_3 + \lambda_5)$, $Y_\Gamma(-3\lambda_1 + \lambda_2 - \lambda_3 + \lambda_5)$, $Y_\Gamma(-3\lambda_1 + \lambda_6)$. The B socle of M is $-\lambda_1$ and so the B socle of F_2 is $-3\lambda_1$. Moreover
$Ext^1_{P_\Gamma}(Y_\Gamma(-2\lambda_1 + \lambda_2 - \lambda_3), Y_\Gamma(-3\lambda_1)) \cong H^1(P_\Gamma, Y_\Gamma(-\lambda_1 + \lambda_2) \otimes -3\lambda_1) \cong$ $H^1(B, -\alpha_1) \cong k$, using (2.1.3) and (1.4.7). Thus F_2 is, by (5.11), the unique non-split extension of $Y_\Gamma(-3\lambda_1)$ by $Y_\Gamma(-2\lambda_1 + \lambda_2 - \lambda_3)$ and hence isomorphic to $Y_\Gamma(\lambda_2 - \lambda_3) \otimes -2\lambda_1$. Thus we have, by the tensor identity and (2.3.1) and (2.1.5), $R^iG(F_2) = 0$ for all $i \geq 0$ and $R^iG(F) \cong R^iG(F/F_2)$. Similarly $R^iG(F_3/F_2) = 0$ for all $i \geq 0$ and $R^iG(F) \cong R^iG(F/F_3)$. By (2.1.1) and (2.1.2), $R^iG(Y_\Gamma(\tau)) \cong R^iG(\tau)$ for $\tau \in X_\Omega$. Thus, by (5.6) and (2.1.5), $R^iG(Y_\Gamma(-3\lambda_1 + \lambda_2 - \lambda_3 + \lambda_5)) = 0$ for all $i \geq 0$ and $R^iG(F) \cong R^iG(-3\lambda_1 + \lambda_6)$. Thus, by (5.6) and (2.3.1), $R^iG(M \otimes -2\lambda_1)$ is isomorphic to $Y_\Sigma(-\lambda_3 + \lambda_6)$ if $i = 2$ and is 0 for $i \neq 2$. By (9.5.2), (9.5.3) we now have (9.5.1) for $j = 3$.

It remains to consider the case $j = 2$. Now let $\Gamma = \Delta\backslash\{\alpha_1, \alpha_2\}$. It is easy to check that

$$\chi_\Omega(-3\lambda_1 + \lambda_2) = \chi_\Gamma(-3\lambda_1 + \lambda_2) + \chi_\Gamma(-2\lambda_1 - \lambda_2 + \lambda_3)$$
$$+ \chi_\Gamma(-\lambda_1 - \lambda_2 + \lambda_6)$$

so that $Y_\Omega(-3\lambda_1 + \lambda_2)$ has a good filtration with quotients $Y_\Gamma(-\lambda_1 - \lambda_2 + \lambda_6)$, $Y_\Gamma(-2\lambda_1 - \lambda_2 + \lambda_3)$, $Y_\Gamma(-3\lambda_1 + \lambda_2)$. Now $N \otimes -\lambda_1$ has a good filtration with quotients $Y_\Omega(-3\lambda_1 + \lambda_2)$, $Y_\Omega(-\lambda_1)$. Moreover $R^iG(-\lambda_1) = 0$ for all $i \geq 0$ by (2.1.5) thus, by (9.5.2), in order to prove (9.5.1) it is enough to prove that $R^iG(Y_\Omega(-3\lambda_1 + \lambda_2))$ has a good filtration for $i = 1$ and is 0 for $i > 1$. Now $R^iG(Y_\Gamma(\tau)) = R^iG(\tau)$ for $\tau \in X_\Gamma$, by (2.1.1) and (2.1.2) so that, by (5.6), (2.1.5) and (2.3.1), $R^iG(Y_\Gamma(\tau)) = 0$ for $i \neq 1$ and $\tau = -\lambda_1 - \lambda_2 + \lambda_6$, $-2\lambda_1 - \lambda_2 + \lambda_3$, $-3\lambda_1 + \lambda_2$. Moreover $RG(Y_\Gamma(-\lambda_1 - \lambda_2 + \lambda_6)) = 0$, $RG(Y_\Gamma(-2\lambda_1 - \lambda_2 + \lambda_3)) = Y_\Sigma(-2\lambda_2 + \lambda_3)$,

$RG(Y_\Gamma(-3\lambda_1 + \lambda_2)) \cong Y_\Sigma(-\lambda_2)$ and so $R^i G(Y_\Omega(-3\lambda_1 + \lambda_2))$ has a good filtration with quotients $Y_\Sigma(-2\lambda_2 + \lambda_3)$, $Y_\Sigma(-\lambda_2)$ if $i = 1$ and is 0 if $i \neq 1$. This completes the proof of (9.5.1).

By construction $Y(\lambda_7)/E$, as a P_Ω module, has a good filtration with quotients $Y_\Omega(\lambda_5 - \lambda_1)$, $Y_\Omega(\lambda_7)$ so we obtain, from (5.2):

(9.5.3) *$R^i \mathrm{Ind}_B^{P_\Sigma}(Y(\lambda_7)/E)$ has a good filtration for $i = 0$ and is 0 for $i > 0$, where $\Sigma = \Delta\backslash\{\alpha_j\}$ and $2 \leq j \leq 7$.*

9.6 Character Formulas

Most of the work involved in showing that $Y(\lambda_7)|_{P_\Sigma}$ has a good filtration for $\Sigma = \Delta\backslash\{\alpha_j\}$, $2 \leq j \leq 7$ is in the analysis of $R^i G(\bar{E}|_B)$, where $G = \mathrm{Ind}_B^{P_\Sigma}$ and $\bar{E} = E/D$. In this section we deal with some character formulas which will be needed in the analysis.

We record:

(9.6.1) *the W_Ω orbit of $\lambda_2 - \lambda_1$ is* $-1\ 1\ \cdot\ \overset{\cdot}{\cdot}\ \cdot\ \cdot$, $\cdot\ -1\ 1\ \overset{\cdot}{\cdot}\ \cdot\ \cdot$,
$\cdot\ \cdot\ -1\ \overset{\cdot}{1}\ \cdot\ \cdot$, $\cdot\ \cdot\ \cdot\ \overset{1}{-1}\ 1\ \cdot$, $\cdot\ \cdot\ \cdot\ \overset{-1}{\cdot}\ 1\ \cdot$, $\cdot\ \cdot\ \cdot\ \overset{1}{\cdot}\ -1\ 1$,
$\cdot\ \cdot\ \cdot\ \overset{-1}{1}\ -1\ 1$, $\cdot\ \cdot\ \cdot\ \overset{1}{\cdot}\ \cdot\ -1$, $\cdot\ \cdot\ 1\ \overset{\cdot}{-1}\ \cdot\ 1$, $\cdot\ \cdot\ \cdot\ \overset{-1}{1}\ \cdot\ -1$,
$1\ -1\ \overset{\cdot}{\cdot}\ \cdot\ 1$, $\cdot\ \cdot\ 1\ \overset{\cdot}{-1}\ 1\ -1$, $1\ -1\ \cdot\ \overset{\cdot}{\cdot}\ \cdot\ 1$, $\cdot\ 1\ -1\ \overset{\cdot}{\cdot}\ 1\ -1$,
$\cdot\ \cdot\ 1\ \overset{\cdot}{\cdot}\ -1\ \cdot$, $1\ -1\ \cdot\ \overset{\cdot}{\cdot}\ 1\ -1$, $\cdot\ 1\ -1\ \overset{\cdot}{1}\ -1\ \cdot$, $1\ -1\ \cdot\ \overset{\cdot}{1}\ -1\ \cdot$,
$1\ \cdot\ \overset{1}{-1}\ \cdot\ \cdot$, $1\ -1\ 1\ \overset{1}{-1}\ \cdot\ \cdot$, $\cdot\ 1\ \cdot\ \overset{-1}{\cdot}\ \cdot\ \cdot$, $1\ \cdot\ -1\ \overset{1}{\cdot}\ \cdot\ \cdot$,
$-1\ 1\ \overset{-1}{\cdot}\ \cdot\ \cdot$, $1\ \cdot\ -1\ \overset{-1}{1}\ \cdot\ \cdot$, $1\ \cdot\ \cdot\ \overset{\cdot}{-1}\ 1\ \cdot$, $1\ \cdot\ \cdot\ \overset{\cdot}{\cdot}\ -1\ 1$, 1
$\cdot\ \cdot\ \overset{\cdot}{\cdot}\ \cdot\ -1$.

For $1 \leq i,j \leq 7$, $i \neq j$ we let $W_{i,j} = W_\Gamma$, $Y_{i,j}(\tau) = Y_\Gamma(\tau)$ nd $\chi_{i,j}(\tau) = \chi_\Gamma(\tau)$ for $\tau \in X$, where $\Gamma = \Delta\backslash\{\alpha_i,\alpha_j\}$. We also rite $P_{i,j}$ for P_Γ .

(9.6.2) The $W_{1,2}$ orbit of $-2\lambda_1 + \lambda_3$ is $-2\ \cdot\ 1\ \overset{\cdot}{\cdot}\ \cdot\ \cdot$,

$-2\ 1\ -1\ \overset{\cdot}{1}\ \cdot\ \cdot$, $-2\ 1\ \cdot\ \overset{1}{-1}\ 1\ \cdot$, $-2\ 1\ \cdot\ \overset{-1}{\cdot}\ 1\ \cdot$, $-2\ 1\ \cdot\ \overset{1}{\cdot}\ -1\ 1$,

$-2\ 1\ \cdot\ \overset{-1}{1}\ -1\ 1$, $-2\ 1\ \cdot\ \overset{1}{\cdot}\ \cdot\ -1$, $-2\ 1\ 1\ \overset{\cdot}{-1}\ \cdot\ 1$, $-2\ 1\ \cdot\ \overset{-1}{1}\ \cdot\ -1$,

$-2\ 2\ -1\ \overset{\cdot}{\cdot}\ \cdot\ 1$, $-2\ 1\ 1\ \overset{\cdot}{-1}\ 1\ -1$, $-2\ 2\ -1\ \overset{\cdot}{\cdot}\ 1\ -1$, $-2\ 1\ 1\ \overset{\cdot}{\cdot}\ -1\ \cdot$,

$-2\ 2\ -1\ \overset{\cdot}{1}\ -1$, $-2\ 2\ \cdot\ \overset{1}{-1}\ \cdot\ \cdot$, $-2\ 2\ \cdot\ \overset{-1}{\cdot}\ \cdot\ \cdot$.

We let $\mathcal{U}$ denote the $W_{1,2}$ orbit of $-2\lambda_1 + \lambda_3$. Let M_π denote the P_π submodule $M_\pi(-2\lambda_1 + \lambda_3)$ (see (5.8)) of $Y_\Omega(-2\lambda_1 + \lambda_3)$, where $\pi = \Delta\backslash\{\alpha_1,\alpha_2\}$ and let $\Gamma = \Delta\backslash\{\alpha_1,\alpha_j\}$ where $2 \le j \le 7$. Then there are non-negative integers a_τ $(\tau \in X_\Gamma)$ such that

$$\chi_\Omega(-2\lambda_1 + \lambda_3) = \sum_\tau a_\tau \chi_\Omega(\tau) .$$

We let χ be the sum of the $a_\tau\chi_\Gamma(\tau)$'s for $\tau \in \mathcal{U} \cap X_\Gamma$ and let χ' be the complementary sum (so that $\chi_\Omega(-2\lambda_1 + \lambda_3) = \chi + \chi'$) . By the argument of the proof of (5.9), M_π is the k-span of weight vectors of weight ξ such that in an expression of $-2\lambda_1 + \lambda_3 - \xi$ as a linear combination of simple roots α_2 appears with positive coefficient. Thus, using (5.8), for $\tau \in \mathcal{U}$, $\dim Y_\Omega(-2\lambda_1 + \lambda_3)^\tau = \dim Y_\pi(-2\lambda_1 + \lambda_3)^\tau$ which is 1 , since $\chi_\pi(-2\lambda_1 + \lambda_3) = s_\pi(-2\lambda_1 + \lambda_3)$ by (9.6.2). Thus $a_\tau \le 1$ for all $\tau \in \mathcal{U}$. Suppose that $\tau \in \mathcal{U}$ and $a_\tau = 0$. Then, since τ is a weight of $Y_\Omega(-2\lambda_1 + \lambda_3)$, τ is a weight of $Y_\Gamma(\xi)$ for some weight ξ satisfying $-2\lambda_1 + \lambda_3 \ge \xi \ge \tau$. Now the coefficient of α_2 in the expression of $-2\lambda_1 + \lambda_3$ as a linear combination of simple roots must be 0 . Thus ξ cannot be a weight of M_π and so $\xi \in \mathcal{U}$. Let $\Lambda = \Gamma \cap \Pi$. Again by the proof of (5.9), τ is a weight of $Y_\Lambda(\xi)$. Since Λ is a subset of π we have, by (9. .2) and (2.2.3), (2.2.5) that $\chi_\Lambda(\xi) = s_\Lambda(\xi)$. Since $\chi_\Lambda(\xi) = \sum_\mu a_\mu \chi_\Lambda(\mu)$, by (2.2.3), the sum running over $W_\Lambda\xi$, and $(\xi,\alpha^\vee) \ge -1$ for all $\alpha \in \Lambda$ by (9.6.2), we must have $\chi_\Lambda(\mu) = 0$ for $\mu \in W_\Lambda\xi$, $\mu \ne \xi$. In particular $\chi_\Lambda(\tau) = 0$ and $(\tau,\alpha^\vee) = -1$ for some $\alpha \in \Lambda$. Thus $\chi_\Gamma(\tau) = 0$ and so

$$\chi = \sum_{\tau \in U} \chi_\Gamma(\tau) .$$

Suppose that $a_\tau \neq 0$ and $\tau \notin U$. We claim that $\tau \not\geq U$ for any $\mu \in U$. The reason being that if $\tau \geq \mu$ then $-2\lambda_1 + \lambda_3 - \mu \geq -2\lambda_1 + \lambda_3 - \tau$ and τ cannot be a weight of M_π and therefore occurs as a weight of $Y_\pi(-2\lambda_1 + \lambda_3)$ and so $\tau \in W_\pi(-2\lambda_1 + \lambda_3) = U$. Thus we obtain

$$\chi_\Omega(-2\lambda_1 + \lambda_3) = \sum_{\tau \in U} \chi_\Gamma(\tau) + \textit{lower terms}..$$

Here "lower terms" denotes a sum of $\chi_\Gamma(\xi)$'s such that for any occurring $\chi_\Gamma(\xi)$, $(\xi,\alpha^\vee) \geq -1$ and for any $\tau \in U$, $\xi \geq \tau$. Specifically, we have from (9.6.2) and (2.2.3), (2.2.5):

(9.6.3) $\chi_\Omega(-2\lambda_1 + \lambda_3) = \chi_{1,2}(-2\lambda_1 + \lambda_3) + \textit{lower terms}$;

$\chi_\Omega(-2\lambda_1 + \lambda_3) = \chi_{1,3}(-2\lambda_1 + \lambda_3) + \chi_{1,3}(-2\lambda_1 + \lambda_2 - \lambda_3 + \lambda_4)$
$+ \chi_{1,3}(-2\lambda_1 + 2\lambda_2 - \lambda_3 + \lambda_6) + \textit{lower terms}$;

$\chi_\Omega(-2\lambda_1 + \lambda_3) = \chi_{1,4}(-2\lambda_1 + \lambda_3) + \chi_{1,4}(-2\lambda_1 + \lambda_2 - \lambda_4 + \lambda_5 + \lambda_7$
$\chi_{1,4}(-2\lambda_1 + \lambda_2 + \lambda_3 - \lambda_4 + \lambda_6) + \chi_{1,4}(-2\lambda_1 + 2\lambda_2 - \lambda_4 + \lambda_7)$
$+ \textit{lower terms}$;

$\chi_\Omega(-2\lambda_1 + \lambda_3) = \chi_{1,5}(-2\lambda_1 + \lambda_3) + \chi_{1,5}(-2\lambda_1 + \lambda_2 - \lambda_5 + \lambda_6 + \lambda_7)$
$+ \chi_{1,5}(-2\lambda_1 + \lambda_2 + \lambda_3 - \lambda_5) + \textit{lower terms}$;

$\chi_\Omega(-2\lambda_1 + \lambda_3) = \chi_{1,6}(-2\lambda_1 + \lambda_3) + \chi_{1,6}(-2\lambda_1 + \lambda_2 - \lambda_6 + \lambda_7)$
$+ \textit{lower terms}$;

$\chi_\Omega(-2\lambda_1 + \lambda_3) = \chi_{1,7}(-2\lambda_1 + \lambda_3) + \chi_{1,7}(-2\lambda_1 + \lambda_2 + \lambda_5 - \lambda_7)$
$+ \chi_{1,7}(-2\lambda_1 + 2\lambda_2 - \lambda_7) + \textit{lower terms}$.

Also, from (9.6.1) and (2.2.3), (2.2.5):

(9.6.4) $\chi_\Omega(-\lambda_1 + \lambda_2) = \chi_{1,2}(-\lambda_1 + \lambda_2) + \chi_{1,2}(-\lambda_2 + \lambda_3)$
$+ \chi_{1,2}(\lambda_1 - \lambda_2 + \lambda_6)$;

$\chi_\Omega(-\lambda_1 + \lambda_2) = \chi_{1,3}(-\lambda_1 + \lambda_2) + \chi_{1,3}(-\lambda_3 + \lambda_4)$
$+ \chi_{1,3}(\lambda_2 - \lambda_3 + \lambda_6) + \chi_{1,3}(\lambda_1 - \lambda_3 + \lambda_7)$;

$\chi_\Omega(-\lambda_1 + \lambda_2) = \chi_{1,4}(-\lambda_1 + \lambda_2) + \chi_{1,4}(-\lambda_4 + \lambda_5 + \lambda_7)$
$+ \chi_{1,4}(\lambda_3 - \lambda_4 + \lambda_6) + \chi_{1,4}(\lambda_2 - \lambda_4 + \lambda_7)$
$+ \chi_{1,4}(\lambda_1 - \lambda_4 + \lambda_5)$;

$\chi_\Omega(-\lambda_1 + \lambda_2) = \chi_{1,5}(-\lambda_1 + \lambda_2) + \chi_{1,5}(-\lambda_5 + \lambda_6 + \lambda_7)$
$+ \chi_{1,5}(\lambda_3 - \lambda_5) + \chi_{1,5}(\lambda_1 - \lambda_5 + \lambda_6)$;

$\chi_\Omega(-\lambda_1 + \lambda_2) = \chi_{1,6}(-\lambda_1 + \lambda_2) + \chi_{1,6}(-\lambda_6 + \lambda_7) + \chi_{1,6}(\lambda_1 - \lambda_6)$;

$\chi_\Omega(-\lambda_1 + \lambda_2) = \chi_{1,7}(-\lambda_1 + \lambda_2) + \chi_{1,7}(\lambda_5 - \lambda_7) + \chi_{1,7}(\lambda_2 - \lambda_7)$.

9.7 $R^i Ind_B(\bar{E})$

We shall now show:

(9.7.1) $R^i Ind_B^{P_\Sigma}(\bar{E})$ *has a good filtration if* $i = 0$ *and is* 0 *if* $i > 0$ *where* $\Sigma = \Delta\backslash\{\alpha_j\}$ *for some* $2 \le j \le 7$.

We prove (9.7.1) by considering each j separately. We give a detailed account of the most complicated case $j = 4$ (which we treat first). In the remaining cases we omit references and leave it to the reader to check minor cohomological calculations. The method is the same in all cases so the reader should have no difficulty in supplying the details by comparison with the case $j = 4$.

Let $j = 4$, $P = P_{1,4}$ and $G = Ind_B^{P_4}$. By (9.6.3) and Proposition 3.2.6 there is a P submodule K of $Y_\Omega(-2\lambda_1 + \lambda_3)$ such that K has a good filtration and $Y_\Omega(-2\lambda_1 + \lambda_3)/K$ has a good filtration with successive quotients $Y_{1,4}(-2\lambda_1 + 2\lambda_2 - \lambda_4 + \lambda_7)$, $Y_{1,4}(-2\lambda_1 + \lambda_2 + \lambda_3 - \lambda_4 + \lambda_6)$, $Y_{1,4}(-2\lambda_1 + \lambda_2 - \lambda_4 + \lambda_5 + \lambda_7)$, $Y_{1,4}(-2\lambda_1 + \lambda_3)$. Moreover, if $Y_{1,4}(\xi)$ is any successive quotient in a good filtration of K then $(\xi,\alpha_1^\vee) \ge -1$ and so, by (5.2), $R^i G(K)$ has a good filtration if $i = 0$ and is 0 if $i > 0$. Thus, identifying $Y_\Omega(-2\lambda_1 + \lambda_3)$, and hence K , with a submodule of $\bar{E}$, it is enough to prove that $R^iG(F)$ has a good filtration for $i = 0$ and is 0 for $i > 0$, where $F = \bar{E}/K$. By (9.6.4) and Proposition 3.2.6, F has a descending filtration $F = F_o, F_1, \ldots, F_8, F_9 = 0$ with successive quotients $Y_{1,4}(-\lambda_1 + \lambda_2)$, $Y_{1,4}(-\lambda_4 + \lambda_5 + \lambda_7)$, $Y_{1,4}(\lambda_3 - \lambda_4 + \lambda_6)$, $Y_{1,4}(\lambda_2 - \lambda_4 + \lambda_7)$, $Y_{1,4}(\lambda_1 - \lambda_4 + \lambda_5)$, $Y_{1,4}(-2\lambda_1 + \lambda_3)$, $Y_{1,4}(-2\lambda_1 + \lambda_2 - \lambda_4 + \lambda_5 + \lambda_7)$, $Y_{1,4}(-2\lambda_1 + \lambda_2 + \lambda_3 - \lambda_4 + \lambda_6)$, $Y_{1,4}(-2\lambda_1 + 2\lambda_2 - \lambda_4 + \lambda_7)$, where $F_5 = Y_\Omega(-2\lambda_1 + \lambda_3)/K$. By (5.8), $F_6 = M_\Gamma(-2\lambda_1 + \lambda_3)/K$, where $\Gamma = \Delta\backslash\{\alpha_1,\alpha_4\}$ and so, by (9.4.10) and (5.1.2) F/F_6 is a non-split extension of F_5/F_6 by F/F_5 . Now $Ext^1_P(Y_{1,4}(\tau), Y_{1,4}(-2\lambda_1 + \lambda_3)) = 0$ for $\lambda = \lambda_1 - \lambda_4 + \lambda_5$, $\lambda_2 - \lambda_4 + \lambda_5$, $\lambda_3 - \lambda_4 + \lambda_6$, $-\lambda_4 + \lambda_5 + \lambda_7$ by Lemma 3.2.1 so that we may rechoose the filtration of F such that F_o/F_1 is a non-split extension of $Y_{1,4}(-2\lambda_1 + \lambda_3)$ by $Y_{1,4}(-\lambda_1 + \lambda_2)$ and the remaining successive quotients are (in order) $Y_{1,4}(-\lambda_4 + \lambda_5 + \lambda_7)$, $Y_{1,4}(\lambda_3 - \lambda_4 + \lambda_6)$, $Y_{1,4}(\lambda_2 - \lambda_4 + \lambda_7)$, $Y_{1,4}(\lambda_1 - \lambda_4 + \lambda_5)$, $Y_{1,4}(-2\lambda_1 + \lambda_2 - \lambda_4 + \lambda_5 + \lambda_7)$, $Y_{1,4}(-2\lambda_1 + \lambda_2 + \lambda_3 - \lambda_4 + \lambda_6)$, $Y_{1,4}(-2\lambda_1 + 2\lambda_2 - \lambda_4 + \lambda_7)$. Now $Ext^1_P(Y_{1,4}(-\lambda_1 + \lambda_2), Y_{1,4}(-2\lambda_1 + \lambda_3)) \cong Ext^1_{P_\Lambda}(Y_\Lambda(-\lambda_1 + \lambda_2), Y_\Lambda(-2\lambda_1 + \lambda_3))$ where $\Lambda = \{\alpha_2\}$, by (5.10). Now $Ext^1_{P_\Lambda}(Y_\Lambda(-\lambda_1 + \lambda_2), Y_\Lambda(-2\lambda_1 + \lambda_3)) \cong H^1(P_\Lambda, Y_\Lambda(\lambda_2 - \lambda_3) \otimes Y_\Lambda(-2\lambda_1 + \lambda_3))$ $\cong H^1(P_\Lambda, Y_\Lambda(-2\lambda_1 + \lambda_2)) \cong H^1(B, -\alpha_1) \cong k$, by (1.4.7). Thus, by (5.11), F_o/F_1 is the unique non-split extension of $Y_{1,4}(-2\lambda_1 + \lambda_3)$ by $Y_{1,4}(-\lambda_1 + \lambda_2)$ and is therefore isomorphic to $Y_\Sigma(\lambda_2) \otimes -\lambda_1$ (where $\Sigma = \Delta\backslash\{\alpha_4\}$) .

Now $Ext^1_P(Y_{1,4}(\tau), Y_{1,4}(-2\lambda_1 + \lambda_2 - \lambda_4 + \lambda_5 + \lambda_7)) = 0$ for $\tau = \lambda_1 - \lambda_4 + \lambda_5$, $\lambda_2 - \lambda_4 + \lambda_7$, $\lambda_3 - \lambda_4 + \lambda_6$ by Lemma 3.2.1 so we may rechoose the filtration leaving F_o/F_1 unchanged, such that F_1/F_2 is an extension of $Y_{1,4}(-2\lambda_1 + \lambda_2 - \lambda_4 + \lambda_5 + \lambda_7)$ by $Y_{1,4}(-\lambda_4 + \lambda_5 + \lambda_7)$, and the remaining quotients, in order, are $Y_{1,4}(\lambda_3 - \lambda_4 + \lambda_6)$, $Y_{1,4}(\lambda_2 - \lambda_4 + \lambda_7)$, $Y_{1,4}(\lambda_1 - \lambda_4 + \lambda_5)$,

$Y_{1,4}(-2\lambda_1 + \lambda_2 + \lambda_3 - \lambda_4 + \lambda_6)$, $Y_{1,4}(-2\lambda_1 + 2\lambda_2 - \lambda_4 + \lambda_7)$. Now $Ext^1_P(F_o/F_1, Y_{1,4}(-2\lambda_1 + \lambda_2 - \lambda_4 + \lambda_5 + \lambda_7) \cong$ $Ext^1_P(Y_\Sigma(\lambda_2) \otimes -\lambda_1, Y_{1,4}(-2\lambda_1 + \lambda_2 - \lambda_4 + \lambda_5 + \lambda_7)) \cong$ $H^1(P, Y_\Sigma(\lambda_2)^* \otimes Y_{1,4}(-\lambda_1 + \lambda_2 - \lambda_4 + \lambda_5 + \lambda_7))$, which is 0 by (2.1.3) and (2.1.6). Thus if the extension F_1/F_2 of $Y_{1,4}(-2\lambda_1 + \lambda_2 - \lambda_4 + \lambda_5 + \lambda_7)$ by $Y_{1,4}(-\lambda_4 + \lambda_5 + \lambda_7)$ were split $Y_{1,4}(-2\lambda_1 + \lambda_2 - \lambda_4 + \lambda_5 + \lambda_7)$ would be an epimorphic image of F and hence of $\bar{E}$. However, by reciprocity,

$$Hom_P(\bar{E}, Y_{1,4}(-2\lambda_1 + \lambda_2 - \lambda_4 + \lambda_5 + \lambda_7)) \cong$$

$$Hom_{P_4}(\bar{E}, Ind_P^{P_4}(Y_{1,4}(-2\lambda_1 + \lambda_2 - \lambda_4 + \lambda_5 + \lambda_7))$$

which is 0 since $Ind_P^{P_4}(Y_{1,4}(-2\lambda_1 + \lambda_2 - \lambda_4 + \lambda_5 + \lambda_7)) = 0$ by the transitivity of induction and (1.7.1). Thus F_1/F_2 is a non-split extension. By (5.10) and (5.11) it is the unique non-split extension and therefore isomorphic to $Y_4(\lambda_1 - \lambda_4 + \lambda_5 + \lambda_7) \otimes -\lambda_1$.

We have $Ext^1_P(Y_{1,4}(\tau), Y_{1,4}(-2\lambda_1 + \lambda_2 + \lambda_3 - \lambda_4 + \lambda_6)) = 0$ for $\tau = \lambda_2 - \lambda_4 + \lambda_7, \lambda_1 - \lambda_4 + \lambda_5$ by Lemma 3.2.1 so that we may rechoose the filtration of F leaving F_o/F_1 , F_1/F_2 unchanged so that F_2/F_3 is an extension of $Y_{1,4}(-2\lambda_1 + \lambda_2 + \lambda_3 - \lambda_4 + \lambda_6)$ by $Y_{1,4}(\lambda_3 - \lambda_4 + \lambda_6)$ and the remaining quotients are $Y_{1,4}(\lambda_2 - \lambda_4 + \lambda_7)$, $Y_{1,4}(\lambda_1 - \lambda_4 + \lambda_5)$, $Y_{1,4}(-2\lambda_1 + 2\lambda_2 - \lambda_4 + \lambda_7)$. Now $Ext^1_P(Y_\Sigma(\tau) \otimes \lambda_1, Y_{1,4}(-2\lambda_1 + \lambda_2 + \lambda_3 - \lambda_4 + \lambda_6)) \cong$ $H^1(P, Y_\Sigma(\tau)^* \otimes Y_{1,4}(-\lambda_1 + \lambda_2 + \lambda_3 - \lambda_4 + \lambda_6)) = 0$ for $\tau \in X_\Sigma$, by (2.1.3) and (2.1.6). Thus if the extension F_2/F_3 of $Y_{1,4}(-2\lambda_1 + \lambda_2 + \lambda_3 - \lambda_4 + \lambda_6)$ by $Y_{1,4}(\lambda_3 - \lambda_4 + \lambda_6)$ were split $Y_{1,4}(-2\lambda_1 + \lambda_2 + \lambda_3 - \lambda_4 + \lambda_6)$ would be an epimorphic image of F , and hence of $\bar{E}$. However, $Hom_P(\bar{E}, Y_{1,4}(-2\lambda_1 + \lambda_2 + \lambda_3 - \lambda_4 + \lambda_6)) = 0$ so this is clearly impossible and F_2/F_3 is a non-split extension. Moreover $Ext^1_P(Y_{1,4}(\lambda_3 - \lambda_4 + \lambda_6), Y_{1,4}(-2\lambda_1 + \lambda_2 + \lambda_3 - \lambda_4 + \lambda_6)) \cong k$

so that F_2/F_3 is the unique non-split extension. By considering $Y_4(\lambda_1 + \lambda_3 - \lambda_4 + \lambda_6)|_P$ we deduce that there is a short exact sequence of P modules

$$0 \to Y_4(\lambda_1 - \lambda_4 + \lambda_6) \otimes -2\lambda_1 \to Y_4(\lambda_1 + \lambda_3 - \lambda_4 + \lambda_6) \otimes -\lambda_1$$
$$\to F_2/F_3 \to 0 .$$

Now $\mathit{Ext}^1_P(Y_{1,4}(\lambda_1 - \lambda_4 + \lambda_5), Y_{1,4}(-2\lambda_1 + 2\lambda_2 - \lambda_4 + \lambda_7)) = 0$ by Lemma 3.2.1 so we may rechoose the filtration leaving $F_o/F_1, F_1/F_2, F_2/F_3$ and F_3/F_4 unchanged with $F_4/F_5 = Y_{1,4}(-2\lambda_1 + 2\lambda_2 - \lambda_4 + \lambda_7)$, $F_5 = Y_{1,4}(\lambda_1 - \lambda_4 + \lambda_5)$, $F_6 = 0$. We have $\mathit{Ext}^1_P(Y_4(\tau) \otimes -\lambda_1, Y_{1,4}(-2\lambda_1 + 2\lambda_2 - \lambda_4 + \lambda_7)) = 0$ for $\tau \in X_\Sigma$ so that $\mathit{Ext}^1_P(F_o/F_2, Y_{1,4}(-2\lambda_1 + 2\lambda_2 - \lambda_4 + \lambda_7)) = 0$. By the above short exact sequence we have

$\mathit{Ext}^1_P(F_2/F_3, Y_{1,4}(-2\lambda_1 + 2\lambda_2 - \lambda_4 + \lambda_7))$

$\cong \mathit{Ext}^2_P(Y_4(\lambda_1 - \lambda_4 + \lambda_6) \otimes -2\lambda_1, Y_{1,4}(-2\lambda_1 + 2\lambda_2 - \lambda_4 + \lambda_7))$

$\cong \mathit{Ext}^2_P(Y_4(\lambda_1 - \lambda_4 + \lambda_6), Y_{1,4}(2\lambda_2 - \lambda_4 + \lambda_7))$. However this is 0 by (1.1.7), (1.1.8) and (2.1.6). Thus $\mathit{Ext}^1_P(F/F_3, Y_{1,4}(-2\lambda_1 + 2\lambda_2 - \lambda_4 + \lambda_7)) = 0$ and it follows that the extension F_3/F_5, of F_4/F_5 by F_3/F_4 is non-split.

To prove (9.7.1) it now suffices to prove that $R^iF(M)$ has a good filtration for $i = 0$ and is 0 for $i > 0$, for $M = F_o/F_1, F_1/F_2, F_2/F_3, F_3/F_5, F_5$. However $R^iG(Y_4(\tau) \otimes -\lambda_1) = 0$ for all $i \geq 0$, $\tau \in X_\Sigma$ by the tensor identity and (2.1.5). Thus it only remains to deal with $M = F_2/F_3, F_3/F_5, F_5$. We have, by the above short exact sequence, $R^iG(F_2/F_3) \cong R^{i+1}G(Y_4(\lambda_1 - \lambda_4 + \lambda_6) \otimes -2\lambda_1) = 0$ for all $i \geq 0$, by the tensor identity, (2.3.1) and (2.1.5). Also $R^iG(Y_{1,4}(\lambda_1 - \lambda_4 + \lambda_5))$ is $Y_4(\lambda_1 - \lambda_4 + \lambda_5)$ if $i = 0$ and 0 if $i > 0$, by (2.1.1) and the vanishing theorem. Finally, one obtains $R^iG(F_3/F_5) = 0$ for all $i \geq 0$ by (5.18). This completes the proof of (9.7.1) in the case $j = 4$.

Now suppose that $j = 2$, let $P = P_{1,2}$, $G = Ind_B^{P_2}$ and $\Gamma = \Delta\backslash\{\alpha_1,\alpha_2\}$. Let $M = M_\Gamma(-2\lambda_1 + \lambda_3)$ (see (5.8)). Then $R^iG(M)$ has a good filtration if $i = 0$ and is 0 if $i > 0$, by (9.6.4). Also $F = \bar{E}/M$ is a non-split extension of $Y_{1,2}(-2\lambda_1 + \lambda_3)$ by $Y_1(-\lambda_1 + \lambda_2)$. By (9.6.4), $Y_1(-\lambda_1 + \lambda_2)|_P$ has a good filtration with quotients $Y_{1,2}(\lambda_1 - \lambda_2 + \lambda_6)$, $Y_{1,2}(-\lambda_2 + \lambda_3)$, $Y_{1,2}(-\lambda_1 + \lambda_2)$. Thus F has a descending filtration $F = F_o,F_1,F_2,F_3,F_4 = 0$ with quotients, in order, $Y_{1,2}(-\lambda_1 + \lambda_2)$, $Y_{1,2}(-\lambda_2 + \lambda_3)$, $Y_{1,2}(\lambda_1 - \lambda_2 + \lambda_6)$, $Y_{1,2}(-2\lambda_1 + \lambda_3)$. Also $Ext^1_P(Y_{1,2}(\lambda_1 - \lambda_2 + \lambda_6),Y_{1,2}(-2\lambda_1 + \lambda_3)) = 0$ and
$Ext^1_P(Y_{1,2}(-\lambda_1 + \lambda_2),Y_{1,2}(-2\lambda_1 + \lambda_3)) \cong H^1(P,Y_{1,2}(-\lambda_1 - \lambda_2 + \lambda_3)) = 0$.
Thus we may rechoose the filtration such that F_o/F_1 is unchanged, F_1/F_2 is a non-split extension of $Y_{1,2}(-2\lambda_1 + \lambda_3)$ by $Y_{1,2}(-\lambda_2 + \lambda_3)$ - and hence isomorphic to $Y_2(\lambda_1 - \lambda_2 + \lambda_3) \otimes -\lambda_1$ - and $F_2 \cong Y_{1,2}(\lambda_1 - \lambda_2 + \lambda_6)$. Now $R^iG(V)$ has a good filtration for $i = 0$ and is 0 for $i > 0$, for $V = F_o/F_1,F_1/F_2,F_2$. Hence $R^iG(F)$ has a good filtration for $i = 0$ and is 0 for $i > 0$. This concludes the analysis of the case $j = 2$.

Now suppose that $j = 3$. By (9.6.3) there is a submodule K of $Y_\Omega(-2\lambda_1 + \lambda_3)$ such that K has a good filtration and $Y_\Omega(-2\lambda_1 + \lambda_3)/K$ has a good filtration with successive quotients
$Y_{1,3}(-2\lambda_1 + 2\lambda_2 - \lambda_3 + \lambda_6)$, $Y_{1,3}(-2\lambda_1 + \lambda_2 - \lambda_3 + \lambda_4)$,
$Y_{1,3}(-2\lambda_1 + \lambda_3)$. Moreover, if $Y_{1,3}(\tau)$ is any successive quotient in a good filtration of K then $(\tau,\check{\alpha}_1) \geq -1$ and so $R^iG(K)$ has a good filtration if $i = 0$ and is 0 if $i > 0$. Thus it is enough to prove that $R^iG(F)$ has a good filtration for $i = 0$ and is 0 for $i > 0$, where $F = \bar{E}/K$. Now $Y_1(-\lambda_1 + \lambda_2)|_P$ has, by (9.6.4), a good filtration with quotients $Y_{1,3}(\lambda_1 - \lambda_3 + \lambda_7)$, $Y_{1,3}(\lambda_2 - \lambda_3 + \lambda_6)$,
$Y_{1,3}(-\lambda_1 + \lambda_2)$. Thus F has a descending filtration
$F = F_o,F_1,\ldots,F_6,F_7 = 0$ with successive quotients $Y_{1,3}(-\lambda_1 + \lambda_2)$,
$Y_{1,3}(-\lambda_3 + \lambda_4)$, $Y_{1,3}(\lambda_2 - \lambda_3 + \lambda_6)$, $Y_{1,4}(\lambda_1 - \lambda_3 + \lambda_7)$,
$Y_{1,3}(-2\lambda_1 + \lambda_3)$, $Y_{1,3}(-2\lambda_1 + \lambda_2 - \lambda_3 + \lambda_4)$, $Y_{1,3}(-2\lambda_1 + 2\lambda_2 - \lambda_3 + \lambda_6)$

Also, F/F_5 is a non-split extension of F_4/F_5 by F/F_4. By Lemma 3.2.1, $Ext^1_P(Y_{1,3}(\tau), Y_{1,3}(-2\lambda_1 + \lambda_3)) = 0$ for $\tau = \lambda_1 - \lambda_3 + \lambda_7$, $\lambda_2 - \lambda_3 + \lambda_6$, $-\lambda_3 + \lambda_4$ so we may rechoose the filtration so that F_o/F_1 is a non-split extension of $Y_{1,3}(-2\lambda_1 + \lambda_3)$ by $Y_{1,3}(-\lambda_1 + \lambda_2)$ and the remaining quotients, in order, are $Y_{1,3}(-\lambda_3 + \lambda_4)$, $Y_{1,3}(\lambda_2 - \lambda_3 + \lambda_6)$, $Y_{1,3}(\lambda_1 - \lambda_3 + \lambda_7)$, $Y_{1,3}(-2\lambda_1 + \lambda_2 - \lambda_3 + \lambda_4)$, $Y_{1,3}(-\lambda_1 + 2\lambda_2 - \lambda_3 + \lambda_6)$. Now F_o/F_1 is the unique non-split P extension of $Y_{1,3}(-2\lambda_1 + \lambda_3)$ by $Y_{1,3}(-\lambda_1 + \lambda_2)$ and hence isomorphic to $Y_3(\lambda_2) \otimes -\lambda_1$.

We have $Ext^1_P(Y_{1,3}(\tau), Y_{1,3}(-2\lambda_1 + \lambda_2 - \lambda_3 + \lambda_4)) = 0$ for $\tau = \lambda_1 - \lambda_3 + \lambda_7$, $\lambda_2 - \lambda_3 + \lambda_6$. Thus we may rechoose the filtration such that F_o/F_1 is unchanged, F_1/F_2 is an extension of $Y_{1,3}(-2\lambda_1 + \lambda_2 - \lambda_3 + \lambda_4)$ by $Y_{1,3}(-\lambda_3 + \lambda_4)$ and the remaining quotients are (in order) $Y_{1,3}(\lambda_2 - \lambda_3 + \lambda_6)$, $Y_{1,3}(\lambda_1 - \lambda_3 + \lambda_7)$, $Y_{1,3}(-2\lambda_1 + 2\lambda_2 - \lambda_3 + \lambda_6)$. We have

$Ext^1_P(Y_3(\lambda_2) \otimes -\lambda_1, Y_{1,3}(-2\lambda_1 + \lambda_2 + \lambda_4)) \cong$
$H^1(P, Y_3(\lambda_2)^* \otimes Y_{1,3}(-\lambda_1 + \lambda_2 - \lambda_3 + \lambda_4))$, which is 0. Moreover F does not have an epimorphic image isomorphic to $Y_{1,3}(-2\lambda_1 + \lambda_2 - \lambda_3 + \lambda_4)$ and so the extension F_1/F_2 of $Y_{1,3}(-2\lambda_1 + \lambda_2 - \lambda_3 + \lambda_4)$ by $Y_{1,3}(-\lambda_3 + \lambda_4)$ is non-split and hence isomorphic to $Y_3(\lambda_1 + \lambda_4) \otimes -\lambda_1$.

Now $Ext^1_P(Y_{1,3}(\lambda_1 - \lambda_3 + \lambda_7), Y_{1,3}(-2\lambda_1 + 2\lambda_2 - \lambda_3 + \lambda_6)) = 0$ so that we may rechoose the filtration leaving $F_o/F_1, F_1/F_2$ unchanged, such that F_2/F_3 is an extension of $Y_{1,3}(-2\lambda_1 + 2\lambda_2 - \lambda_3 + \lambda_6)$ by $Y_{1,3}(\lambda_2 - \lambda_3 + \lambda_6)$ and $F_3 = Y_{1,3}(\lambda_1 - \lambda_3 + \lambda_7)$. Also $Ext^1_P(S, Y_{1,3}(-2\lambda_1 + 2\lambda_2 - \lambda_3 + \lambda_6)) = 0$ for $S = F_o/F_1, F_1/F_2$ and since $Y_{1,3}(-2\lambda_1 + 2\lambda_2 - \lambda_3 + \lambda_6)$ is not an image of F, F_2/F_3 is the unique non-split extension of $Y_{1,3}(-2\lambda_1 + 2\lambda_2 - \lambda_3 + \lambda_6)$ by $Y_{1,3}(\lambda_2 - \lambda_3 + \lambda_6)$.

We have $R^iG(V) = 0$ for all $i \geq 0$ for $V = F_o/F_1, F_1/F_2, F_2/F_3$ and $R^iG(F_3)$ is $Y_3(\lambda_1 - \lambda_3 + \lambda_7)$ for $i = 0$ and is 0 for $i > 0$. Thus $R^iG(F)$ has a good filtration if $i = 0$ and is 0 if $i > 0$.

This concludes the proof of (9.7.1) in the case $j = 3$.

We now examine the case $j = 5$. Let $P = P_{1,5}$ and $G = Ind_B^{P_5}$. By (9.6.3), (9.6.4) we need only show that $R^iG(F)$ has a good filtrat- for $i = 0$ and is 0 for $i > 0$, where F is an epimorphic image of $\overline{E}$ which has a descending filtration $F = F_o, F_1, \ldots, F_7 = 0$ with successive quotients $Y_{1,5}(-\lambda_1 + \lambda_2)$, $Y_{1,5}(-\lambda_5 + \lambda_6 + \lambda_7)$, $Y_{1,5}(\lambda_3 - \lambda_5)$, $Y_{1,5}(\lambda_1 - \lambda_5 + \lambda_6)$, $Y_{1,5}(-2\lambda_1 + \lambda_3)$, $Y_{1,5}(-2\lambda_1 + \lambda_2 - \lambda_5 + \lambda_6 + \lambda_7)$, $Y_{1,5}(-2\lambda_1 + \lambda_2 + \lambda_3 - \lambda_5)$ and such that F/F_5 is a non-split extension of F_4/F_5 by F/F_4. We may re-choose the filtration such that F_1/F_1 is a non-split extension of $Y_{1,5}(-2\lambda_1 + \lambda_3)$ by $Y_{1,5}(-\lambda_1 + \lambda_2)$ - and hence isomorphic to $Y_5(\lambda_2) \otimes -\lambda_1$ - F_1/F_2 is a non-split extension of $Y_{1,5}(-2\lambda_1 + \lambda_2 - \lambda_5 + \lambda_6 + \lambda_7)$ by $Y_{1,5}(-\lambda_5 + \lambda_6 + \lambda_7)$ - and hence isomorphic to $Y_5(\lambda_1 - \lambda_5 + \lambda_6 + \lambda_7) \otimes -\lambda_1$ - F_2/F_3 is a non-split extension of $Y_{1,5}(-2\lambda_1 + \lambda_2 + \lambda_3 - \lambda_5)$ by $Y_{1,5}(\lambda_3 - \lambda_5)$, $F_3 = Y_{1,5}(\lambda_1 - \lambda_5 + \lambda_6)$ and $F_4 = 0$. Now $R^iG(Y_5(\tau) \otimes -\lambda_1) = 0$ for all $i \geq 0$, $\tau \in X_\Sigma$, where $\Sigma = \Delta\backslash\{\alpha_5\}$, so that $R^iG(F_o/F_1) = R^iG(F_1/F_2) = 0$ for all $i \geq 0$. Also $R^iG(F_2/F_3) = 0$ for all $i \geq 0$ and $R^iG(Y_{1,5}(\lambda_1 - \lambda_5 + \lambda_6))$ is $Y_5(\lambda_1 - \lambda_5 + \lambda_6)$ if $i = 0$ and is 0 if $i > 0$. Thus $R^iG(F)$ has a good filtration for $i = 0$ and is 0 for $i > 0$. This completes the case $j = 5$.

Now suppose that $j = 6$. Let $P = P_{1,6}$ and $G = Ind_B^{P_6}$. By (9.6.3), (9.6.4) we need only show that $R^iG(F)$ has a good filtration for $i = 0$ and is 0 for $i > 0$, where F is a P epimorphic image of $\overline{E}$ with a descending filtration $F = F_o, F_1, \ldots, F_4, F_5 = 0$ with successive quotients $Y_{1,6}(-\lambda_1 + \lambda_2)$, $Y_{1,6}(-\lambda_6 + \lambda_7)$, $Y_{1,6}(\lambda_1 - \lambda_6)$, $Y_{1,6}(-2\lambda_1 + \lambda_3)$, $Y_{1,6}(-2\lambda_1 + \lambda_2 - \lambda_6 + \lambda_7)$ and such that F/F_4 is a non-split extension of F_4/F_5 by F/F_4. We may rechoose the filtration such that F_o/F_1 is a non-split extension of $Y_{1,6}(-2\lambda_1 + \lambda_3)$ by $Y_{1,6}(-\lambda_1 + \lambda_2)$, $F_1/F_2 = Y_{1,6}(-\lambda_6 + \lambda_7)$, $F_2/F_3 = Y_{1,6}(\lambda_1 - \lambda_6)$, $F_3 = Y_{1,6}(-2\lambda_1 + \lambda_2 - \lambda_6 + \lambda_7)$ and $F_4 = 0$. By considering $Y_6(\lambda_2)|_P$ we find that there is a short exact sequence of P modules

$$0 \to Y_6(\lambda_1 + \lambda_6) \otimes -2\lambda_1 \to Y_6(\lambda_2) \otimes -\lambda_1 \to F_o/F_1 \to 0 .$$

Thus $Ext_P^1(F_o/F_1, Y_{1,6}(-2\lambda_1 + \lambda_2 - \lambda_6 + \lambda_7)) =$
$Ext_P^2(Y_6(\lambda_1 + \lambda_6) \otimes -2\lambda_1, Y_{1,6}(-2\lambda_1 + \lambda_2 - \lambda_6 + \lambda_7))$
$= Ext_P^2(Y_6(\lambda_1 + \lambda_6), Y_{1,6}(\lambda_2 - \lambda_6 + \lambda_7))$
$= Ext_{P_6}^2(Y_6(\lambda_1 + \lambda_6), Y_6(\lambda_2 - \lambda_6 + \lambda_7)) = 0$ and so we may rechoose the filtration so that F_o/F_1 is unchanged, F_1/F_2 is a non-split extension of $Y_{1,6}(-2\lambda_1 + \lambda_2 - \lambda_6 + \lambda_7)$ by $Y_{1,6}(-\lambda_6 + \lambda_7)$, $F_2 = Y_{1,6}(\lambda_1 - \lambda_6)$ and $F_3 = 0$. We have
$R^iG(F_o/F_1) \cong R^{i+1}(Y_6(\lambda_1 + \lambda_6) \otimes -2\lambda_1) \cong R^iG(Y_6(\lambda_1 + \lambda_6) \otimes -\lambda_2) = 0$ for all $i \geq 0$. Also $R^iG(F_1/F_2) = 0$ for all $i \geq 0$, and $R^iG(F_2)$ is $Y_6(\lambda_1 - \lambda_6)$ if $i = 0$ and 0 if $i > 0$. Hence $R^iG(F)$ has a good filtration for $i = 0$ and is 0 for $i > 0$, as required.

We now come to the final case $j = 7$. Let $P = P_{1,7}$ and $G = Ind_B^{P_7}$. By (9.6.3), (9.6.4) we need only show that $R^iG(F)$ has a good filtration for $i = 0$ and is 0 for $i > 0$ where F is a P epimorphic image of $\tilde{E}$ which has a descending filtration $F = F_o, F_1, \ldots, F_6 = 0$ with successive quotients $Y_{1,7}(-\lambda_1 + \lambda_2)$, $Y_{1,7}(\lambda_5 - \lambda_7)$, $Y_{1,7}(\lambda_2 - \lambda_7)$, $Y_{1,7}(-2\lambda_1 + \lambda_3)$, $Y_{1,7}(-2\lambda_1 + \lambda_2 + \lambda_5 - \lambda_7)$, $Y_{.,7}(-2\lambda_1 + 2\lambda_2 - \lambda_7)$ and F/F_4 is a non-split extension of F_3/F_4 by F/F_3. We may rechoose the filtration such that F_o/F_1 is a non-split extension of $Y_{1,7}(-2\lambda_1 + \lambda_2 + \lambda_5 - \lambda_7)$ by $Y_{1,7}(-\lambda_1 + \lambda_2)$ - and hence isomorphic to $Y_7(\lambda_2) \otimes -\lambda_1$ - F_1/F_2 is a non-split extension of $Y_{1,7}(-2\lambda_1 + \lambda_2 + \lambda_5 - \lambda_7)$ by $Y_{1,7}(\lambda_5 - \lambda_7)$, $F_2/F_3 = Y_{1,7}(\lambda_2 - \lambda_7)$, $F_3 = F_{1,7}(-2\lambda_1 + 2\lambda_2 - \lambda_7)$ and $F_4 = 0$. By considering $Y_7(\lambda_1 + \lambda_5 - \lambda_7)|_P$ we find that there is a short exact sequence

$$0 \to K \to Y_7(\lambda_1 + \lambda_5 - \lambda_7) \otimes -\lambda_1 \to F_1/F_2 \to 0$$

where K is a P module extension of $Y_{1,7}(-3\lambda_1 + \lambda_2 + \lambda_6)$ by $Y_{1,7}(-\lambda_1 + \lambda_6)$. Thus

$Ext^1_P(F_1/F_2, Y_{1,7}(-2\lambda_1 + 2\lambda_2 - \lambda_7)) \cong Ext^2_P(K, Y_{1,7}(-2\lambda_1 + 2\lambda_2 - \lambda_7))$ which is 0 since $Ext^2_P(Y_{1,7}(\tau), Y_{1,7}(-2\lambda_1 + 2\lambda_2 - \lambda_7)) = 0$ for $\tau = -\lambda_1 + \lambda_6$, $-3\lambda_1 + \lambda_2 + \lambda_6$. Also $Ext^1_P(F_o/F_1, Y_{1,7}(-2\lambda_1 + 2\lambda_2 - \lambda_7)) = 0$ and it follows that F_2 is a non-split extension of F_3 by F_2/F_3. We have $R^iG(F_o/F_1) = R^iG(F_1/F_2) = R^iG(F_2/F_3) = 0$ for all $i \geq 0$. Hence $R^iG(F) = 0$ for all $i \geq 0$ and the proof of (9.7.1) is complete.

9.8 $Y(\lambda_7)$

We now have by (9.4.11), (9.5.2), (9.5.3), (9.7.1) and the remarks at the beginning of section 9.5 that $Y(\lambda_7)|_{P_\Sigma}$ has a good filtration for $\Sigma = \Delta\backslash\{\alpha_j\}$, $1 \leq j \leq 7$. Since $\underline{H}(P_\Gamma)$ is true for every proper subset Γ of Δ we obtain:

(9.8.1) $Y(\lambda_7)|_{P_\Sigma}$ *has a good filtration for every* $\Sigma \subseteq \Delta$.

Since $(\lambda_7, \beta_o^\vee) = 2$ (where β_o is the highest short root) we have $(\tau, \alpha^\vee) \geq -2$ for any weight τ of $Y(\lambda_7)$ and $\alpha \in \Delta$. Thus Proposition 3.5.2 yields the following.

(9.8.2) *For every* G-*module* V *with a good filtration,* $Y(\lambda_7) \otimes V$ *has a good filtration.*

We now obtain, as in section 9.3, that $\underline{H}(G)$ is true if the characteristic of k is at least 5. Combining this with (9.3.4) we obtain the following.

(9.8.3) *If the characteristic of* k *is not* 2 *then* $\underline{H}(G)$ *is true.*

10. $\underline{E}_8$

10.1 $Y(\lambda_1)$

Let G be semisimple, simply connected group of type E_8 over an algebraically closed field k . We assume whenever convenient in this chapter that $\underline{H}(E_7)$ is true. By chapter 9 this certainly holds if the characteristic of k is anything but 2 (of course it is true also in characteristic 2 but we can't prove it). In this chapter we show that, provided that the characteristic of k is not 2 , $\underline{H}(G)$ is true. The proof follows the proof, given in the last chapter, for E_7 . In this section we analyse $Y(\lambda_1)$, in 10.2 we use some block theory to obtain the desired result in characteristic not equal to 2 or 7 . The remainder of the chapter consists of the detailed analysis of $Y(\lambda_7)$ necessary to obtain the result in characteristic 7 .

By p.69 of [34] (with suitable relabelling) we have:

$$(10.1.1)\quad \lambda_1 = \langle 2\,3\,4\,5\,\overset{3}{6}\,4\,2\rangle , \quad \lambda_2 = \langle 3\;6\;8\;10\;\overset{6}{12}\;8\;4\rangle ,$$

$$\lambda_3 = \langle 4\;8\;12\;15\;\overset{9}{18}\;12\;6\rangle , \quad \lambda_4 = \langle 5\;10\;15\;20\;\overset{12}{24}\;16\;8\rangle ,$$

$$\lambda_5 = \langle 6\;12\;18\;24\;\overset{15}{30}\;20\;10\rangle , \quad \lambda_6 = \langle 4\;8\;12\;16\;\overset{10}{20}\;14\;7\rangle ,$$

$$\lambda_7 = \langle 2\;4\;6\;8\;\overset{5}{10}\;7\;4\rangle , \quad \lambda_8 = \langle 3\;6\;9\;12\;\overset{8}{15}\;10\;5\rangle .$$

Thus λ_1 is the unique minimal non-zero dominant weight from which it follows that $\chi(\lambda_1)$ is the character of the adjoint representation of G so that

$$(10.1.2)\quad \chi(\lambda_1) = s(\lambda_1) + 8s(0) .$$

Let $\Omega = \Delta\backslash\{\alpha_1\}$. We claim:

$$(10.1.3)\quad s(\lambda_1) = s_\Omega(\lambda_1) + s_\Omega(-\lambda_1+\lambda_2) + s_\Omega(-\lambda_1+\lambda_7) + s_\Omega(-2\lambda_1+\lambda_2) + s_\Omega(-\lambda_1) .$$

et U be the set of roots α which belong to X_Ω . Thus (10.1.3) s equivalent to the assertion $U = \{\lambda_1, -\lambda_1+\lambda_2, -\lambda_1+\lambda_7, -2\lambda_1+\lambda_2, -\lambda_1\}$ nich we now prove. Suppose $\alpha \in U$ and $i > 1$. Either $(\alpha,\alpha_i^\vee) = 0$, $= \alpha_i$ or $(\alpha,\alpha_i^\vee) = 1$ and α together with α_i generates a root system of type A_2. Actually, no α_i, for $i > 1$, belongs to $\mathcal{U}$. If $(\alpha,\alpha_i^\vee) = 0$ for $2 \le i \le 8$ then $\alpha = m\lambda_1$ for some integer m and, since α is conjugate to λ_1, we must have $\alpha = \pm\lambda_1$. It remains to show that if $\alpha \in \mathcal{U}$ and $(\alpha,\alpha_i^\vee) = 1$ for some $i > 1$ then α is $-\lambda_1 + \lambda_2$, $-\lambda_1 + \lambda_7$ or $-2\lambda_1 + \lambda_2$. Suppose that $(\alpha,\alpha_i^\vee) = (\alpha,\alpha_j^\vee) = 1$ for $i,j > 1$, $i \neq j$. Let $\beta = \beta_1 + \beta_2 + \dots + \beta_r$ where $\beta_1,\beta_2,\dots,\beta_r$ are the simple roots indexing the nodes in the path from α_i to α_j in the Dynkin-Diagram. Then $(\alpha,\beta^\vee) \ge (\alpha,\alpha_i^\vee) + (\alpha,\alpha_j^\vee) = 2$ so that $\alpha = \beta$. However no such β lies in X_Ω so that $(\alpha,\alpha_i^\vee) = 1$ for precisely one $2 \le i \le 8$. If $(\alpha,\alpha_1^\vee) = -2$ then $\alpha = -\alpha_1 = -2\lambda_1 + \lambda_2$. Otherwise $\alpha = -\lambda_1 + \lambda_i$ for some $2 \le i \le 8$. The condition $(\alpha,\beta_0^\vee) \le 2$, where β_0 is the highest root $\langle 2345\overset{3}{6}42\rangle$ now implies that $i = 2$, 7 or 8. It is easy to check that $-\lambda_1 + \lambda_2$ and $-\lambda_1 + \lambda_7$ are conjugate to λ_1 but $-\lambda_1 + \lambda_8$ is not, completing the proof of (10.1.3).

It is easy to deduce from (10.1.2) and (10.1.3):

(10.1.4) $$\chi(\lambda_1) = \chi_\Omega(\lambda_1) + \chi_\Omega(-\lambda_1 + \lambda_2) + \chi_\Omega(-\lambda_1 + \lambda_7) + \chi_\Omega(0) + \chi_\Omega(-2\lambda_1 + \lambda_2) + \chi_\Omega(-\lambda_1) .$$

Let $K = F_\Omega^{-1}Y$ where $Y = Y(\lambda_1)$. Then $ch K = \chi_\Omega(-2\lambda_1 + \lambda_2) + \chi_\Omega(-\lambda_1)$, moreover $\chi_\Omega(-2\lambda_1 + \lambda_2) = s_\Omega(-2\lambda_1 + \lambda_2)$ and $\chi_\Omega(-\lambda_1) = s_\Omega(-\lambda_1)$ so that $Y_\Omega(-2\lambda_1 + \lambda_2)$ and $Y_\Omega(-\lambda_1)$ are simple and so any composition series of K is a good filtration. By the usual argument $(Y/K)^{P_\Omega} \cong H^1(B,K)$. Moreover the long exact sequence of cohomology, applied to the short exact sequence

$$0 \to Y_\Omega(-\lambda_1) \to K \to Y_\Omega(-2\lambda_1 + \lambda_2) \to 0$$

gives $H^1(B,K) \cong H^1(B,Y_\Omega(-2\lambda_1 + \lambda_2))$. By (1.4.7) and (2.1.3) this is isomorphic to k. Hence there is a P_Ω submodule M of Y containing K such that M/K is isomorphic to k. Now from (10.1.4), $ch F_\Omega^0(Y/M) = \chi_\Omega(-\lambda_1 + \lambda_7)$ so that, by (1.5.4), to show that $F_\Omega^0(Y/M)$ is isomorphic to $Y_\Omega(-\lambda_1 + \lambda_7)$ it suffices to show that $((F_\Omega^0(Y/M)) \otimes \tau)^B = 0$ for any $\tau \in X_\Omega$ different from $-\lambda_1 + \lambda_7$ such that $-\tau$ is a weight of $Y_\Omega(-\lambda_1 + \lambda_7)$. Thus it suffices to show that $(F_\Omega^0(Y/M))^B = 0$ and so it is enough to prove that $(Y/M)^B = 0$. Let

$N/M = (Y/M)^B$. Then N/K is a trivial B module by Lemma 3.2.1 and so $N/K = (Y/K)^B = M/K$ and $N = M$. Thus $(Y/M)^B = 0$ and $F_\Omega^0(Y/M) \cong Y_\Omega(-\lambda_1 + \lambda_7)$. Now, by (10.1.4), $chY/F_\Omega^0 Y = \chi_\Omega(\lambda_1) + \chi_\Omega(-\lambda_1 + \lambda_2)$. Moreover $\chi_\Omega(\lambda_1) = s_\Omega(\lambda_1)$ and $\chi_\Omega(-\lambda_1 + \lambda_2) = s_\Omega(-\lambda_1 + \lambda_2)$ so that $Y_\Omega(\lambda_1)$ and $Y_\Omega(-\lambda_1 + \lambda_2)$ are simple P_Ω modules. Thus $Y/F_\Omega^0 Y$ has a good filtration (any composition series is a good filtration), we have shown that $F_\Omega^0 Y$ has a good filtration and therefore:

(10.1.5) $Y(\lambda_1)|_{P_\Omega}$ *has a good filtration.*

We shall need, in the later sections,

(10.1.6) Y/M *has a simple* B *socle* $\lambda_1 - \lambda_7$

and it is convenient to observe this now. The quotients in a good filtration of Y/M are $Y_\Omega(-\lambda_1 + \lambda_7)$, $Y_\Omega(-\lambda_1 + \lambda_2)$ and $Y_\Omega(\lambda_1)$. The B socle of $Y_\Omega(-\lambda_1 + \lambda_7)$ is $\lambda_1 - \lambda_7$, the B socle of $Y_\Omega(-\lambda_1 + \lambda_2)$ is $2\lambda_1 - \lambda_2$ and the B-socle of $Y_\Omega(\lambda_1)$ is λ_1 . Thus it suffices to prove that $(Y/M \otimes \tau)^B = 0$ for $\tau = -\lambda_1 , -2\lambda_1 + \lambda_2$. If $(Y/M \otimes -\lambda_1)^B \neq 0$ then Y/M contains a copy, say M'/M of $Y_\Omega(\lambda_1)$. By Proposition 3.2.4, Y/M' has a good filtration with quotients $Y_\Omega(-\lambda_1 + \lambda_7)$, $Y_\Omega(-\lambda_1 + \lambda_2)$. Thus for $\tau = -\lambda_1 + \lambda_7$ or $\tau = -\lambda_1 + \lambda_2$, $Hom_{P_\Omega}(Y, Y_\Omega(\tau)) \neq 0$. But this is impossible, by the reciprocity of induction. Similarly, if $(Y/M \otimes -2\lambda_1 + \lambda_2)^B \neq 0$, Y/M contains a P_Ω submodule, say M''/M , isomorphic to $Y_\Omega(-\lambda_1 + \lambda_2)$ and Y/M'' has a good filtration with quotients $Y_\Omega(-\lambda_1 + \lambda_7)$ and $Y_\Omega(\lambda_1)$. However $Ext^1_{P_\Omega}(Y((\lambda_1), Y_\Omega(-\lambda_1 + \lambda_7)) \cong H^1(P_\Omega, Y_\Omega(-2\lambda_1 + \lambda_7)) \cong H^1(B, -2\lambda_1 + \lambda_7)$ which is 0 by Corollary 2.3.5. Thus $Hom_{P_\Omega}(Y, Y_\Omega(-\lambda_1 + \lambda_7)) = 0$ which, by the reciprocity of induction, is impossible. Thus Y/M does indeed have B socle $\lambda_1 - \lambda_7$.

We shall now show:

(10.1.7) $Y(\lambda_1)|_{P_\Sigma}$ *has a good filtration for every* $\Sigma \subseteq \Delta$.

Since $\underline{H}(\Phi_\Gamma)$ is true for proper subsets Γ of Δ it suffices to prove (10.1.7) with Σ maximal and, by (10.1.5) we may suppose $\Sigma = \Delta\backslash\{\alpha_j\}$ where $2 \le j \le 8$. By (5.1) it suffices to show that $R^i F(Z)$ has a good filtration for $i = 0$ and is 0 for $i > 0$ for $Z = F_\Omega^{-2}Y, V, Y/M$ where $V = M/F_\Omega^{-2}Y$ and $F = Ind_{P_\Gamma}^{P_\Sigma}$ with $\Gamma = \Omega \cap \Sigma$. By (5.2) it suffices to prove this for $Z = V$.

We claim:

(10.1.8) *the short exact sequence* $0 \to Y_\Omega(-2\lambda_1 + \lambda_2) \to V \to k \to 0$ *is not split.*

If the sequence were split we would have, in side Y, a B extension of $-\lambda_1$ by k. However $H^1(B,-\lambda_1) = 0$ so that Y would contain a copy of the trivial B module k. However Y^B is 0 by (1.5.2).

Let $S = M_\Gamma(-2\lambda_1 + \lambda_2)$ (see (5.8)), identified with a submodule of V, and $V' = V/S$. By (5.12), V' is a non-split extension of $Y_\Gamma(-2\lambda_1 + \lambda_2)$ by k. To prove (10.1.7) it now suffices to prove that $R^i F(Z)$ has a good filtration for $i = 0$ and is 0 for $i > 0$, for $Z = S$ and $Z = V'$. Let τ be a non-zero weight of S. Then τ is a root so that $(\tau,\alpha_1^\vee) = 0$ or τ and α_1 generate a root system of type A_1 or A_2. Hence $(\tau,\alpha_1^\vee) \ge -1$ or $(\tau,\alpha_1^\vee) = -2$ and $\tau = -\alpha_1$. However each root occurs with multiplicity one as a weight of $Y(\lambda_1)$ and $-\alpha_1$ is a weight of V'. Hence $(\tau,\alpha_1^\vee) \ge -1$ for all weights τ of S. Choosing a good filtration $0 = S_0, S_1, \ldots$ of S we see that $R^i F(S_m/S_{m-1}) = 0$ for all $i > 0$, $m > 0$ by (2.3.1) and (2.1.5) and each $F(S_m/S_{m-1})$ has a good filtration. Thus, by the long exact sequence of cohomology, $R^i F(S)$ has a good filtration if $i = 0$ and is 0 if $i > 0$.

It remains to show that $R^i F(V')$ has a good filtration for $i = 0$ and is 0 for $i > 0$. The non-split short exact sequence

$$0 \to Y_\Gamma(-2\lambda_1 + \lambda_2) \to V' \to k \to 0 \qquad (\dagger)$$

gives rise, via the long exact sequence of cohomology, to an exact sequence

$$0 \to 0 \to F(V') \to k \to k \to RF(V') \to 0$$

and $R^i F(V') = 0$ for $i > 1$ (using (1.5.1), (2.3.1) and the vanishing theorem). Thus $F(V')$ is isomorphic to a submodule of k and $RF(V')$

is isomorphic to $F(V')$. Thus it suffices to prove that $F(V')^{P_\Sigma} = 0$. However, by reciprocity, $F(V')^{P_\Sigma} \cong (V')^{P_\Gamma}$ which is 0 since (†) is non split. This completes the proof of (10.1.7).

We now have by Proposition 3.5.3:

(10.1.9) *For every* G-*module* V *with a good filtration,* $V \otimes Y(\lambda_1)$ *has a good filtration.*

10.2 $p \neq 2$ or 7

To make progress with the hypothesis $\underline{H}(G)$ when the characteristic of k is not too small we shall need the following formulas:

(10.2.1) $\chi(\lambda_1)^2 = \chi(2\lambda_1) + \chi(\lambda_7) + \chi(\lambda_2) + \chi(\lambda_1) + \chi(0)$

and

(10.2.2) $\chi(\lambda_1)\chi(\lambda_7) = \chi(\lambda_1 + \lambda_7) + \chi(\lambda_8) + \chi(\lambda_7) + \chi(\lambda_2) + \chi(\lambda_1)$.

By (10.1.2) and "Brauer's Formula" (2.2.3),
$\chi(\lambda_1)^2 = \Sigma\, \chi(\lambda_1 + \alpha) + 8\, \chi(\lambda_1)$, where the sum runs over the roots α .
Of course if $(\lambda_1 + \alpha, \alpha_j^\vee) = -1$ for some $1 \leq j \leq 8$ then $\chi(\lambda_1 + \alpha) = 0$ so that

$$\chi(\lambda_1)^2 = \sum_{\alpha \in U} \chi(\lambda_1 + \alpha) + 8\, \chi(\lambda_1)$$

where U is the set of roots α such that for each $2 \leq j \leq 8$, $(\alpha,\alpha_j^\vee) \neq -1$. If $(\alpha,\alpha_j^\vee) = -2$ then $\alpha = -\alpha_j$ so that $\alpha \in U$ is either $-\alpha_j$ for some $2 \leq j \leq 8$ or $(\alpha,\alpha_j^\vee) \geq 0$ for $2 \leq j \leq 8$. In the latter case, by (10.1.3), α is λ_1 , $-\lambda_1 + \lambda_2$, $-\lambda_1 + \lambda_7$, $-2\lambda_1 + \lambda_2$ or $-\lambda_1$. Thus we obtain

$$\chi(\lambda_1)^2 = \sum_{j=2}^{8} \chi(\lambda_1 - \alpha_j) + 8\, \chi(\lambda_1) + \chi(2\lambda_1) + \chi(\lambda_2) + \chi(\lambda_7) + \chi(-\lambda_1 + \lambda_2) + \chi(0) .$$

However $\chi(\lambda_1 - \alpha_j) = -\chi(\lambda_1)$ by (2.2.5), for $2 \leq j \leq 8$ and $\chi(-\lambda_1 + \lambda_2) = 0$ giving (10.2.1). The proof of (10.2.2) is similar.

The characters of $\Lambda^2 Y(\lambda_1)$ and $Symm^2 Y(\lambda_1)$ are independent of the characteristic of k and must, therefore, each be a sum of $\chi(\tau)$'s for various $\tau \in X^+$. Moreover the sum of the character of $\Lambda^2 Y(\lambda_1)$ and of $Symm^2 Y(\lambda_1)$ is $\chi(\lambda_1)^2$ and λ_2 is the highest weight of $\Lambda^2 Y(\lambda_1)$. By (10.2.1) and dimensions (see p.50 of [51]) we must have that the character of $\Lambda^2 Y(\lambda_1)$ is given by:

(10.2.3) $\Lambda^2 \chi(\lambda_1) = \chi(\lambda_2) + \chi(\lambda_1)$.

Let S denote the set of rational G-modules V such that $V \otimes V'$ has a good filtration for every G-module V' with a good filtration and $V|_{P_\Sigma}$ has a good filtration for every $\Sigma \subseteq \Delta$. We have, by (10.1.7) and (10.1.8):

(10.2.4) $Y(\lambda_1) \in S$.

From (10.2.3) and (10.2.4) we deduce in the usual manner (cf. section 7.4):

(10.2.5) $Y(\lambda_2) \in S$ *provided that* $p \neq 2$.

We now wish to show that, except possibly for small primes, $Y(\lambda_7) \in S$. As in sections 7.13, 9.3 and chapter 8 we write $C = \{\lambda \in X: 0 \leq (\lambda,\alpha^\vee) \leq p \text{ for all } \alpha \in \Phi^+\}$ and, for $\lambda \in X$, denote by $\tilde{\lambda}$ the element of C to which λ is conjugate under the affine Weyl group W_p. Note that a weight λ belongs to C if and only if $\lambda \in X^+$ and $(\lambda,\beta_0^\vee) \leq p$ where β_0 is the highest root $\langle 2345\overset{3}{6}42\rangle$.

We leave it to the reader to check the following calculations. For $p > 31$, $\widetilde{2\lambda_1} = 2\lambda_1 + \rho$, $\tilde{\lambda}_7 = \lambda_7 + \rho$; for $p = 31$, $\widetilde{2\lambda_1} = \rho$, $\tilde{\lambda}_7 = \lambda_7 + \rho$; for $p = 29$, $\widetilde{2\lambda_1} = 1\cdot 11\overset{1}{1}11$, $\tilde{\lambda}_7 = 1\cdot 11\overset{1}{1}12$; for $p = 23$, $\widetilde{2\lambda_1} = 111\cdot\overset{1}{1}\cdot 1$, $\tilde{\lambda}_7 = 111\cdot\overset{1}{1}1\cdot$; for $p = 19$, $\widetilde{2\lambda_1} = 1\cdot 1\cdot\overset{1}{1}\cdot 1$, $\tilde{\lambda}_7 = 11\cdot 1\overset{1}{\cdot}11$; for $p = 17$, $\widetilde{2\lambda_1} = 111\cdot\overset{\cdot}{1}\cdot 1$, $\tilde{\lambda}_7 = \cdot 1\cdot 1\overset{1}{\cdot}1\cdot$; for $p = 13$, $\widetilde{2\lambda_1} = 1\cdot 1\cdot\overset{1}{\cdot}1\cdot$, $\tilde{\lambda}_7 = \cdot 1\cdot\cdot\overset{\cdot}{1}\cdot 1$; for $p = 11$, $\widetilde{2\lambda_1} = 1\cdot 1\cdot\overset{1}{\cdot}\cdot 1$, $\tilde{\lambda}_7 = \cdot\cdot 1\cdot\overset{\cdot}{1}\cdot\cdot$; for $p = 7$, $\widetilde{2\lambda_1} = \cdot\cdot\cdot\cdot\overset{\cdot}{1}\cdot\cdot$,

$\tilde{\lambda}_7 = \cdot\cdot\cdot\cdot\overset{\cdot}{1}\cdot\cdot$; for $p = 5$, $\widetilde{2\lambda_1} = \cdot\cdot\cdot\cdot\overset{\cdot}{\cdot}1\cdot$, $\tilde{\lambda}_7 = 1\cdot\cdot\cdot\overset{1}{\cdot}\cdot\cdot$; for $p = 3$, $\widetilde{2\lambda_1} = \cdot 1\cdot\cdot\overset{\cdot}{\cdot}\cdot\cdot$, $\tilde{\lambda}_7 = \cdot\cdot\cdot\cdot\overset{1}{\cdot}\cdot\cdot$; for $p = 2$, $\widetilde{2\lambda_1} = \lambda_7$, $\tilde{\lambda}_7 = \lambda_7$.

In particular we have:

(10.2.6) *If* $p = 2$ *or* 7 *then* $\widetilde{2\lambda_1} \neq \tilde{\lambda}_7$.

We now assume that $p \neq 2$ or 7 . Let $M = Y(\lambda_1) \otimes Y(\lambda_1)$ and $N = M_{\tilde{\lambda}_7}$ (see section 7.13). Then N is a G component of $Y(\lambda_1) \otimes Y(\lambda_1)$ and N has, by (10.2.1) and (10.2.6), highest weight λ_7 . We deduce (cf. section 9.3) with the help of (5.15):

(10.2.7) *If* $p \neq 2$ *or* 7 *then* $Y(\lambda_7) \in S$.

We now wish to exploit (10.2.2) in order to prove that $Y(\lambda_8) \in S$. To do so we shall need the following calculations. For $p > 31$, $\widetilde{\lambda_1 + \lambda_7} = \lambda_1 + \lambda_7 + \rho$, $\tilde{\lambda}_8 = \lambda_8 + \rho$; for $p = 31$, $\widetilde{\lambda_1 + \lambda_7} = \cdot 111\overset{1}{1}12$, $\tilde{\lambda}_8 = \cdot 111\overset{2}{1}11$; for $p = 29$, $\widetilde{\lambda_1 + \lambda_7} = 11\cdot 1\overset{1}{1}12$, $\tilde{\lambda}_8 = 11\cdot 1\overset{2}{1}11$; for $p = 23$, $\widetilde{\lambda_1 + \lambda_7} = 11\cdot 1\overset{1}{1}\cdot 1$, $\tilde{\lambda}_8 = 111\cdot\overset{1}{1}\cdot 2$; for $p = 19$, $\widetilde{\lambda_1 + \lambda_7} = 2\cdot 1\cdot\overset{1}{1}\cdot 1$, $\tilde{\lambda}_8 = 1\cdot 1\cdot\overset{1}{1}\cdot 2$; for $p = 17$, $\widetilde{\lambda_1 + \lambda_7} = \cdot 11\cdot\overset{\cdot}{1}\cdot 1$, $\tilde{\lambda}_8 = 1\cdot 1\cdot\overset{\cdot}{1}1$; for $p = 13$, $\widetilde{\lambda_1 + \lambda_7} = 1\cdot\cdot 1\overset{1}{\cdot}\cdot 1$, $\tilde{\lambda}_8 = \cdot\cdot 1\cdot\overset{\cdot}{1}\cdot 1$; for $p = 11$, $\widetilde{\lambda_1 + \lambda_7} = \cdot 1\cdot\cdot\overset{\cdot}{\cdot}1\cdot$, $\tilde{\lambda}_8 = \cdot 1\cdot\cdot\overset{\cdot}{1}\cdot\cdot$; for $p = 7$, $\widetilde{\lambda_1 + \lambda_7} = \lambda_3 + \lambda_8$, $\tilde{\lambda}_8 = \lambda_2 + \lambda_6$; for $p = 5$, $\widetilde{\lambda_1 + \lambda_7} = \lambda_4$, $\tilde{\lambda}_8 = \lambda_6$; for $p = 3$, $\widetilde{\lambda_1 + \lambda_7} = \lambda_2$, $\tilde{\lambda}_8 = \lambda_8$; for $p = 2$, $\widetilde{\lambda_1 + \lambda_7} = \lambda_7$, $\tilde{\lambda}_8 = \lambda_7$.

Thus we have:

(10.2.8) *If* $p \neq 2$ *then* $\widetilde{\lambda_1 + \lambda_7} \neq \tilde{\lambda}_8$.

We obtain, from (10.2.2) that $Y(\lambda_1) \otimes Y(\lambda_7)$ has a G component with highest weight λ_8. We deduce (cf. section 9.3) with the help of (5.15):

(10.2.9) *If* $p \neq 2$ *or* 7 *then* $Y(\lambda_8) \in S$.

Suppose now that $p \neq 2, 3$ or 7. As in section 9.3 we can now show, by considering $\Lambda^i Y(\lambda_1)$ for $2 \leq i \leq 4$, $\Lambda^2 Y(\lambda_7)$ and $\Lambda^2 Y(\lambda_8)$, that $Y(\tau) \in S$ for all $\tau \in X^+$ and so $\underline{H}(G)$. Thus we have:

(10.2.10) *If* $p \neq 2, 3$ *or* 7 *then* $\underline{H}(G)$ *is true.*

We now do some more work for $p = 3$ to obtain that $\underline{H}(G)$ is true in this case also. Assume, for a contradiction, that $\underline{H}(G)$ is false. Let $\underline{H}(G;i)$, for $i \geq 0$, be the assertion that $Y(\tau) \in S$ for all $\tau \in X^+$ such that $(\tau, \beta_0^\vee) \leq i$. Now choose i to be minimal such that $\underline{H}(G;i)$ is false and let $\tau \in X^+$ be minimal with respect to the partial order $\prec$ such that $(\tau, \beta_0^\vee) = i$ and $Y(\tau) \notin S$. It follows from the argument of Proposition 3.5.4 (ii) that $\tau = \lambda_j$ for some $1 \leq j \leq 8$ (and therefore $i \leq 6$ since $(\lambda_j, \beta_0^\vee) \leq 6$ for all j). If $(\tau, \beta_0^\vee) \leq 3$ then τ is $0, \lambda_1, \lambda_7, \lambda_2$ or λ_8 so that, by (10.2.4), (10.2.5), (10.2.7) and (10.2.9), $\underline{H}(G;3)$ is true and $i > 3$. Suppose that $\underline{H}(G;4)$ is false. Then we must have $\tau = \lambda_3$ or λ_6.

In order to prove that $Y(\lambda_3) \in S$ we shall need

(10.2.11) $\widetilde{2\lambda_7} = \lambda_7$, $\tilde{\lambda}_6 = \lambda_2$ *and* $\tilde{\lambda}_3 = \lambda_8$

which may be easily checked. We also claim:

(10.2.12) $Y(\lambda_3)$ *occurs as a successive quotient in a good filtration of* $Y(\lambda_7) \otimes Y(\lambda_7)$.

Of course $Y(\lambda_7) \otimes Y(\lambda_7)$ has a good filtration since $Y(\lambda_7) \in S$. The number of times $Y(\lambda_3)$ occurs in a good filtration is equal to the coefficient of $\chi(\lambda_3)$ in an expression of $\chi(\lambda_7)^2$ as a linear combination of $\chi(\lambda)$'s for $\lambda \in X^+$. This coefficient is independent of characteristic so that it suffices to show that, in characteristic zero $\mathrm{Hom}_G(Y(\lambda_7) \otimes Y(\lambda_7), Y(\lambda_3)) \neq 0$. To this end we assume for the moment that k has characteristic zero. We let $\Sigma = \{\alpha_4, \alpha_5, \alpha_6, \alpha_7, \alpha_8\}$. By the reciprocity of induction $Hom_G(Y(\lambda_7) \otimes Y(\lambda_7), Y(\lambda_3))$ is isomorphic to $Hom_{P_\Sigma}(Y(\lambda_7) \otimes Y(\lambda_7), Y_\Sigma(\lambda_3))$. Moreover $Y(\lambda_3)$ has a P_Σ epimorphic image isomorphic to $Y_\Sigma(\lambda_7)$ so it suffices to show that $Hom_{P_\Sigma}(Y_\Sigma(\lambda_7) \otimes Y_\Sigma(\lambda_7), Y_\Sigma(\lambda_3)) \neq 0$. Since the unipotent radical U^Σ acts trivially on $Y_\Sigma(\lambda_7) \otimes Y_\Sigma(\lambda_7)$ this is equivalent to showing $Hom_{G_\Sigma}(Y_\Sigma(\lambda_7) \otimes Y_\Sigma(\lambda_7), Y_\Sigma(\lambda_3)) \neq 0$. The dimension of this space is, by complete reducibility of rational modules for reductive groups in characteristic 0 , the composition multiplicity of $Y_\Sigma(\lambda_3)$ in $Y_\Sigma(\lambda_7) \otimes Y_\Sigma(\lambda_7)$. This multiplicity is also the coefficient of $\chi_\Sigma(\lambda_3)$ in an expression of $\chi_\Sigma(\lambda_7)^2$ as a linear combination of $\chi_\Sigma(\lambda)$'s for $\lambda \in X_\Sigma$. However using (2.2.3) and (2.2.5) one easily obtains $\chi_\Sigma(\lambda_7)^2 = \chi_\Sigma(2\lambda_7) + \chi_\Sigma(\lambda_6) + \chi_\Sigma(\lambda_3)$ which completes the proof of (10.2.12).

Let $M = Y(\lambda_7) \otimes Y(\lambda_7)$ and $N = M_{\tilde{\lambda}_3}$. Since N is a component of M , N has a good filtration. Suppose $Y(\mu)$ occurs as a successive quotient in a good filtration of N . Then $\mu \leq 2\lambda_7$ and so $(\mu,\beta_0^\vee) \leq 3$ or μ is $2\lambda_1$, λ_3 , λ_6 or $2\lambda_7$. However, by (10.2.11), μ cannot be $2\lambda_7$ or λ_6 . It follows that λ_3 is a maximal weight of N and by Lemma 3.2.3 there is s submodule N' such that N' has a good filtration, λ_3 is not a weight of N' and N/N' is a direct sum of copies of $Y(\lambda_3)$. If $Y(\mu)$ occurs in a good filtration of N' then $(\mu,\beta_0^\vee) \leq 3$ or $\mu = 2\lambda_1$. Hence if $\tau = \lambda_3$ we have, by the minimality of τ , that $N' \in S$. However $M = Y(\lambda_7) \otimes Y(\lambda_7) \in S$ and so N - a component of M - belongs to S . Hence N/N' belongs to S . But N/N' is a direct sum of copies of $Y(\lambda_3)$ so that $Y(\lambda_3) \in S$.

We now rule out the possibility that $\tau = \lambda_6$. Since $\Lambda^2 Y(\lambda_7)$ is a direct summand of M and $M \in S$, $\Lambda^2 Y(\lambda_7) \in S$. However λ_6 is the highest weight of $\Lambda^2 Y(\lambda_7)$ so that there is a submodule, say K , of $\Lambda^2 Y(\lambda_7)$ such that K has a good filtration and $(\Lambda^2 Y(\lambda_7))/K$ is isomorphic to $Y(\lambda_6)$. If $\tau = \lambda_6$ then, by minimality, $K \in S$ and so $Y(\lambda_6) \cong (\Lambda^2 Y(\lambda_7))/K$ belongs to S - a contradiction. Thus both $\tau = \lambda_3$ and $\tau = \lambda_6$ are impossible and $\underline{H}(G;4)$ is true.

Now suppose that $\underline{H}(G;5)$ is false. We must have $\tau = \lambda_4$. We shall need

(10.2.13) $\tilde{\lambda}_4 = \lambda_8$ *and* $\widetilde{\lambda_7 + \lambda_8} = \lambda_2$

which is easily checked, and

(10.2.14) $Y(\lambda_4)$ *occurs as a successive quotient in a good filtration of* $Y(\lambda_7) \otimes Y(\lambda_8)$.

The details of the proof of (10.2.14) will be ommitted. It is obtained in the same way as (10.2.12) taking $\Sigma = \{\alpha_5, \alpha_6, \alpha_7, \alpha_8\}$.

We now take $M = Y(\lambda_7) \otimes Y(\lambda_8)$ and $N = M_{\tilde{\lambda}_4}$. Since N is a component of M , N has a good filtration. Suppose that $Y(\mu)$ occurs as a successive quotient. Then $\mu \leq \lambda_7 + \lambda_8$ and so $(\mu, \beta_0^\vee) \leq 4$ or μ is λ_4 , $\lambda_1 + \lambda_2$, $\lambda_1 + \lambda_8$, $\lambda_7 + \lambda_2$ or $\lambda_7 + \lambda_8$. However, μ cannot be $\lambda_7 + \lambda_8$ by (10.2.13) and it follows that λ_4 is a maximal weight of N . By Lemma 3.2.3 there is a submodule N' such that N' has a good filtration, λ_4 is not a weight of N' and N/N' is a direct sum of copies of $Y(\lambda_4)$. If $Y(\mu)$ occurs in a good filtration of N' then $(\mu, \beta_0^\vee) \leq 4$ or μ is $\lambda_1 + \lambda_2$, $\lambda_1 + \lambda_8$ or $\lambda_7 + \lambda_2$. By the minimality of $\tau = \lambda_4$ we have $N' \in S$. Also N is a component of M and $M \in S$ so that $N \in S$ and hence $N/N' \in S$. But N/N' is a direct sum of copies of $Y(\lambda_4)$ so that $Y(\lambda_4) \in S$. Thus $\underline{H}(G;5)$ is true. The remaining case $\underline{H}(G;6)$ and $\tau = \lambda_5$ is ruled out by considering $\Lambda^2 Y(\lambda_8)$. Thus $\underline{H}(G;r)$ is true for all $r \geq 0$ and we have shown $\underline{H}(G)$ to be true in characteristic 3 . Combining this with (10.2.10) we obtain:

(10.2.15) *If* $p \neq 2$ *or* 7 *then* $\underline{H}(G)$ *is true.*

10.3 The restriction of $Y(\lambda_7)$ to P_Ω

As usual $\Omega = \Delta \backslash \{\alpha_1\}$. It is convenient to exhibit here the following orbit.

(10.3.1) *The* W_Ω *orbit of* $-\lambda_1 + \lambda_2$ *is:*

```
        •               •               •               •
-1 1 • • • • • , • -1 1 • • • • • , • • -1 1 • • • , • • • • -1 1 • • ,

        1               1               -1              -1
• • • • -1 1 • , • • • • • • -1 1 , • • • • • • 1 • , • • • • • 1 -1 1 ,

       1                •               -1              •
• • • • • • • -1 , • • • 1 -1 • 1 , • • • • • 1 • -1 , • • 1 -1 • • 1 ,

        •                •               •               •
• • • 1 -1 1 -1 , • 1 -1 • • • 1 , • • 1 -1 • 1 -1 , • • • • 1 • -1 • ,

        •               •               •               •
1 -1 • • • • 1 , • 1 -1 • • 1 -1 , • • 1 -1 1 -1 • , 1 -1 • • • 1 -1 ,

        •                1               •               1
• 1 -1 • 1 -1 • , • • • 1 • -1 • • , 1 -1 • • 1 -1 • , • 1 -1 1 -1 • • ,

     -1                 1               1               -1
• • 1 • • • • , 1 -1 • 1 -1 • • , • 1 • -1 • • • , • 1 -1 1 • • • ,

       1                -1              -1              1
1 -1 1 -1 • • • , 1 -1 • 1 • • • , • 1 • -1 1 • • , 1 • -1 • • • • ,

      -1                •               -1              •
1 -1 1 -1 1 • • , • 1 • • -1 1 • , 1 • -1 • 1 • • , 1 -1 1 • -1 1 • ,

       •                •               •               •
• 1 • • • -1 1 , 1 • -1 1 -1 1 • , 1 -1 1 • • -1 1 , • 1 • • • • -1 ,

        •               •               •               •
1 • • -1 • 1 • , 1 • -1 1 • -1 1 , 1 -1 1 • • • -1 , 1 • • -1 1 -1 1 ,

        •                1               •               -1
• • -1 1 • • -1 , 1 • • • -1 • 1 , 1 • • -1 1 • -1 , 1 • • • • • 1 ,

        1               -1              1               -1
• • • • -1 1 -1 , 1 • • • • 1 -1 , 1 • • • • -1 • , 1 • • • 1 -1 • ,

        •               •               •               •
• • • 1 -1 • • , 1 • 1 -1 • • • , 1 1 -1 • • • • , 2 -1 • • • • .
```

We now begin our analysis of $Y(\lambda_7)$. We first note:

(10.3.2) $\mathit{Hom}_{P_\Omega}(Y(\lambda_1) \otimes -\lambda_1, Y(\lambda_7)) \cong k$.

This is proved in the same way as (9.4.1), with the help of (10.2.2).

Let θ be a non-zero element of $\mathit{Hom}_{P_\Omega}(Y(\lambda_1) \otimes -\lambda_1, Y(\lambda_7))$ and let D be the image of θ . The B socle of $Y(\lambda_7)$ is $-\lambda_7$ and $-\lambda_7$ is not a weight of $M \otimes -\lambda_1$ (see 10.1 for the definition of M) hence θ induces a non-zero map $\overline{\theta}: (Y(\lambda_1) \otimes -\lambda_1)/(M \otimes -\lambda_1) \to D$.

Since $\overline{\theta}$ is non-zero and the B socle of $Y(\lambda_7)$ is $-\lambda_7$ we have that $\overline{\theta}(\nu) \neq 0$ for some weight ν of $(Y(\lambda_1) \otimes -\lambda_1)/(M \otimes -\lambda_1)$ of weight $-\lambda_7$. Now $\lambda_1 - \lambda_7$ is conjugate to λ_1 and so occurs with multiplicity one as a weight of $Y(\lambda_7)$. It follows that the k-span of v is, by (10.1.6), the B socle of $(Y(\lambda_1) \otimes -\lambda_1)/(M \otimes -\lambda_1)$. Thus $\bar{\theta}$ is injective on the B socle and therefore injective. Thus:

(10.3.3) *there is a short exact sequence of* P_Ω*-modules*

$$0 \to M \otimes -\lambda_1 \to Y(\lambda_1) \otimes -\lambda_1 \to D \to 0 .$$

We now show:

(10.3.4) $((Y/D) \otimes (-\lambda_1 + \lambda_2))^B = 0$

where $Y = Y(\lambda_7)$. By the long exact sequence of cohomology it suffices to prove that $H^1(B, D \otimes (-\lambda_1 + \lambda_2)) = 0$. Moreover, since $H^1(B, Y(\lambda_1) \otimes (-2\lambda_1 + \lambda_2)) \cong Y(\lambda_1)^B = 0$ it suffices, by (10.3.3), to show that $H^2(B, M \otimes (-2\lambda_1 + \lambda_2)) = 0$. We also have that $H^i(B, -3\lambda_1 + \lambda_2) = 0$ for all i so that it suffices to show:

(10.3.5) $H^2(B, \bar{M} \otimes (-2\lambda_1 + \lambda_2)) = 0$

where $\bar{M} = M/F_\Omega^{-2} Y(\lambda_1)$. By (10.1.8), $\bar{M}$ is a non-split extension of $Y_\Omega(-2\lambda_1 + \lambda_2)$ by k. By (10.1.5), (10.1.4), Proposition 3.2.4 and Proposition 3.2.6 there is a submodule, say N, of $Y(\lambda_1)$ containing M such that N has a good filtration and $Y(\lambda_1)/N$ has a good filtration with quotients $Y_\Omega(-\lambda_1 + \lambda_2)$ and $Y_\Omega(\lambda_1)$. Since $Y(\lambda_1)$ does have $Y_\Omega(-\lambda_1 + \lambda_2)$ as a quotient $Y(\lambda_1)/N$ is a non-split extension of $Y_\Omega(-\lambda_1 + \lambda_2)$. It now follows from (5.10) and (5.11) that

(10.3.6) $\bar{M} \cong (Y(\lambda_1)/N) \otimes -\lambda_1$.

Using (10.3.5) and the long exact sequence of cohomology we obtain that $H^2(B, \bar{M} \otimes (-2\lambda_1 + \lambda_2))$ is isomorphic to $H^3(B, N \otimes (-3\lambda_1 + \lambda_2))$. Now the quotients in a good filtration of N have the form $Y_\Omega(\tau)$ for $\tau \in S = \{-\lambda_1, -2\lambda_1 + \lambda_2, 0, -\lambda_1 + \lambda_7\}$ so by (2.1.3), it suffices to prove that $H^3(B, Y_\Omega(\tau) \otimes Y_\Omega(-3\lambda_1 + \lambda_2)) = 0$ for $\tau \in U$. For $\tau = -\lambda_1$ or 0 this follows from (2.1.3) and (5.4). It is easy to deduce from (10.3.1), (2.2.3) and (2.2.5), that

$$\chi_\Omega(-2\lambda_1 + \lambda_2)\chi_\Omega(-3\lambda_1 + \lambda_2) = \chi_\Omega(-5\lambda_1 + 2\lambda_2)$$

$$+ \chi_\Omega(-4\lambda_1 + \lambda_3) + \chi_\Omega(-3\lambda_1 + \lambda_7)$$

$$+ \chi_\Omega(-2\lambda_1) .$$

Hence $Y_\Omega(-2\lambda_1 + \lambda_2) \otimes Y_\Omega(-3\lambda_1 + \lambda_2)$ has a good filtration with quotients of the form $Y_\Omega(\tau)$ for $\tau \in V = \{-5\lambda_1 + 2\lambda_2, -4\lambda_1 + \lambda_3, -3\lambda_1 + \lambda_7, -2\lambda_1\}$. However, it follows from (2.1.3) and (5.4) that $H^3(B,Y_\Omega(\tau)) = 0$ for $\tau \in V$ and so $H^3(B,Y_\Omega(-2\lambda_1 + \lambda_2) \otimes Y_\Omega(-3\lambda_1 + \lambda_2))$. It follows from (10.3.1), (2.2.3) and (2.2.5) that

$$\chi_\Omega(-\lambda_1 + \lambda_7)\chi_\Omega(-2\lambda_1 + \lambda_2) = \chi_\Omega(-3\lambda_1 + \lambda_2 + \lambda_7)$$

$$+ \chi_\Omega(-2\lambda_1 + \lambda_8) + \chi_\Omega(-2\lambda_1 + \lambda_2) .$$

Hence $Y_\Omega(-\lambda_1 + \lambda_7) \otimes Y_\Omega(-2\lambda_1 + \lambda_2)$ has a good filtration with quotients $Y_\Omega(\tau)$ for $\tau \in W = \{-3\lambda_1 + \lambda_2 + \lambda_7, -2\lambda_1 + \lambda_8, -2\lambda_1 + \lambda_2\}$. However, it follows from (2.1.3) and (5.4) that $H^3(B,Y_\Omega(\tau)) = 0$ for $\tau \in W$ and so $H^3(B,Y_\Omega(-\lambda_1 + \lambda_7) \otimes Y_\Omega(-2\lambda_1 + \lambda_2)) = 0$. We have now shown that $H^3(B,Y_\Omega(\tau) \otimes Y_\Omega(-3\lambda_1 + \lambda_2)) = 0$ for all $\tau \in U$ and so the proof of (10.3.5) is complete.

We now claim:

(10.3.6) $F_\Omega^{-1}Y/D$ *is isomorphic to* $Y_\Omega(-2\lambda_1 + \lambda_8)$.

By p.230 of [27] the character of $F_\Omega^{(2)}Y$ is $\chi_\Omega(\lambda_7)$. Since *ch*Y s equal to *ch*Y* , the character of the dual module Y* , $hF_\Omega^{(-2)}Y = chF_\Omega^{-2}Y$ is $\chi_\Omega(-2\lambda_1 + \lambda_8)$. However D contains a copy of $_\Omega(-2\lambda_1 + \lambda_8)$ so we have $F_\Omega^{-2}Y \subseteq D$. Thus any composition factor $_\Omega(\tau)$ of $V = F_\Omega^{-1}Y/D$ satisfies $f_\Omega(\tau) = -1$. Since τ is a weight f $Y(\lambda_7)$ we have $(\tau,\alpha_1^\vee) \geq -2$ and so τ must be $-2\lambda_1 + \lambda_2$ or $2\lambda_1 + \lambda_8$. In characteristic zero V is a completely reducible G_Ω-odule. If $L_\Omega(-2\lambda_1 + \lambda_2) = Y_\Omega(-2\lambda_1 + \lambda_2)$ were a composition factor n that case it would follow that $((Y/D) \otimes (-\lambda_1 + \lambda_2))^B \neq 0$ contrary o (10.3.4). Thus in characteristic zero only $_\Omega(-2\lambda_1 + \lambda_8) = Y_\Omega(-2\lambda_1 + \lambda_8)$ can be involved. However $-2\lambda_1 + \lambda_8$ ccurs precisely once as a weight of Y so that in charactersitic zero,

$V \cong Y_\Omega(-2\lambda_1 + \lambda_8)$. However the character is independent of the characteristic so that, for k of arbitrary characteristic, $chV = \chi_\Omega(-2\lambda_1 + \lambda_8)$. Now (10.3.6) follows from (10.3.4) and (1.5.4).

Our next claim is:

(10.3.7) $H^i(B, D \otimes (-\lambda_1 + \lambda_7)) = 0$ *for all* i .

By (10.3.3), and the long exact sequence of cohomology, it is enough to show that $H^i(B, M \otimes (-2\lambda_1 + \lambda_7)) = 0$ for all $i > 0$. Since M has a good filtration with quotients $-\lambda_1$, $Y_\Omega(-2\lambda_1 + \lambda_2)$, $Y_\Omega(0)$ it is enough, by (2.1.3), to show that $H^i(B, Y_\Omega(\tau) \otimes Y_\Omega(-2\lambda_1 + \lambda_7)) = 0$ for $\tau = -\lambda_1$, $-2\lambda_1 + \lambda_2$, 0 . For $\tau = -\lambda_1$ or 0 this follows from (2.1.3) and (5.4). Also $Y_\Omega(-2\lambda_1 + \lambda_2) \otimes Y_\Omega(-2\lambda_1 + \lambda_7)$ has a good filtration with quotients $Y_\Omega(-3\lambda_1 + \lambda_2)$, $Y_\Omega(-3\lambda_1 + \lambda_8)$ and $Y_\Omega(-4\lambda_1 + \lambda_2 + \lambda_7)$ so that one obtains $H^i(B, Y_\Omega(-2\lambda_1 + \lambda_2) \otimes Y_\Omega(-2\lambda_1 + \lambda_7)) = 0$ from (2.1.3) and (5.4). This completes the proof of (10.3.7).

It is convenient to note at this point:

(10.3.8) $H^i(B, F_\Omega^{-1}Y) = 0$ *for all* i .

Since $H^i(B, Y_\Omega(-2\lambda_1 + \lambda_8)) = 0$ for all i by (2.1.3), (2.3.3) and (2.1.6) it is enough to prove that $H^i(B, D) = 0$ for all i . By (10.3.3) and the long exact sequence of cohomology it suffices to show that $H^i(B, M \otimes -\lambda_1) = 0$ for all $i > 0$. But M has a good filtration with quotients $Y_\Omega(-\lambda_1)$, $Y_\Omega(-2\lambda_1 + \lambda_2)$, $Y_\Omega(0)$ so the result follows from (2.1.3), (2.1.6), (2.3.3) and (5.6). By the long exact sequence of cohomology we have $((Y/F_\Omega^{-1}Y) \otimes (-\lambda_1 + \lambda_7))^B \cong H^1(B, (F_\Omega^{-1}Y) \otimes (-\lambda_1 + \lambda_7))$. By the long exact sequence of cohomology, (10.3.7) and (10.3.6) this is isomorphic to $H^1(B, Y_\Omega(-2\lambda_1 + \lambda_8) \otimes (-\lambda_1 + \lambda_7))$. By (2.1.3), $H^1(B, Y_\Omega(-2\lambda_1 + \lambda_8) \otimes (-\lambda_1 + \lambda_7))$ is isomorphic to $H^1(B, Y_\Omega(-2\lambda_1 + \lambda_8) \otimes Y_\Omega(-\lambda_1 + \lambda_7))$. The possible composition factors of $Y_\Omega(-\lambda_1 + \lambda_7)$ are $L_\Omega(-\lambda_1 + \lambda_7)$ and $L_\Omega(0) = k$. However

$H^i(B, Y_\Omega(-2\lambda_1 + \lambda_8)) = 0$ for all i, by (5.4), so we now have that $((Y/F_\Omega^{-1}Y) \otimes (-\lambda_1 + \lambda_7))^B$ is isomorphic to $H^1(B, Y_\Omega(-2\lambda_1 + \lambda_8) \otimes L_\Omega(-\lambda_1 + \lambda_7))$. Since $L_\Omega(-\lambda_1 + \lambda_7)$ is self dual this is $Ext^1_B(L_\Omega(-\lambda_1 + \lambda_7), Y_\Omega(-2\lambda_1 + \lambda_8))$. Again, since $H^i(B, Y_\Omega(-2\lambda_1 + \lambda_8)) = 0$ for all i, this is $Ext^1_B(Y_\Omega(-\lambda_1 + \lambda_7), Y_\Omega(-2\lambda_1 + \lambda_8))$ which, by (2.1.3), is equal to $Ext^1_{P_\Omega}(Y_\Omega(-\lambda_1 + \lambda_7), Y_\Omega(-2\lambda_1 + \lambda_8))$. Now, by (5.10), this is $Ext^1_{P_\Gamma}(Y_\Gamma(-\lambda_1 + \lambda_7), Y_\Gamma(-2\lambda_1 + \lambda_8))$ where $\Gamma = \Delta\backslash\{\alpha_1, \alpha_8\}$. However the dual of $Y_\Gamma(-\lambda_1 + \lambda_7)$ is $Y_\Gamma(\lambda_2 - \lambda_8)$ so that $Ext^1_{P_\Gamma}(Y_\Gamma(-\lambda_1 + \lambda_7), Y_\Gamma(-2\lambda_1 + \lambda_8))$ is isomorphic to $H^1(P_\Gamma, Y_\Gamma(-2\lambda_1 + \lambda_2))$ which, by (2.1.3) and (1.4.7), is k. We have shown:

(10.3.9) $((Y/F_\Omega^{-1}Y) \otimes (-\lambda_1 + \lambda_7))^B \cong k$

and so, by reciprocity, that $((Y/F_\Omega^{-1}Y) \otimes Y_\Omega(-\lambda_1 + \lambda_7))^{P_\Omega} \cong k$. It follows that $L_\Omega(-\lambda_1 + \lambda_7)$ or $L_\Omega(0)$ embeds in $Y/F_\Omega^{-1}Y$. However $(Y/F_\Omega^{-1}Y)^B \cong H^1(B, F_\Omega^{-1}Y)$ which is 0 by (10.3.8) so that $L_\Omega(-\lambda_1 + \lambda_7)$ embeds in $Y/F_\Omega^{-1}Y$. Now $Y_\Omega(-\lambda_1 + \lambda_7)/L_\Omega(-\lambda_1 + \lambda_7)$ is a trivial P_Ω-module so applying $Hom_{P_\Omega}(-, Y/F_\Omega^{-1}Y)$ to the short exact sequence $0 \to L_\Omega(-\lambda_1 + \lambda_7) \to Y_\Omega(-\lambda_1 + \lambda_7) \to Y_\Omega(-\lambda_1 + \lambda_7)/L_\Omega(-\lambda_1 + \lambda_7) \to 0$ yields, via (10.3.9), an embedding of $Y_\Omega(-\lambda_1 + \lambda_7)$ in $Y/F_\Omega^{-1}Y$. Thus there is a P_Ω submodule, say E, of Y containing $F_\Omega^{-1}Y$ such that $E/F_\Omega^{-1}Y$ is isomorphic to $Y_\Omega(-\lambda_1 + \lambda_7)$.

We note:

(10.3.10) E/D *is a non-split extension of* $Y_\Omega(-2\lambda_1 + \lambda_8)$ *by* $Y_\Omega(-\lambda_1 + \lambda_7)$.

If the extension were split we would have $\lambda_1 - \lambda_7$ in the B socle of Y/D but $((Y/D) \otimes (-\lambda_1 + \lambda_7))^B$ is isomorphic to $H^1(B, D \otimes (-\lambda_1 + \lambda_7))$ which is 0 by (10.3.7).

Our next claim is:

(10.3.11) $(F_\Omega^0 Y)/E$ *is isomorphic to* $Y_\Omega(-2\lambda_1 + \lambda_3)$.

Suppose that $L_\Omega(\tau)$ is a composition factor of $(F_\Omega^0 Y)/E$. Then $f_\Omega(\tau) = 0$ and, for any root α , $|(\tau,\alpha^\vee)| \le (\lambda_7,\beta_0^\vee) = 2$. Taking $\alpha = \alpha_1$ limits the choice of τ to one of the following: 0 , $-\lambda_1 + \lambda_7$, $-2\lambda_1 + 2\lambda_7$, $-2\lambda_1 + \lambda_6$ and $-2\lambda_1 + \lambda_3$. Taking $\alpha = \langle \cdot 1\overset{2}{2}3432 \rangle$ limits the choice of τ to 0 , $-\lambda_1 + \lambda_7$ and $-2\lambda_1 + \lambda_3$, moreover these are all weights of $Y(\lambda_7)$. Maximal among these is $-2\lambda_1 + \lambda_3$ so that $L_\Omega(-2\lambda_1 + \lambda_3)$ must occur. Thus in characteristic zero we now know that $Y_\Omega(\tau)$, for $\tau = -2\lambda_1 + \lambda_7$, $-2\lambda_1 + \lambda_2$, 0 , $-2\lambda_1 + \lambda_8$, $-\lambda_1 + \lambda_7$, $-2\lambda_1 + \lambda_3$, occurs as a composition factor. Moreover, Y is self dual (in characteristic zero) so that $Y_\Omega(\tau)^*$ occurs whenever $Y_\Omega(\tau)$ occurs. Thus $Y(\lambda_7)$ has as a composition factor $Y_\Omega(\tau)$ for $\tau \in \mathcal{U} = \{-2\lambda_1 + \lambda_7, -2\lambda_1 + \lambda_2, 0, -2\lambda_1 + \lambda_8, -\lambda_1 + \lambda_7, -2\lambda_1 + \lambda_3, -\lambda_1 + \lambda_2, -\lambda_1 + \lambda_8, \lambda_7\}$. However, one may easily check rom the information on p.47 and p.50 of [51] that $\dim Y(\lambda_7) = \sum_{\tau\in\mathcal{U}} \dim Y_\Omega(\tau)$. Hence $\chi(\lambda_7) = \sum_{\tau\in\mathcal{U}} \chi_\Omega(\tau)$. However $\chi(\lambda_7)$ and $\chi_\Omega(\tau)$ are independent of characteristic so we have, in arbitrary characteristic:

(10.3.12) $\chi(\lambda_7) = \chi_\Omega(-2\lambda_1 + \lambda_7) + \chi_\Omega(-2\lambda_1 + \lambda_2) + \chi_\Omega(0)$

$+ \chi_\Omega(-2\lambda_1 + \lambda_8) + \chi_\Omega(-\lambda_1 + \lambda_7) + \chi_\Omega(-2\lambda_1 + \lambda_3) + \chi_\Omega(-\lambda_1 + \lambda_2)$

$+ \chi_\Omega(-\lambda_1 + \lambda_8) + \chi_\Omega(\lambda_7)$.

Thus $ch(F_\Omega^0 Y)/E = \chi_\Omega(-2\lambda_1 + \lambda_3)$ and, by (1.5.4), to prove (10.3.11), it suffices to show that $(((F_\Omega^0 Y)/E) \otimes \tau)^B = 0$ for $\tau = 0$ and $\tau = -\lambda_1 + \lambda_7$. Hence it suffices to show that $((Y/E) \otimes \tau)^B = 0$ for $\tau = 0$ and $\tau = -\lambda_1 + \lambda_7$ and so, by the long exact sequence of cohomology, it suffices to show that $H^1(B,E \otimes \tau) = 0$. For $\tau = 0$ by (10.3.8), we need only observe that $H^1(B,Y_\Omega(-\lambda_1 + \lambda_7)) = 0$, which is so by (2.1.6). Suppose, for a contradiction, that $(Y/E \otimes (-\lambda_1 + \lambda_7))^B \neq 0$. We have $Ext_B^1(\lambda_1 - \lambda_7, Y_\Omega(-\lambda_1 + \lambda_7)) = H^1(B,Y_\Omega(-\lambda_1 + \lambda_7) \otimes (-\lambda_1 + \lambda_7))$ which is

0 by (1.4.6) so that $Y/F_{\Omega}^{-1}Y$ would contain a split extension of $Y_{\Omega}(-\lambda_1 + \lambda_7)$ by $\lambda_1 - \lambda_7$ and hence two copies of $\lambda_1 - \lambda_7$ in the B socle of $Y/F_{\Omega}^{-1}Y$. But this contradicts (10.3.9). The proof of (10.3.11) is now complete.

We have thus constructed a good filtration of $F_{\Omega}^{0}Y$ and so, by the argument of Proposition 4.7.1:

(10.3.13) $Y(\lambda_7)|_{P_{\Omega}}$ *has a good filtration.*

10.4 $R^i Ind_B^{P_\Sigma}(D)$

By (10.3.13) and Proposition 3.2.4, Y/E has a good filtration and it follows from Proposition 3.2.6 and (10.3.12) that Y/E contains a P_{Ω} submodule, say F/E, which is an extension of $Y_{\Omega}(-2\lambda_1 + \lambda_3)$ by $Y_{\Omega}(-\lambda_1 + \lambda_2)$. Let $\Sigma = \Delta\backslash\{\alpha_j\}$ where $2 \le j \le 8$. In order to show that $Y(\lambda_7)|_{P_\Sigma} \cong Ind_B^{P_\Sigma}(Y/B)$ has a good filtration we prove that $F(D)$, $F(E/D)$, $F(F/E)$, $F(Y/F)$ each has a good filtration and

$RF(D) = RF(E/D) = RF(F/E) = 0$, where $F = Ind_B^{P_\Sigma}$.

We first show:

(10.4.1) $R^i Ind_B^{P_\Sigma}(D)$ *has a good filtration if* $i = 0$ *and is* 0 *if* $i > 0$, *where* $\Sigma = \Delta\backslash\{\alpha_j\}$ *and* $2 \le j \le 8$.

We have $R^i F(Y(\lambda_1) \otimes -\lambda_1) = 0$ for all i by (2.1.5) and the tensor identity. Thus by (10.3.2):

(10.4.2) $R^i F(D) \cong R^{i+1}F(M \otimes -\lambda_1)$.

Let $\Gamma = \Omega \cap \Sigma$. By (1.1.6) and the vanishing theorem $R^i F(M \otimes -\lambda_1) \cong R^i G(M \otimes -\lambda_1)$ where $G = Ind_{P_\Gamma}^{P_\Sigma}$. Hence it suffices to show:

(10.4.3) $R^i G(M \otimes -\lambda_1)$ *has a good filtration if* $i = 0$ *and is* 0 *if* $i > 0$.

Now $M \otimes -\lambda_1$ has a good filtration with quotients $Y_{\Omega}(\tau)$ for $\tau \in U = \{-2\lambda_1, -3\lambda_1 + \lambda_2, -\lambda_1\}$. Thus it suffices to show that, for

$\tau \in \mathcal{U}$, $R^i G(Y_\Omega(\tau))$ has a good filtration for $i = 0$ and is 0 for $i > 0$. For $\tau = -\lambda_1$, $-2\lambda_1$ this is true by (5.7). Hence it remains to prove that $R^i G(Y_\Omega(-3\lambda_1 + \lambda_2))$ has a good filtration for $i = 1$ and is 0 if $i \neq 1$. We have $G(Y_\Gamma(-3\lambda_1 + \lambda_2)) = 0$ by transitivity and (1.7.1). For $i > 0$, by (2.1.2) and (5.5), $R^i G(Y_\Gamma(-3\lambda_1 + \lambda_2)) \cong R^{i-1}F(\lambda_1 - \lambda_2)$ which for $i = 1$ is $Y_\Sigma(\lambda_1 - \lambda_2)$ if $j = 2$ and 0 otherwise and is 0 for $i > 1$ by (2.1.5) and the vanishing theorem. Thus we need only show that $R^i G(M_\Gamma(-3\lambda_1 + \lambda_2))$ is 0 if $i \neq 1$ and has a good filtration if $i = 1$.

It follows from (10.3.1) that, unless $j = 2$, for any successive quotient $Y_\Gamma(\xi)$ in a good filtration of $M_\Gamma(-3\lambda_1 + \lambda_2)$, $-1 \geq (\xi, \alpha_1^\vee) \geq -2$ and, by (5.7), $R^i G(M_\Gamma(-3\lambda_1 + \lambda_2))$ is 0 if $i \neq 1$ and has a good filtration when $i = 1$.

Now suppose $j = 2$ and let $V_1 < V_2$ be P_Ω submodules of $M \otimes -\lambda_1$ with $V_1 \cong -2\lambda_1$, $V_2/V_1 \cong Y_\Omega(-3\lambda_1 + \lambda_2)$, $(M \otimes -\lambda_1)/V_2 \cong -\lambda_1$. Now $M_\Gamma(-3\lambda_1 + \lambda_2)$ contains a copy of $Y_\Gamma(-\lambda_2)$ so that V_2 contains an extension, say V_1' , of $-2\lambda_1$ by $-\lambda_2$. Since the B socle of $Y(\lambda_1)$, and hence $M \otimes -\lambda_1$, is simple V_1' is a non split extension and therefore isomorphic to $Y_\Sigma(\lambda_1 - \lambda_2) \otimes -\lambda_1$. We now have an ascending P_Γ module filtration of $M \otimes -\lambda_1$, 0 , V_1' , V_2' , V_3' , $M \otimes -\lambda_1$ where $V_3' = V_2$, $V_2'/V_1' \cong M_\Gamma(-3\lambda_1 + \lambda_2)/Y_\Gamma(-\lambda_2)$, $V_3'/V_2' \cong Y_\Gamma(-3\lambda_1 + \lambda_2)$, $(M \otimes -\lambda_1)/V_3' \cong -\lambda_1$. However, for each successive quotient Q , $R^i G(Q)$ is 0 for $i = 1$ and has a good filtration if $i = 1$. We have seen this already for $Q = (M \otimes -\lambda_1)/V_3'$, $Y_\Gamma(-3\lambda_1 + \lambda_2)$. It is true by (5.7) for $Q = V_2'/V_1'$ and true by (2.1.5) for $Q = V_1'$. This completes the proof of (10.4.1).

The P_Ω module Y/F has a good filtration with quotients $Y_\Omega(-\lambda_1 + \lambda_8)$ and $Y_\Omega(\lambda_7)$ so by (5.2) we have:

(10.4.4) $R^i \mathrm{Ind}_B^{P_\Sigma}(Y/F)$ *has a good filtration if* $i = 0$ *and is* 0 *if* $i > 0$, *where* $\Sigma = \Delta\backslash\{\alpha_j\}$.

10.5 Orbits and characters

We present here the character formulas needed in our proof that $Y(\lambda_7)|_{P_\Sigma}$ has a good filtration for all $\Sigma \subseteq \Delta$. It is convenient to exhibit here the following orbits.

(10.5.1) *The* W_Ω *orbit of* $-\lambda_1 + \lambda_7$ *is:*

```
              ·                       ·                       ·                       1
 -1  ·  ·  ·  ·  ·  1 ,  -1  ·  ·  ·  ·  1 -1 ,  -1  ·  ·  ·  1 -1  · ,  -1  ·  ·  1 -1  ·  · ,

              1                      -1                       1                      -1
 -1  ·  1 -1  ·  ·  · ,  -1  ·  ·  1  ·  ·  · ,  -1  1 -1  ·  ·  ·  · ,  -1  ·  1 -1  1  ·  · ,

              1                      -1                       ·                      -1
  · -1  ·  ·  ·  ·  · ,  -1  1 -1  ·  1  ·  · ,  -1  ·  1  · -1  1  · ,   · -1  ·  ·  1  ·  · ,

              ·                       ·                       ·                       ·
 -1  1 -1  1 -1  1  · ,  -1  ·  1  ·  · -1  1 ,   · -1  ·  1 -1  1  · ,  -1  1  · -1  ·  1  · ,

              ·                       ·                       ·                       ·
 -1  1 -1  1  · -1  1 ,  -1  ·  1  ·  ·  · -1 ,   · -1  1 -1  ·  1  · ,   · -1  ·  1  · -1  1 ,

              ·                       ·                       ·                       ·
 -1  1  · -1  1 -1  1 ,  -1  1 -1  1  ·  · -1 ,   ·  · -1  ·  ·  1  · ,   · -1  1 -1  1 -1  1 ,

              ·                       1                       ·                       ·
  · -1  ·  1  ·  · -1 ,  -1  1  ·  · -1  ·  1 ,  -1  1  · -1  1  · -1 ,   ·  · -1  ·  1 -1  1 ,

              1                       ·                      -1                       1
  · -1  1  · -1  ·  1 ,   · -1  1 -1  1  · -1 ,  -1  1  ·  ·  ·  ·  1 ,  -1  1  ·  · -1  1 -1 ,

              1                       ·                      -1                       1
  ·  · -1  1 -1  ·  1 ,   ·  · -1  ·  1  · -1 ,   · -1  1  ·  ·  ·  1 ,   · -1  1  · -1  1 -1 ,

             -1                       1                       1                      -1
 -1  1  ·  ·  ·  1 -1 ,  -1  1  ·  ·  · -1  · ,   ·  ·  · -1  ·  ·  1 ,   ·  · -1  1  ·  ·  1 ,

              1                      -1                       1                      -1
  ·  · -1  1 -1  1 -1 ,   · -1  1  ·  ·  1 -1 ,   · -1  1  ·  · -1  · ,  -1  1  ·  ·  1 -1  · ,

          -1                       1                      -1                       1
  ·  · -1  1  ·  1 ,   ·  ·  · -1  ·  1 -1 ,   ·  · -1  1  ·  1 -1 ,   ·  · -1  1  · -1  · ,

          -1                       ·                       ·                      -1
 -1  1  ·  1 -1  · ,  -1  1  ·  1 -1  ·  · ,   ·  ·  ·  · -1  1  1 ,   ·  ·  · -1  1  1 -1 ,

           1                      -1                       ·                       ·
  ·  · -1  1 -1  · ,   ·  · -1  1  1 -1  · ,   · -1  1  1 -1  ·  · ,  -1  1  1 -1  ·  ·  · ,

           ·                       ·                      -1                       2
  ·  ·  ·  · -1  2 ,   ·  ·  ·  · -1  2 -1 ,   ·  ·  · -1  2 -1  · ,   ·  ·  ·  · -1  ·  · ,

           ·                       ·                       ·                       ·
  · -1  2 -1  ·  · ,   · -1  2 -1  ·  ·  · ,  -1  2 -1  ·  ·  ·  · ,   ·  ·  ·  ·  ·  1 -2 ,

           ·                       1                      -2                       ·
  ·  ·  ·  1 -2  1 ,   ·  ·  ·  1 -2  1  · ,   ·  ·  ·  ·  1  ·  · ,   ·  ·  1 -2  1  ·  · ,
```

```
              ·                       ·                       ·
  ·  1 -2  1  ·  ·  · ,   1 -2  1  ·  ·  ·  · ,   ·  ·  ·  ·  1 -1 -1 ,   ·  ·  ·  1 -1 -1  1 ,

             -1                       1                       ·                       ·
  ·  ·  ·  1 -1  1  · ,   ·  ·  1 -1 -1  1  · ,   ·  1 -1 -1  1  ·  · ,   1 -1 -1  1  ·  ·  · ,

              1                      -1                       1                      -1
  ·  ·  ·  1 -1  · -1 ,   ·  ·  ·  1  · -1  1 ,   ·  ·  1 -1  · -1  1 ,   ·  ·  1 -1  ·  1  · ,

              1                       ·                      -1                       1
  ·  1 -1  · -1  1  · ,   1 -1  · -1  1  ·  · ,   ·  ·  ·  1  ·  · -1 ,   ·  ·  1 -1  ·  · -1 ,

             -1                       1                      -1                       1
  ·  ·  1 -1  1 -1  1 ,   ·  1 -1  ·  · -1  1 ,   ·  1 -1  ·  ·  1  · ,   1 -1  ·  · -1  1  · ,

             -1                       ·                       1                      -1
  ·  ·  1 -1  1  · -1 ,   ·  ·  1  · -1  ·  1 ,   ·  1 -1  ·  ·  · -1 ,   ·  1 -1  ·  1 -1  1 ,

              1                      -1                       ·                      -1
  1 -1  ·  ·  · -1  1 ,   1 -1  ·  ·  ·  1  · ,   ·  ·  1  · -1  1 -1 ,   ·  1 -1  ·  1  · -1 ,

              ·                       1                      -1                       ·
  ·  1 -1  1 -1  ·  1 ,   1 -1  ·  ·  ·  · -1 ,   1 -1  ·  ·  1 -1  1 ,   ·  ·  1  ·  · -1  · ,

              ·                       ·                      -1                       ·
  ·  1 -1  1 -1  1 -1 ,   ·  1  · -1  ·  ·  1 ,   1 -1  ·  ·  1  · -1 ,   1 -1  ·  1 -1  ·  1 ,

              ·                       ·                       ·                       ·
  ·  1 -1  1  · -1  · ,   1 -1  ·  1 -1  1 -1 ,   ·  1  · -1  ·  1 -1 ,   1 -1  1 -1  ·  ·  1 ,

              ·                       ·                       ·                       ·
  1 -1  ·  1  · -1  · ,   ·  1  · -1  1 -1  · ,   1 -1  1 -1  ·  1 -1 ,   1  · -1  ·  ·  ·  1 ,

              ·                       1                       ·                       ·
  1 -1  1 -1  1 -1  · ,   ·  1  ·  · -1  ·  · ,   1  · -1  ·  ·  1 -1 ,   1  · -1  ·  1 -1  · ,

              1                      -1                       1                      -1
  1 -1  1  · -1  ·  · ,   ·  1  ·  ·  ·  ·  · ,   1  · -1  1 -1  ·  · ,   1 -1  1  ·  ·  ·  · ,

              1                      -1                      -1                       ·
  1  ·  · -1  ·  ·  · ,   1  · -1  1  ·  ·  · ,   1  ·  · -1  1  ·  · ,   1  ·  ·  · -1  1  · ,

              ·                       ·
  1  ·  ·  ·  · -1  1 ,   1  ·  ·  ·  ·  · -1 .
```

We adopt the notation of section 9.6, in particular $W_{1,2}$ is W_Σ for $\Sigma = \Delta\setminus\{\alpha_1,\alpha_2\}$.

(10.5.2) *The* $W_{1,2}$ *orbit of* $-2\lambda_1 + \lambda_8$ *is:*

```
              1                      -1                       ·                       ·
 -2  ·  ·  ·  ·  ·  · ,  -2  ·  ·  ·  1  ·  · ,  -2  ·  ·  1 -1  1  · ,  -2  ·  1 -1  ·  1  · ,

              ·                       ·                       ·                       ·
 -2  ·  ·  1  · -1  1 ,  -2  1 -1  ·  ·  1  · ,  -2  ·  1 -1  1 -1  1 ,  -2  ·  ·  1  ·  · -1 ,

              ·                       1                       ·
 -2  1 -1  ·  1 -1  1 ,  -2  ·  1  · -1  ·  1 ,  -2  ·  1 -1  1  · -1 ,
```

```
             1                     -1                      1                      ·
-2  1 -1  1 -1  ·  1 , -2  ·  1  ·  ·  ·  1 , -2  ·  1  · -1  1 -1 , -2  1 -1  ·  1  · -1 ,

             1                     -1                      1                     -1
-2  1  · -1  ·  ·  1 , -2  1 -1  1  ·  ·  1 , -2  1 -1  1 -1  1 -1 , -2  ·  1  ·  ·  1 -1 ,

            -1                      1                     -1
-2  1  · -1  1  ·  1 , -2  1  · -1  ·  1 -1 , -2  1 -1  1  ·  1 -1 ,

             1                     -1                      ·                     -1
-2  1 -1  1  · -1  · , -2  ·  1  ·  1 -1  · , -2  1  ·  · -1  1  1 , -2  1  · -1  1  1 -1 ,

             1                     -1                      ·                      ·
-2  1  · -1  1 -1  · , -2  1 -1  1  1 -1  · , -2  ·  1  1 -1  ·  · , -2  1  ·  ·  · -1  2 ,

             ·                     -1                      2
-2  1  ·  · -1  2 -1 , -2  1  · -1  2 -1  · , -2  1  ·  · -1  ·  · ,

             ·                      ·                      ·                      ·
-2  1 -1  2 -1  ·  · , -2  ·  2 -1  ·  ·  · , -2  1  ·  ·  ·  1 -2 , -2  1  ·  ·  1 -2  1 ,

             1                     -2                      ·                      ·
-2  1  ·  1 -2  1  · , -2  1  ·  ·  1  ·  · , -2  1  1 -2  1  ·  · , -2  2 -2  1  ·  ·  · ,

             ·                      1                      1
-2  1  ·  ·  1 -1 -1 , -2  1  ·  1 -1 -1  1 , -2  1  1 -1 -1  1  · ,

            -1                      ·                     -1                     -1
-2  1  ·  1 -1  1  · , -2  2 -1 -1  1  ·  · , -2  1  1 -1  ·  1  · , -2  1  ·  1  · -1  1 ,

             1                      1                      1
-2  2 -1  · -1  1  · , -2  1  ·  1 -1  · -1 , -2  1  1 -1  · -1  1 ,

            -1                     -1                     -1                      1
-2  2 -1  ·  ·  1  · , -2  1  1 -1  1 -1  1 , -2  1  ·  1  ·  · -1 , -2  2 -1  ·  · -1  1 ,

             1                     -1                      ·
-2  1  1 -1  ·  · -1 , -2  2 -1  ·  1 -1  1 , -2  1  1  · -1  ·  1 ,

            -1                      1                      ·
-2  1  1 -1  1  · -1 , -2  2 -1  ·  ·  · -1 , -2  2 -1  1 -1  ·  1 ,

            -1                      ·                      ·
 2  2 -1  ·  1  · -1 , -2  1  1  · -1  1 -1 , -2  2  · -1  ·  ·  1 ,

             ·                      ·                      ·
 2  2 -1  1 -1  1 -1 , -2  1  1  ·  · -1  · , -2  2  · -1  ·  1 -1 ,

             ·                      ·                      1                     -1
 2  2 -1  1  · -1  · , -2  2  · -1  1 -1  · , -2  2  ·  · -1  ·  · , -2  2  ·  ·  ·  ·  · .
```

10.5.3) *The* $W_{1,2}$ *orbit of* $-2\lambda_1 + \lambda_3$ *is:*

```
             ·                      ·                      ·                      1
 2  ·  1  ·  ·  ·  · , -2  1 -1  1  ·  ·  · , -2  1  · -1  1  ·  · , -2  1  ·  · -1  1  · ,

            -1                      1                     -1                      1
 2  1  ·  ·  ·  1  · , -2  1  ·  ·  · -1  1 , -2  1  ·  ·  1 -1  1 , -2  1  ·  ·  ·  · -1 ,

             ·                     -1                      ·                      ·
 2  1  ·  1 -1  ·  1 , -2  1  ·  ·  1  · -1 , -2  1  1 -1  ·  ·  1 , -2  1  ·  1 -1  1 -1 ,
```

```
        •                •                 •
-2 2 -1 • • • 1 , -2 1 1 -1 • 1 -1 , -2 1 • 1 • -1 • ,

        •                •                 •
-2 2 -1 • • 1 -1 , -2 1 1 -1 1 -1 • , -2 2 -1 • 1 -1 • ,

        1                1                 -1                 1
-2 1 1 • -1 • • , -2 2 -1 1 -1 • • , -2 1 1 • • • • , -2 2 • -1 • • • ,

        -1               -1                •                  •
-2 2 -1 1 • • • , -2 2 • -1 1 • • , -2 2 • • -1 1 • , -2 2 • • • -1 1 ,

        •
-2 2 • • • • -1 .
```

It is easy to deduce the following from the above and (2.2.3), (2.2.5).

(10.5.4) $\chi_{\Omega}(-\lambda_1 + \lambda_7) = \delta_{\Omega}(-\lambda_1 + \lambda_7) + 7\delta_{\Omega}(0)$.

(10.5.5) $\chi_{1,2}(-2\lambda_1 + \lambda_8) = \delta_{1,2}(-2\lambda_1 + \lambda_8) + 6\delta_{1,2}(-2\lambda_1 + \lambda_2)$

(10.5.6) $\chi_{1,2}(-2\lambda_1 + \lambda_3) = \delta_{1,2}(-2\lambda_1 + \lambda_3)$.

The following formulas are obtained from (10.3.1), (10.5.4), (2.2.3) and (2.2.5).

(10.5.7) $\chi_{\Omega}(-\lambda_1 + \lambda_7)$ *is equal to the following expressions:*

$\chi_{1,2}(-\lambda_1 + \lambda_7) + \chi_{1,2}(-\lambda_2 + \lambda_8) + \chi_{1,2}(0) + \chi_{1,2}(\lambda_1 - 2\lambda_2 + \lambda_3)$;

$\chi_{1,3}(-\lambda_1 + \lambda_7) + \chi_{1,3}(-\lambda_1 + \lambda_2 - \lambda_3 + \lambda_8) + \chi_{1,3}(-\lambda_3 + \lambda_6)$
$+ \chi_{1,3}(-\lambda_1 + 2\lambda_2 - \lambda_3) + \chi_{1,3}(0) + \chi_{1,3}(\lambda_2 - 2\lambda_3 + \lambda_4)$
$+ \chi_{1,3}(\lambda_1 - \lambda_3 + \lambda_3)$;

$\chi_{1,4}(-\lambda_1 + \lambda_7) + \chi_{1,4}(-\lambda_1 + \lambda_3 - \lambda_4 + \lambda_8) + \chi_{1,4}(-\lambda_1 + \lambda_2 - \lambda_4 + \lambda_6$
$+ \chi_{1,4}(-\lambda_4 + \lambda_7 + \lambda_8) + \chi_{1,4}(-\lambda_1 + \lambda_2 + \lambda_3 - \lambda_4) + \chi_{1,4}(0)$
$+ \chi_{1,4}(\lambda_3 - 2\lambda_4 + \lambda_5) + \chi_{1,4}(\lambda_2 - \lambda_4 + \lambda_7) + \chi_{1,4}(\lambda_1 - \lambda_4 + \lambda_8)$;

$\chi_{1,5}(-\lambda_1 + \lambda_7) + \chi_{1,5}(-\lambda_1 + \lambda_4 - \lambda_5 + \lambda_8) + \chi_{1,5}(-\lambda_1 + \lambda_3 - \lambda_5 + \lambda_6$
$+ \chi_{1,5}(-\lambda_1 + \lambda_2 - \lambda_5 + \lambda_7 + \lambda_8) + \chi_{1,5}(-\lambda_1 + \lambda_2 + \lambda_4 - \lambda_5)$
$+ \chi_{1,5}(-\lambda_5 + \lambda_6 + \lambda_7) + \chi_{1,5}(-\lambda_5 + 2\lambda_8) + \chi_{1,5}(0)$

$+ \chi_{1,5}(\lambda_4 - 2\lambda_5 + \lambda_6 + \lambda_8) + \chi_{1,5}(\lambda_3 - \lambda_5 + \lambda_7) + \chi_{1,5}(\lambda_2 - \lambda_5 + \lambda_8)$

$+ \chi_{1,5}(\lambda_1 - \lambda_5 + \lambda_6)$;

$\chi_{1,6}(-\lambda_1 + \lambda_7) + \chi_{1,6}(-\lambda_1 + \lambda_5 - \lambda_6) + \chi_{1,6}(-\lambda_1 + \lambda_3 - \lambda_6 + \lambda_7)$

$+ \chi_{1,6}(-\lambda_1 + \lambda_2 - \lambda_6 + \lambda_8) + \chi_{1,6}(-\lambda_6 + 2\lambda_7) + \chi_{1,6}(0)$

$+ \chi_{1,6}(\lambda_5 - 2\lambda_6 + \lambda_7) + \chi_{1,6}(\lambda_3 - \lambda_6) + \chi_{1,6}(\lambda_1 - \lambda_6 + \lambda_7)$;

$\chi_{1,7}(-\lambda_1 + \lambda_7) + \chi_{1,7}(-\lambda_1 + \lambda_6 - \lambda_7) + \chi_{1,7}(-\lambda_1 + \lambda_3 - \lambda_7)$

$+ \chi_{1,7}(0) + \chi_{1,7}(\lambda_6 - 2\lambda_1) + \chi_{1,7}(\lambda_1 - \lambda_7)$;

$\chi_{1,8}(-\lambda_1 + \lambda_7) + \chi_{1,8}(-\lambda_1 + \lambda_4 - \lambda_8) + \chi_{1,8}(-\lambda_1 + \lambda_2 + \lambda_7 - \lambda_8)$

$+ \chi_{1,8}(0) + \chi_{1,8}(\lambda_5 - 2\lambda_8) + \chi_{1,8}(\lambda_2 - \lambda_8)$.

(10.5.8) $\chi_\Omega(-\lambda_1 + \lambda_2)$ *is equal to the following expressions:*

$\chi_{1,2}(-\lambda_1 + \lambda_2) + \chi_{1,2}(-\lambda_2 + \lambda_3) + \chi_{1,2}(2\lambda_1 - \lambda_2)$;

$\chi_{1,3}(-\lambda_1 + \lambda_2) + \chi_{1,3}(-\lambda_3 + \lambda_4) + \chi_{1,3}(\lambda_2 - \lambda_3 + \lambda_7)$

$+ \chi_{1,3}(\lambda_1 - \lambda_3 + \lambda_8)$;

$\chi_{1,4}(-\lambda_1 + \lambda_2) + \chi_{1,4}(-\lambda_4 + \lambda_5) + \chi_{1,4}(\lambda_3 - \lambda_4 + \lambda_7)$

$+ \chi_{1,4}(\lambda_2 - \lambda_4 + \lambda_8) + \chi_{1,4}(\lambda_1 - \lambda_4 + \lambda_6) + \chi_{1,4}(\lambda_1 + \lambda_3 - \lambda_4)$;

$\chi_{1,5}(-\lambda_1 + \lambda_2) + \chi_{1,5}(-\lambda_5 + \lambda_6 + \lambda_8) + \chi_{1,5}(\lambda_4 - \lambda_5 + \lambda_7)$

$\chi_{1,5}(\lambda_3 - \lambda_5 + \lambda_8) + \chi_{1,5}(\lambda_2 - \lambda_5 + \lambda_6) + \chi_{1,5}(\lambda_1 - \lambda_5 + \lambda_7 + \lambda_8)$

$+ \chi_{1,5}(\lambda_1 + \lambda_4 - \lambda_5)$;

$\chi_{1,6}(-\lambda_1 + \lambda_2) + \chi_{1,6}(-\lambda_6 + \lambda_7 + \lambda_8) + \chi_{1,6}(\lambda_4 - \lambda_6)$

$+ \chi_{1,6}(\lambda_2 - \lambda_6 + \lambda_7) + \chi_{1,6}(\lambda_1 - \lambda_6 + \lambda_8)$;

$\chi_{1,7}(-\lambda_1 + \lambda_2) + \chi_{1,7}(-\lambda_7 + \lambda_8) + \chi_{1,7}(\lambda_2 - \lambda_7)$;

$\chi_{1,8}(-\lambda_1 + \lambda_2) + \chi_{1,8}(\lambda_6 - \lambda_8) + \chi_{1,8}(\lambda_3 - \lambda_8)$

$+ \chi_{1,8}(\lambda_1 + \lambda_7 - \lambda_8)$.

Let $\lambda = -2\lambda_1 + \lambda_3$ or $-2\lambda_1 + \lambda_8$, let $\mathcal{U}$ be the set of weights of $Y_\Sigma(\lambda)$ and $a_\tau = \dim Y_\Sigma(\lambda)^\tau$ for $\tau \in \mathcal{U}$. Let $\Gamma = \Delta\backslash\{\alpha_1, \alpha_j\}$, where $2 \le j \le 7$ and $\Pi = \Sigma \cap \Gamma$. It is our contention that

$$(10.5.9) \quad \chi_\Omega(\lambda) = \sum_{\tau\in\mathcal{U}} a_\tau \chi_\Gamma(\tau) + \textit{lower terms} .$$

Here, as in section 9.6, "lower terms" denotes a sum of $\chi_\Gamma(\xi)$'s for $\xi \in X_\Gamma$ such that for any occurring $\chi_\Gamma(\xi)$, $(\xi,\alpha_1^\vee) \ge -1$ and for any $\tau \in \mathcal{U}$, $\xi \not\ge \tau$. Now $\chi_\Omega(\lambda)$ is a sum of $\chi_\Gamma(\tau)$'s for various $\tau \in X_\Gamma$ so we may express $\chi_\Omega(\lambda)$ as $\Sigma b_\tau \chi_\Gamma(\tau) + \Sigma c_\tau \chi_\Gamma(\tau)$ where b_τ, $c_\tau \ge 0$, $\tau \in X_\Gamma$, $b_\tau \neq 0$ implies that $\lambda - \tau \in \mathbb{Z}\Sigma$ and $c_\tau \neq 0$ implies that $\lambda - \tau \notin \mathbb{Z}\Sigma$. If $c_\tau \neq 0$ then, since τ is a weight of $Y_\Omega(\lambda)$, $\tau = \lambda - \sum_{i=2}^{8} m_i\alpha_i$ with all $m_i \ge 0$ and $m_2 \neq 0$. Thus $(\tau,\alpha_1^\vee) = (\lambda,\alpha_1^\vee) + m_1 \ge -1$. If $c_\xi \neq 0$ and $\xi \ge \tau$ for some $\tau \in \mathcal{U}$ then $\lambda - \tau \ge 0$, $\lambda - \xi \ge 0$, $\lambda - \tau \in \mathbb{Z}\Sigma$ and $\lambda - \tau \ge \lambda - \xi$ imply that $\lambda - \xi \in \mathbb{Z}\Sigma$ which is a contradiction. Hence $c_\xi \neq 0$ implies that, for every $\tau \in \mathcal{U}$, $\xi \not\ge \tau$. Thus to prove (10.5.9) it suffices to show that $\Sigma a_\tau \chi_\Gamma(\tau)$ is equal to $\Sigma b_\tau \chi_\Gamma(\tau)$. Let $\mathfrak{f}: \mathbb{Z}X \to \mathbb{Z}X$ be the $\mathbb{Z}$-linear map such that $\mathfrak{f}(e(\mu))$ is $e(\mu)$ if $\lambda - \mu \in \mathbb{Z}\Sigma$ and 0 if $\lambda - \mu \notin \mathbb{Z}\Sigma$. It follows from p.230 of [27] that $\mathfrak{f}(\chi_\Omega(\lambda)) = \chi_\Sigma(\lambda)$ and that, for $\tau \in X_\Gamma$, $\mathfrak{f}(\chi_\Gamma(\tau)) = \chi_\Pi(\tau)$ if $\lambda - \tau \in \mathbb{Z}\Sigma$ and 0 otherwise. Thus we obtain $\chi_\Sigma(\lambda) = \Sigma b_\tau \chi_\Pi(\tau)$. Let χ be $\chi_{\Gamma,E}$ (see section 2.2) where E is the empty set. Applying χ to the equation $\Sigma a_\tau e(\tau) = \Sigma b_\tau \chi_\Gamma(\tau)$ we obtain, with the help of (2.2.1) and (2.2.2), $\Sigma a_\tau \chi_\Gamma(\tau) = \Sigma b_\tau \chi_\Gamma(\tau)$. This completes the proof of (10.5.9).

The next result is deduced from (10.5.9), (10.5.3) and (10.5.6).

(10.5.10) $\chi_\Omega(-2\lambda_1 + \lambda_3)$ *is equal to the following expressions:*

$$\chi_{1,2}(-2\lambda_1 + \lambda_3) + \textit{lower terms} ;$$

$$\chi_{1,3}(-2\lambda_1 + \lambda_3) + \chi_{1,3}(-2\lambda_1 + \lambda_2 - \lambda_3 + \lambda_4)$$
$$+ \chi_{1,3}(-2\lambda_1 + 2\lambda_2 - \lambda_3 + \lambda_7) + \textit{lower terms} ;$$

$\chi_{1,4}(-2\lambda_1 + \lambda_3) + \chi_{1,4}(-2\lambda_1 + \lambda_2 - \lambda_4 + \lambda_5)$

$+ \chi_{1,4}(-2\lambda_1 + \lambda_2 + \lambda_3 - \lambda_4 + \lambda_7) + \chi_{1,4}(-2\lambda_1 + 2\lambda_2 - \lambda_4 + \lambda_8)$

$+$ *lower terms* ;

$\chi_{1,5}(-2\lambda_1 + \lambda_3) + \chi_{1,5}(-2\lambda_1 + \lambda_2 - \lambda_5 + \lambda_6 + \lambda_8)$

$+ \chi_{1,5}(-2\lambda_1 + \lambda_2 + \lambda_4 - \lambda_5 + \lambda_7) + \chi_{1,5}(-2\lambda_1 + \lambda_2 + \lambda_3 - \lambda_5 + \lambda_8)$

$+ \chi_{1,5}(-2\lambda_1 + 2\lambda_2 - \lambda_5 + \lambda_6)$ + *lower terms* ;

$\chi_{1,6}(-2\lambda_1 + \lambda_3) + \chi_{1,6}(-2\lambda_1 + \lambda_2 - \lambda_6 + \lambda_7 + \lambda_8)$

$+ \chi_{1,6}(-2\lambda_1 + \lambda_2 + \lambda_4 - \lambda_6) + \chi_{1,6}(-2\lambda_1 + 2\lambda_2 - \lambda_6 + \lambda_7)$

$+$ *lower terms* ;

$\chi_{1,7}(-2\lambda_1 + \lambda_3) + \chi_{1,7}(-2\lambda_1 + \lambda_2 - \lambda_7 + \lambda_8) + \chi_{1,7}(-2\lambda_1 + 2\lambda_2 - \lambda_7)$

$+$ *lower terms* ;

$\chi_{1,8}(-2\lambda_1 + \lambda_3) + \chi_{1,8}(-2\lambda_1 + \lambda_2 + \lambda_6 - \lambda_8)$

$+ \chi_{1,8}(-2\lambda_1 + \lambda_2 + \lambda_3 - \lambda_8)$ + *lower terms* .

n the same way we obtain the corresponding result for $\chi_\Omega(-2\lambda_1 + \lambda_8)$ rom (10.5.9), (10.5.5) and (10.5.2).

10.5.11) $\chi_\Omega(-2\lambda_1 + \lambda_8)$ *is equal to the following expressions:*

$\chi_{1,2}(-2\lambda_1 + \lambda_8)$ + *lower terms* ;

$\chi_{1,3}(-2\lambda_1 + \lambda_8) + \chi_{1,3}(-2\lambda_1 + \lambda_2 - \lambda_3 + \lambda_6) + \chi_{1,3}(-2\lambda_1 + \lambda_2)$

$+ \chi_{1,3}(-2\lambda_1 + 2\lambda_2 - 2\lambda_3 + \lambda_4)$ + *lower terms* ;

$\chi_{1,4}(-2\lambda_1 + \lambda_8) + \chi_{1,4}(-2\lambda_1 + \lambda_3 - \lambda_4 + \lambda_6)$

$+ \chi_{1,4}(-2\lambda_1 + \lambda_2 - \lambda_4 + \lambda_7 + \lambda_8) + \chi_{1,4}(-2\lambda_1 + 2\lambda_3 - \lambda_4)$

$+ \chi_{1,4}(-2\lambda_1 + \lambda_2) + \chi_{1,4}(-2\lambda_1 + \lambda_2 + \lambda_3 - 2\lambda_4 + \lambda_5)$

$+ \chi_{1,4}(-2\lambda_1 + 2\lambda_2 - \lambda_4 + \lambda_7)$ + *lower terms* ;

$$\chi_{1,5}(-2\lambda_1 + \lambda_8) + \chi_{1,5}(-2\lambda_1 + \lambda_4 - \lambda_5 + \lambda_6)$$
$$+ \chi_{1,5}(-2\lambda_1 + \lambda_3 - \lambda_5 + \lambda_7 + \lambda_8) + \chi_{1,5}(-2\lambda_1 + \lambda_2 - \lambda_5 + \lambda_6 + \lambda_7)$$
$$+ \chi_{1,5}(-2\lambda_1 + \lambda_3 + \lambda_4 - \lambda_5) + \chi_{1,5}(-2\lambda_1 + \lambda_2 - \lambda_5 + 2\lambda_8)$$
$$+ \chi_{1,5}(-2\lambda_1 + \lambda_2) + \chi_{1,5}(-2\lambda_1 + \lambda_2 + \lambda_4 - 2\lambda_5 + \lambda_6 + \lambda_8)$$
$$+ \chi_{1,5}(-2\lambda_1 + \lambda_2 + \lambda_3 - \lambda_5 + \lambda_7) + \chi_{1,5}(-2\lambda_1 + 2\lambda_2 - \lambda_5 + \lambda_8)$$
$$+ \textit{lower terms} \;;$$

$$\chi_{1,6}(-2\lambda_1 + \lambda_8) + \chi_{1,6}(-2\lambda_1 + \lambda_4 - \lambda_6 + \lambda_7)$$
$$+ \chi_{1,6}(-2\lambda_1 + \lambda_2 - \lambda_6 + 2\lambda_7) + \chi_{1,6}(-2\lambda_1 + \lambda_2)$$
$$+ \chi_{1,6}(-2\lambda_1 + \lambda_2 + \lambda_5 - 2\lambda_6 + \lambda_7) + \chi_{1,6}(-2\lambda_1 + \lambda_2 + \lambda_3 - \lambda_6)$$
$$+ \textit{lower terms} \;;$$

$$\chi_{1,7}(-2\lambda_1 + \lambda_8) + \chi_{1.7}(-2\lambda_1 + \lambda_4 - \lambda_7) + \chi_{1,7}(-2\lambda_1 + \lambda_2)$$
$$+ \chi_{1,7}(-2\lambda_1 + \lambda_2 + \lambda_6 - 2\lambda_7) + \textit{lower terms} \;;$$

$$\chi_{1,8}(-2\lambda_1 + \lambda_8) + \chi_{1,8}(-2\lambda_1 + \lambda_5 - \lambda_8) + \chi_{1,8}(-2\lambda_1 + \lambda_3 + \lambda_7 - \lambda_8)$$
$$+ \chi_{1,8}(-2\lambda_1 + \lambda_2) + \chi_{1,8}(-2\lambda_1 + \lambda_2 + \lambda_5 - 2\lambda_8)$$
$$+ \chi_{1,8}(-2\lambda_1 + 2\lambda_2 - \lambda_8) + \textit{lower terms} \;.$$

10.6 $R^i Ind_B^{P_\Sigma}(\bar{E})$

Let $\bar{E} = E/D$. It is the purpose of this section to show:

(10.6.1) $R^i Ind_B^{P_\Sigma}(\bar{E})$ *has a good filtration for* $i = 0$ *and is* 0 *for* $i > 0$, *where* $\Sigma = \Delta\backslash\{\alpha_j\}$ *for some* $2 \le j \le 8$.

The method of proof is the same as that employed in proving (9.7.1). We shall therefore usually omit references and minor cohomological calculations in this section, except where a point of some subtlety or complexity is involved. In particular we shall constantly use (10.5.7) and (10.5.11) without reference. It is recommended that readers refresh their memories with a perusal of section 9.7 before proceeding.

The module $\bar{E}$ is a non-split extension of $Y_\Omega(-2\lambda_1 + \lambda_8)$ by

$Y_\Omega(-\lambda_1 + \lambda_7)$ and we identify $Y_\Omega(-2\lambda_1 + \lambda_8)$ with a submodule of $\bar{E}$. Let $j = 2$, $P = P_{1,2}$, $G = Ind_B^{P_2}$ and $\Gamma = \Delta\backslash\{\alpha_1,\alpha_2\}$. Let $K = M_\Gamma(-2\lambda_1 + \lambda_8)$. Then $R^iG(K)$ has a good filtration if $i = 0$ and is 0 if $i > 0$. Thus, to prove (10.6.1) for $j = 2$, it suffices to to show that the non-split extension $J = \bar{E}/K$ of $Y_{1,2}(-2\lambda_1 + \lambda_8)$ by $Y_1(-\lambda_1 + \lambda_7)$ has the property that $R^iG(J)$ has a good filtration for $i = 0$ and is 0 for $i > 0$. Now $Y_1(-\lambda_1 + \lambda_7)|_P$ has a good filtration with quotients $Y_{1,2}(-\lambda_1 + \lambda_7)$, $Y_{1,2}(-\lambda_2 + \lambda_8)$, $Y_{1,2}(0)$, $Y_{1,2}(\lambda_1 - 2\lambda_2 + \lambda_3)$. Hence J has a descending filtration $J = J_o,J_1,J_2,J_3,J_4,J_5 = 0$ with quotients, in order, $Y_{1,2}(-\lambda_1 + \lambda_7)$, $Y_{1,2}(-\lambda_2 + \lambda_8)$, $Y_{1,2}(0)$, $Y_{1,2}(\lambda_1 - 2\lambda_2 + \lambda_3)$, $Y_{1,2}(-2\lambda_1 + \lambda_8)$ and such that J is a non-split extension J_4 by J/J_4. We have $Ext_P^1(Y_{1,2}(\tau),Y_{1,2}(-2\lambda_1 + \lambda_8)) = 0$ for $\tau = \lambda_1 - 2\lambda_2 + \lambda_3, 0$. Also $Ext_P^1(Y_{1,2}(-\lambda_1 + \lambda_7),Y_{1,2}(-2\lambda_1 + \lambda_8))$ is isomorphic to $Ext_{P_\Pi}^1(Y_\Pi(-\lambda_1 + \lambda_7),Y_\Pi(-2\lambda_1 + \lambda_8))$, where $\Pi = \Delta\backslash\{\alpha_1,\alpha_2,\alpha_8\}$, and hence to $H^1(P_\Pi,Y_\Pi(\lambda_1 - \lambda_2 + \lambda_3 - \lambda_8) \otimes (-2\lambda_1 + \lambda_8))$. However, this is isomorphic to $H^1(B,-\lambda_1 - \lambda_2 + \lambda_3)$ which is 0. Thus we may rechoose the filtration such that J_o/J_1 is unchanged, J_1/J_2 is a non-split extension of $Y_{1,2}(-2\lambda_1 + \lambda_8)$ by $Y_{1,2}(-\lambda_2 + \lambda_8)$, J_2/J_3 is $Y_{1,2}(0)$, J_3 is $Y_{1,2}(\lambda_1 - 2\lambda_2 + \lambda_3)$ and $J_4 = 0$. It is easy to see that $R^iG(J_r/J_{r+1})$ has a good filtration if $i = 0$ and is 0 if $i > 0$, for $r = 0$, 2 and 3; for $r = 1$ this is so by (5.18). Thus $R^iG(J)$ has a good filtration if $i = 0$ and is 0 if $i > 0$ and the proof of (10.6.1), for $r = 2$, is complete.

Now let $j = 3$, $P = P_{1,3}$ and $G = Ind_B^{P_3}$. There is a submodule, say K, of $Y_1(-2\lambda_1 + \lambda_8)$ such that K has a good filtration and $Y_1(-2\lambda_1 + \lambda_8)/K$ has a good filtration with quotients $Y_{1,3}(-2\lambda_1 + \lambda_8)$, $Y_{1,3}(-2\lambda_1 + \lambda_2 - \lambda_3 + \lambda_6)$, $Y_{1,3}(-2\lambda_1 + \lambda_2)$ and $Y_{1,3}(-2\lambda_1 + 2\lambda_2 - 2\lambda_3 + \lambda_4)$. Moreover $R^iG(K)$ has a good filtration if $i = 0$ and is 0 if $i > 0$. Thus it suffices to show that $R^iG(J)$ has a good filtration if i 0 and is 0 if $i > 0$ where $J = \bar{E}/K$. Now J has a descending filtration $J = J_o,J_1,\ldots,J_{11} = 0$ with quotients, in order, $Y_{1,3}(-\lambda_1 + \lambda_7)$, $Y_{1,3}(-\lambda_1 + \lambda_2 - \lambda_3 + \lambda_8)$, $Y_{1,3}(-\lambda_3 + \lambda_6)$, $Y_{1,3}(-\lambda_1 + 2\lambda_2 - \lambda_3)$, $Y_{1,3}(0)$, $Y_{1,3}(\lambda_2 - 2\lambda_3 + \lambda_4)$, $Y_{1,3}(\lambda_1 - \lambda_3 + \lambda_7)$, $Y_{1,3}(-2\lambda_1 + \lambda_8)$, $Y_{1,3}(-2\lambda_1 + \lambda_2 - \lambda_3 + \lambda_6)$, $Y_{1,3}(-2\lambda_1 + \lambda_2)$, $Y_{1,3}(-2\lambda_1 + 2\lambda_2 - 2\lambda_3 + \lambda_4)$ such that J/J_8 is a non-split extension of $Y_{1,3}(-2\lambda_1 + \lambda_8)$ by J/J_7 . We have that $Ext^1_P(Y_{1,3}(-\lambda_1 + \lambda_7), Y_{1,3}(-2\lambda_1 + \lambda_8))$ is isomorphic to $Ext^1_{P_\Pi}(Y_\Pi(-\lambda_1 + \lambda_7), Y_\Pi(-2\lambda_1 + \lambda_8))$, where $\Pi = \Delta\backslash\{\alpha_1, \alpha_3, \alpha_8\}$, and hence isomorphic to $H^1(B, Y_\Pi(\lambda_1 - \lambda_3 + \lambda_4 + \lambda_7 - \lambda_8) \otimes (-2\lambda_1 + \lambda_8))$. This is isomorphic to $H^1(B, -\lambda_1 - \lambda_3 + \lambda_4 + \lambda_7)$ and therefore 0 . Thus $Ext^1_P(Y_{1,3}(-\lambda_1 + \lambda_7), Y_{1,3}(-2\lambda_1 + \lambda_8)) = 0$ and this, together with applications of Lemma 3.2.1 shows that the filtration may be rechosen so that J_1/J_1 is unchanged, J_1/J_2 is a non-split extension of $Y_{1,3}(-2\lambda_1 + \lambda_8)$ by $Y_{1,3}(-\lambda_1 + \lambda_2 - \lambda_3 + \lambda_8)$, J_2/J_3 is an extension of $Y_{1,3}(-2\lambda_1 + \lambda_2 - \lambda_3 + \lambda_6)$ by $Y_{1,3}(-\lambda_3 + \lambda_6)$, J_3/J_4 is isomorphic to $Y_{1,3}(-\lambda_1 + 2\lambda_2 - \lambda_3)$, J_4/J_5 is an extension of $Y_{1,3}(-2\lambda_1 + \lambda_2)$ by $Y_{1,2}(0)$, J_5 is an extension of $Y_{1,3}(-2\lambda_1 + 2\lambda_2 - 2\lambda_3 + \lambda_4)$ by $Y_{1,3}(\lambda_2 - 2\lambda_3 + \lambda_4)$ and $J_6 = 0$. There is only one non-split extension of $Y_{1,3}(-2\lambda_1 + \lambda_8)$ by $Y_{1,3}(-\lambda_1 + \lambda_2 - \lambda_3 + \lambda_8)$ so that J_1/J_2 is isomorphic to $Y_3(\lambda_2 - \lambda_3 + \lambda_8) \otimes -\lambda_1$. We claim that the extension J_2/J_3 , of $Y_{1,3}(-2\lambda_1 + \lambda_2 - \lambda_3 + \lambda_6)$ by $Y_{1,3}(-\lambda_3 + \lambda_6)$, is non-split. Suppose, for a contradiction, that this is not the case. Since $Ext^1_P(J_1/J_2, Y_{1,3}(-2\lambda_1 + \lambda_2 - \lambda_3 + \lambda_6)) = Ext^1_P(Y_3(\lambda_2 - \lambda_3 + \lambda_8) \otimes -\lambda_1 , Y_{1,3}(-2\lambda_1 + \lambda_2 - \lambda_3 + \lambda_6))$ is 0 and $Hom_P(\bar{E}, Y_{1,3}(-2\lambda_1 + \lambda_2 - \lambda_3 + \lambda_6)) = 0$ we must have $Ext^1_P(Y_{1,3}(-\lambda_1 + \lambda_7), Y_{1,3}(-2\lambda_1 + \lambda_2 - \lambda_3 + \lambda_6)) \neq 0$. This is isomorphic to $H^1(P, Y_{1,3}(\lambda_1 - \lambda_3 + \lambda_4) \otimes Y_{1,3}(-2\lambda_1 + \lambda_2 - \lambda_3 + \lambda_6))$ and hence to $H^1(P, Y_{1,3}(-\lambda_1 + \lambda_2 - 2\lambda_3) \otimes Y_3(\lambda_4) \otimes Y_3(\lambda_6))$. However $R^i Ind_P^{P_3}(Y_{1,3}(-\lambda_1 + \lambda_2 - 2\lambda_3)) = 0$ for all i so that (1.1.8) implies that $H^1(P, Y_{1,3}(-\lambda_1 + \lambda_2 - 2\lambda_3) \otimes Y_3(\lambda_4) \otimes Y_3(\lambda_6))$ is 0 . We have now obtained a contradiction and so J_2/J_3 is a non-split extension of

$Y_{1,3}(-2\lambda_1 + \lambda_2 - \lambda_3 + \lambda_6)$ by $Y_{1,3}(-\lambda_3 + \lambda_6)$ and so isomorphic to $Y_3(\lambda_1 - \lambda_3 + \lambda_6) \otimes -\lambda_1$. We now claim that the extension J_4/J_5 , of $Y_{1,3}(-2\lambda_1 + \lambda_2)$ by $Y_{1,3}(0)$, is non-split. Since $Ext^1_P(J_r/J_{r+1}, Y_{1,3}(-2\lambda_1 + \lambda_2)) = 0$ for $r = 1,2,3$ and $Y_{1,3}(-2\lambda_1 + \lambda_2)$ is not an image of $\bar{E}$ the alternative is that $Ext^1_P(Y_{1,3}(-\lambda_1 + \lambda_7), Y_{1,3}(-2\lambda_1 + \lambda_2)) \neq 0$. However this Ext group is isomorphic to $H^1(P, Y_{1,3}(\lambda_1 - \lambda_3 + \lambda_4) \otimes Y_{1,3}(-2\lambda_1 + \lambda_2))$ and so isomorphic to $H^1(P, Y_{1,3}(-\lambda_1 + \lambda_2 - \lambda_3 + \lambda_4))$ which is 0 . Thus J_4/J_5 is a non-split extension and therefore isomorphic to $Y_3(\lambda_1) \otimes -\lambda_1$. We now claim that the extension J_5 of $Y_{1,3}(-2\lambda_1 + 2\lambda_2 - 2\lambda_3 + \lambda_4)$ by $Y_{1,3}(\lambda_2 - 2\lambda_3 + \lambda_4)$. Once more, since $Ext^1_P(J_r/J_{r+1}, Y_{1,3}(-2\lambda_1 + 2\lambda_2 - 2\lambda_3 + \lambda_4)) = 0$ for $r = 1,2,3,4$ and $Y_{1,3}(-2\lambda_1 + 2\lambda_2 - 2\lambda_3 + \lambda_4)$ is not an image of $\bar{E}$ the alternative is that $Ext^1_P(Y_{1,3}(-\lambda_1 + \lambda_7), Y_{1,3}(-2\lambda_1 + 2\lambda_2 - 2\lambda_3 + \lambda_4)) \neq 0$. This Ext group is isomorphic to $H^1(P, Y_{1,3}(\lambda_1 - \lambda_3 + \lambda_4) \otimes Y_{1,3}(-2\lambda_1 + 2\lambda_2 - 2\lambda_3 + \lambda_4))$ and so isomorphic to $H^1(P, Y_{1,3}(-\lambda_1 + 2\lambda_2) \otimes Y_3(-\lambda_3 + \lambda_4) \otimes Y_3(-2\lambda_3 + \lambda_4))$. Since $R^i Ind_P^{P_3}(Y_{1,3}(-\lambda_1 + 2\lambda_2)) = 0$ for all i , (1.1.8) implies that the above cohomology group is 0 . Hence J_5 is a non-split extension of $Y_{1,3}(-2\lambda_1 + 2\lambda_2 - 2\lambda_3 + \lambda_4)$ by $Y_{1,3}(\lambda_2 - 2\lambda_3 + \lambda_4)$. It follows that there is a short exact sequence of P-modules

$$0 \to A \to Y_3(\lambda_1 + \lambda_2 - 2\lambda_3 + \lambda_4) \otimes -\lambda_1 \to J_5 \to 0$$

where A is an extension of $Y_{1,3}(-3\lambda_1 + \lambda_2 - \lambda_3 + \lambda_4)$ by $Y_{1,3}(-\lambda_1 - \lambda_3 + \lambda_4)$. To show that $R^i G(J)$ has a good filtration for $i = 0$ and is 0 for $i > 0$ it suffices to show that, for $0 \le r \le 5$, the same is true with J replaced by J_r/J_{r+1} . However, the description of J_r/J_{r+1} obtained so far, reveals that $R^i G(J_r/J_{r+1}) = 0$ for all i and $0 \le r \le 4$ and it only remains to deal with the case $r = 5$. By the above short exact sequence $R^i G(J_5)$ is isomorphic to $R^{i+1} G(A)$. However $R^i G(Y_{1,3}(\tau)) = 0$ for all i and $\tau = -3\lambda_1 + \lambda_2 - \lambda_3 + \lambda_4, -\lambda_1 - \lambda_3 + \lambda_4$ so that $R^i G(A) = 0$ for all i .

This completes the proof of (10.6.1) in the case $j = 3$.

Now let $j = 4$, $P = P_{1,4}$ and $G = Ind_B^{P_4}$. We find, in the usual way, a submodule, say K , of $Y_1(-2\lambda_1 + \lambda_8)$ such that K has a good filtration and $Y_1(-2\lambda_1 + \lambda_8)/K$ has a good filtration with quotients
$Y_{1,4}(-2\lambda_1 + \lambda_8)$, $Y_{1,4}(-2\lambda_1 + \lambda_3 - \lambda_4 + \lambda_6)$,
$Y_{1,4}(-2\lambda_1 + \lambda_2 - \lambda_4 + \lambda_7 + \lambda_8)$, $Y_{1,4}(-2\lambda_1 + 2\lambda_3 - \lambda_4)$,
$Y_{1,4}(-2\lambda_1 + \lambda_2)$, $Y_{1,4}(-2\lambda_1 + \lambda_2 + \lambda_3 - 2\lambda_4 + \lambda_5)$,
$Y_{1,4}(-2\lambda_1 + 2\lambda_2 - \lambda_4 + \lambda_7)$ and $R^iG(K)$ has a good filtration for $i = 0$ and is 0 if $i > 0$. The module $J = \bar{E}/K$ has a descending filtration $J = J_o, J_1, \ldots, J_{16} = 0$ with quotients, in order,
$Y_{1,4}(-\lambda_1 + \lambda_7)$, $Y_{1,4}(-\lambda_1 + \lambda_3 - \lambda_4 + \lambda_8)$, $Y_{1,4}(-\lambda_1 + \lambda_2 - \lambda_4 + \lambda_6)$,
$Y_{1,4}(-\lambda_4 + \lambda_7 + \lambda_8)$, $Y_{1,4}(-\lambda_1 + \lambda_2 + \lambda_3 - \lambda_4)$, $Y_{1,4}(0)$,
$Y_{1,4}(\lambda_3 - 2\lambda_4 + \lambda_5)$, $Y_{1,4}(\lambda_2 - \lambda_4 + \lambda_7)$, $Y_{1,4}(\lambda_1 - \lambda_4 + \lambda_8)$,
$Y_{1,4}(-2\lambda_1 + \lambda_8)$, $Y_{1,4}(-2\lambda_1 + \lambda_3 - \lambda_4 + \lambda_6)$,
$Y_{1,4}(-2\lambda_1 + \lambda_2 - \lambda_4 + \lambda_7 + \lambda_8)$, $Y_{1,4}(-2\lambda_1 + 2\lambda_3 - \lambda_4)$,
$Y_{1,4}(-2\lambda_1 + \lambda_2)$, $Y_{1,4}(-2\lambda_1 + \lambda_2 + \lambda_3 - 2\lambda_4 + \lambda_5)$,
$Y_{1,4}(-2\lambda_1 + 2\lambda_2 - \lambda_4 + \lambda_7)$ and such that J/J_{10} is a non-split extension of J_9/J_{10} by J/J_9 . We have that
$Ext_P^1(Y_{1,4}(-\lambda_1 + \lambda_7), Y_{1,4}(-2\lambda_1 + \lambda_8))$ is isomorphic to
$Ext_{P_\Pi}^1(Y_\Pi(-\lambda_1 + \lambda_7), Y_\Pi(-2\lambda_1 + \lambda_8))$, where $\Pi = \Delta\backslash\{\alpha_1,\alpha_4,\alpha_8\}$, and so isomorphic to $H^1(P_\Pi, Y_\Pi(\lambda_1 - \lambda_4 + \lambda_5 - \lambda_8) \otimes Y_\Pi(-2\lambda_1 + \lambda_8))$. However, this is isomorphic to $H^1(P_\Pi, Y_\Pi(-\lambda_1 - \lambda_4 + \lambda_5))$ which is 0 . Thus $Ext_P^1(Y_{1,4}(-\lambda_1 + \lambda_7), Y_{1,4}(-2\lambda_1 + \lambda_8)) = 0$ and this, together with Lemma 3.2.1, shows that the filtration may be rechosen so that J_o/J_1 is unchanged, J_1/J_2 is a non-split extension of $Y_{1,4}(-2\lambda_1 + \lambda_8)$ by $Y_{1,4}(-\lambda_1 + \lambda_3 - \lambda_4 + \lambda_8)$, J_2/J_3 is an extension of $Y_{1,4}(-2\lambda_1 + \lambda_3 - \lambda_4 + \lambda_6)$ by $Y_{1,4}(-\lambda_1 + \lambda_2 - \lambda_4 + \lambda_6)$, J_3/J_4 is an extension of $Y_{1,4}(-2\lambda_1 + \lambda_2 - \lambda_4 + \lambda_7 + \lambda_8)$ by $Y_{1,4}(-\lambda_4 + \lambda_7 + \lambda_8)$, J_4/J_5 is an extension of $Y_{1,4}(-2\lambda_1 + 2\lambda_3 - \lambda_4)$ by $Y_{1,4}(-\lambda_1 + \lambda_2 + \lambda_3 - \lambda_4)$, J_5/J_6 is an extension of $Y_{1,4}(-2\lambda_1 + \lambda_2)$

by $Y_{1,4}(0)$, J_6/J_7 is an extension of $Y_{1,4}(-2\lambda_1 + \lambda_2 + \lambda_3 - 2\lambda_4 + \lambda_5)$ by $Y_{1,4}(\lambda_3 - 2\lambda_4 + \lambda_5)$, J_7/J_8 is an extension of $Y_{1,4}(-2\lambda_1 + 2\lambda_2 - \lambda_4 + \lambda_7)$ by $Y_{1,4}(\lambda_2 - \lambda_4 + \lambda_7)$, $J_8 \cong Y_{1,4}(\lambda_1 - \lambda_4 + \lambda_8)$ and $J_9 = 0$.

There is a unique non-split extension of $Y_{1,4}(-2\lambda_1 + \lambda_8)$ by $Y_{1,4}(-\lambda_1 + \lambda_3 - \lambda_4 + \lambda_8)$ so that J_1/J_2 is isomorphic to $Y_4(\lambda_3 - \lambda_4 + \lambda_8) \otimes -\lambda_1$. We now note that the extension J_2/J_3 of $Y_{1,4}(-2\lambda_1 + \lambda_3 - \lambda_4 + \lambda_6)$ by $Y_{1,4}(-\lambda_1 + \lambda_2 - \lambda_4 + \lambda_6)$, is non-split. Suppose for a contradiction that this is not the case. Since $Y_{1,4}(-2\lambda_1 + \lambda_3 - \lambda_4 + \lambda_6)$ is not an image of $\bar{E}$ we must have $Ext^1_P(J/J_2, Y_{1,4}(-2\lambda_1 + \lambda_3 - \lambda_4 + \lambda_6)) \neq 0$. However $Ext^1_P(J_1/J_2, Y_{1,4}(-2\lambda_1 + \lambda_3 - \lambda_4 + \lambda_6))$ is isomorphic to $Ext^1_P(Y_4(\lambda_3 - \lambda_4 + \lambda_8) \otimes -\lambda_1 , Y_{1,4}(-2\lambda_1 + \lambda_3 - \lambda_4 + \lambda_6))$ which is 0 . ʌlso, $Ext^1_P(Y_{1,4}(-\lambda_1 + \lambda_7), Y_{1,4}(-2\lambda_1 + \lambda_3 - \lambda_4 + \lambda_6))$ is isomorphic to $^1(P, Y_{1,4}(\lambda_1 - \lambda_4 + \lambda_8) \otimes Y_{1,4}(-2\lambda_1 + \lambda_3 - \lambda_4 + \lambda_6))$ and so isomorphic o $H^1(P, Y_{1,4}(-\lambda_1 + \lambda_3 - 2\lambda_4) \otimes Y_4(\lambda_8) \otimes Y_4(\lambda_6))$. Since $^i Ind_P^{P_4}(Y_{1,4}(-\lambda_1 + \lambda_3 - 2\lambda_4)) = 0$ for all i we obtain, from (1.1.8), hat the above cohomology group is 0 and so $xt^1_P(Y_{1,4}(-\lambda_1 + \lambda_7), Y_{1,4}(-2\lambda_1 + \lambda_3 - \lambda_4 + \lambda_6)) = 0$. Hence $xt^1_P(J/J_2, Y_{1,4}(-2\lambda_1 + \lambda_3 - \lambda_4 + \lambda_6)) = 0$. This contradiction shows hat J_2/J_3 is a non-split extension of $Y_{1,4}(-2\lambda_1 + \lambda_3 - \lambda_4 + \lambda_6)$ by $_{1,4}(-\lambda_1 + \lambda_2 - \lambda_4 + \lambda_6)$ and hence isomorphic to $Y_4(\lambda_2 - \lambda_4 + \lambda_6) \otimes -$ $_4(\lambda_2 - \lambda_4 + \lambda_6) \otimes -\lambda_1$.

Our next claim is that J_3/J_4 is a non-split extension of $_{1,4}(-2\lambda_1 + \lambda_2 - \lambda_4 + \lambda_7 + \lambda_8)$ by $Y_{1,4}(-\lambda_4 + \lambda_7 + \lambda_8)$. The altern- tive is that $Ext^1_P(J/J_3, Y_{1,4}(-2\lambda_1 + \lambda_2 - \lambda_4 + \lambda_7 + \lambda_8)) \neq 0$ since $_{1,4}(-2\lambda_1 + \lambda_2 - \lambda_4 + \lambda_7 + \lambda_8)$ is not an image of $\bar{E}$. However $xt^1_P(J_r/J_{r+1}, Y_{1,4}(-2\lambda_1 + \lambda_2 - \lambda_4 + \lambda_7 + \lambda_8)) = 0$ for $r = 1,2$ by the escription of J_r/J_{r+1} so far obtained. Also

$Ext_P^1(J_o/J_1, Y_{1,4}(-2\lambda_1 + \lambda_2 - \lambda_4 + \lambda_7 + \lambda_8))$ is isomorphic to $H^1(P, Y_{1,4}(\lambda_1 - \lambda_4 + \lambda_8) \otimes Y_{1,4}(-2\lambda_1 + \lambda_2 - \lambda_4 + \lambda_7 + \lambda_8))$ and so isomorphic to $H^1(P, Y_{1,4}(-\lambda_1 + \lambda_2 - 2\lambda_4) \otimes Y_4(\lambda_8) \otimes Y_4(\lambda_7 + \lambda_8))$. This cohomology group is 0 however, by (1.1.8), since $R^i Ind_P^{P_4}(Y_{1,4}(-\lambda_1 + \lambda_2 - 2\lambda_4)) = 0$ for all i. Thus we have $Ext_P^1(J/J_3, Y_{1,4}(-2\lambda_1 + \lambda_2 - \lambda_4 + \lambda_7 + \lambda_8)) = 0$. Hence J_3/J_4 is a non-split extension of $Y_{1,4}(-2\lambda_1 + \lambda_2 - \lambda_4 + \lambda_7 + \lambda_8)$ by $Y_{1,4}(-\lambda_4 + \lambda_7 + \lambda_8)$ and therefore isomorphic to $Y_4(\lambda_1 - \lambda_4 + \lambda_7 + \lambda_8) \otimes -\lambda_1$.

We now treat J_4/J_5. If this is a split extension of $Y_{1,4}(-2\lambda_1 + 2\lambda_3 - \lambda_4)$ by $Y_{1,4}(-\lambda_1 + \lambda_2 + \lambda_3 - \lambda_4)$ we must have $Ext_P^1(J/J_4, Y_{1,4}(-2\lambda_1 + 2\lambda_3 - \lambda_4)) \neq 0$. However, $Ext_P^1(J_r/J_{r+1}, Y_{1,4}(-2\lambda_1 + 2\lambda_3 - \lambda_4)) = 0$ for $r = 1,2,3$ by the description so far obtained of J_r/J_{r+1}. Also $Ext_P^1(J_o/J_1, Y_{1,4}(-2\lambda_1 + 2\lambda_3 - \lambda_4))$ is isomorphic to $H^1(P, Y_{1,4}(\lambda_1 - \lambda_4 + \lambda_8) \otimes Y_{1,4}(-2\lambda_1 + 2\lambda_3 - \lambda_4))$ and so isomorphic to $H^1(P, Y_{1,4}(-\lambda_1 + 2\lambda_3 - 2\lambda_4) \otimes Y_4(\lambda_8))$, which is 0. Thus $Ext_P^1(J/J_4, Y_{1,4}(-2\lambda_1 + 2\lambda_3 - \lambda_4)) = 0$ and J_4/J_5 is a non-split extension of $Y_{1,4}(-2\lambda_1 + 2\lambda_3 - \lambda_4)$ by $Y_{1,4}(-\lambda_1 + \lambda_2 + \lambda_3 - \lambda_4)$. It follows that there is a short exact sequence

$$0 \to A \to Y_4(\lambda_2 + \lambda_3 - \lambda_4) \otimes -\lambda_1 \to J_4/J_5 \to 0 \tag{1}$$

where A is an extension of $Y_{1,4}(-3\lambda_1 + \lambda_3)$ by $Y_{1,4}(-2\lambda_1 + \lambda_2)$. Since the B-socle of $Y_4(\lambda_2 + \lambda_3 - \lambda_4)$, and therefore A, is simple, A is a non-split extension of $Y_{1,4}(-3\lambda_1 + \lambda_3)$ by $Y_{1,4}(-2\lambda_1 + \lambda_2)$. It is easy to check that $Ext_P^1(Y_{1,4}(-2\lambda_1 + \lambda_2), Y_{1,4}(-3\lambda_1 + \lambda_3))$ is isomorphic to k and it follows that A is isomorphic to $Y_4(\lambda_2) \otimes -2\lambda_1$.

We now wish to show that J_5/J_6 is a non-split extension of $Y_{1,4}(-2\lambda_1 + \lambda_2)$ by $Y_{1,4}(0)$. As usual the alternative is that $Ext_P^1(J/J_5, Y_{1,4}(-2\lambda_1 + \lambda_2)) \neq 0$. Moreover familiar arguments reveal that $Ext^1_P(J_r/J_{r+1},Y_{1,4}(-2\lambda_1 + \lambda_2)) = 0$ for $r = 0,1,2,3$ so that $Ext^1_P(J/J_4,Y_{1,4}(-2\lambda_1 + \lambda_2)) = 0$. Also, (1) reveals that $Ext^1_P(J_4/J_5,Y_{1,4}(-2\lambda_1 + \lambda_2))$ is isomorphic to $Ext^2_P(A,Y_{1,4}(-2\lambda_1 + \lambda_2))$. Since A is isomorphic to $Y_4(\lambda_2) \otimes -2\lambda_1$, the Ext^2 cohomology group is isomorphic to $H^2(P,Y_4(\lambda_2 - \lambda_4) \otimes Y_{1,4}(\lambda_2))$. By (1.1.10) this is isomorphic to $H^2(P_4,Y_4(\lambda_2 - \lambda_4) \otimes Y_4(\lambda_2))$. However, $Y_4(\lambda_2 - \lambda_4) \otimes Y_4(\lambda_2)$ has a filtration with quotients $Y_4(2\lambda_2 - \lambda_4)$, $Y_4(\lambda_1 + \lambda_3 - \lambda_4)$, $Y_4(0)$ and $H^2(P_4,Y_4(\tau)) = 0$ for $\tau = 2\lambda_2 - \lambda_4,\lambda_1 + \lambda_3 - \lambda_4,0$ so that $H^2(P_4,Y_4(\lambda_2 - \lambda_4) \otimes Y_4(\lambda_2))$ and hence $Ext^1_P(J_4/J_5,Y_{1,4}(-2\lambda_1 + \lambda_2))$ is 0 . Hence $Ext^1_P(J/J_5,Y_{1,4}(-2\lambda_1 + \lambda_2)) = 0$ and J_5/J_6 is a non-split extension of $Y_{1,4}(-2\lambda_1 + \lambda_2)$ by $Y_{1,4}(0)$ and therefore isomorphic to $Y_4(\lambda_1) \otimes -\lambda_1$.

We now move on to J_6/J_7 . If this is a split extension of $Y_{1,4}(-2\lambda_1 + \lambda_2 + \lambda_3 - 2\lambda_4 + \lambda_5)$ by $Y_{1,4}(\lambda_3 - 2\lambda_4 + \lambda_5)$ we obtain, as usual, $Ext^1_P(J/J_6,Y_{1,4}(-2\lambda_1 + \lambda_2 + \lambda_3 - 2\lambda_4 + \lambda_5)) \neq 0$. Familiar arguments reveal that $Ext^1_P(J_r/J_{r+1},Y_{1,4}(-2\lambda_1 + \lambda_2 + \lambda_3 - 2\lambda_4 + \lambda_5)) = 0$ for $r = 0,1,2,3,5$. Also, (1) reveals that $Ext^1_P(J_4/J_5,Y_{1,4}(-2\lambda_1 + \lambda_2 + \lambda_3 - 2\lambda_4 + \lambda_5))$ is isomorphic to $Ext^2_P(A,Y_{1,4}(-2\lambda_1 + \lambda_2 + \lambda_3 - 2\lambda_4 + \lambda_5))$ and hence to $H^2(P,Y_4(\lambda_2 - \lambda_4) \otimes Y_{1,4}(\lambda_2 + \lambda_3 - 2\lambda_4 + \lambda_5))$. By (1.1.10) this is isomorphic to $H^2(P_4,Y_4(\lambda_2 - \lambda_4) \otimes Y_4(\lambda_2 + \lambda_3 - 2\lambda_4 + \lambda_5))$. However, $Y_4(\lambda_2 - \lambda_4) \otimes Y_4(\lambda_2 + \lambda_3 - 2\lambda_4 + \lambda_5)$ has a filtration with quotients $Y_4(2\lambda_2 + \lambda_3 - 3\lambda_4 + \lambda_5)$, $Y_4(\lambda_1 - 2\lambda_3 - 3\lambda_4 + \lambda_5)$, $Y_4(\lambda_1 + \lambda_2 - 2\lambda_4 + \lambda_5)$, $Y_4(\lambda_3 - 2\lambda_4 + \lambda_5)$ and $H^2(P_4,Y_4(\tau) = 0$ for $\tau = 2\lambda_2 + \lambda_3 - 3\lambda_4 + \lambda_5,\lambda_1 + 2\lambda_3 - 3\lambda_4 + \lambda_5,\lambda_1 + \lambda_2 - 2\lambda 4 +\lambda_5,$ $\lambda_3 - 2\lambda_4 + \lambda_5$. Hence $H^2(P_4,Y_4(\lambda_2 - \lambda_4) \otimes Y_4(\lambda_2 + \lambda_3 - 2\lambda_4 + \lambda_5)) = 0$ and so $Ext^1_P(J_4/J_5,Y_{1,4}(-2\lambda_1 + \lambda_2 + \lambda_3 - 2\lambda_4 + \lambda_5)) = 0$. Therefore $Ext^1_P(J/J_6,Y_{1,4}(-2\lambda_1 + \lambda_2 + \lambda_3 - 2\lambda_4 + \lambda_5)) = 0$ and J_6/J_7 is a non-split extension of $Y_{1,4}(-2\lambda_1 + \lambda_2 + \lambda_3 - 2\lambda_4 + \lambda_5)$ by

$Y_{1,4}(\lambda_3 - 2\lambda_4 + \lambda_5)$. There is only one non-split extension of $Y_{1,4}(-2\lambda_1 + \lambda_2 + \lambda_3 - 2\lambda_4 + \lambda_5)$ by $Y_{1,4}(\lambda_3 - 2\lambda_4 + \lambda_5)$ and it follows that there is a short exact sequence

$$0 \to A' \to Y_4(\lambda_1 + \lambda_3 - 2\lambda_4 + \lambda_5) \otimes -\lambda_1 \to J_6/J_7 \to 0 \qquad (2)$$

where A' is an extension of $Y_{1,4}(-3\lambda_1 + \lambda_2 - \lambda_4 + \lambda_5)$ by $Y_{1,4}(-\lambda_1 - \lambda_4 + \lambda_5)$. Since the B socle of $Y_4(\lambda_1 + \lambda_3 - 2\lambda_4 + \lambda_5)$ is simple A' is a non-split extension and hence A' is isomorphic to $Y_4(\lambda_1 - \lambda_4 + \lambda_5) \otimes -2\lambda_1$.

As usual, if the extension J_7/J_8 of $Y_{1,4}(-2\lambda_1 + 2\lambda_2 - \lambda_4 + \lambda_7)$ by $Y_{1,4}(\lambda_2 - \lambda_4 + \lambda_7)$ is split then $Ext^1_P(J/J_7, Y_{1,4}(-2\lambda_1 + 2\lambda_2 - \lambda_4 + \lambda_7)) \neq 0$. Familiar arguments give that $Ext^1_P(J_r/J_{r+1}, Y_{1,4}(-2\lambda_1 + 2\lambda_2 - \lambda_4 + \lambda_7)) = 0$ for $r = 0,1,2,3,5$. Also, (1) reveals that $Ext^1_P(J_4/J_5, Y_{1,4}(-2\lambda_1 + 2\lambda_2 - \lambda_4 + \lambda_7))$ is isomorphic to $H^2(P, Y_4(\lambda_2 - \lambda_4) \otimes Y_{1,4}(2\lambda_2 - \lambda_4 + \lambda_7))$ which is isomorphic to $H^2(P, Y_4(\lambda_2 - \lambda_4) \otimes Y_4(2\lambda_2 - \lambda_4 + \lambda_7))$. However, $Y_4(\lambda_2 - \lambda_4) \otimes Y_4(2\lambda_2 - \lambda_4 + \lambda_7)$ has a filtration with quotients $Y_4(3\lambda_2 - 2\lambda_4 + \lambda_7)$, $Y_4(\lambda_1 + \lambda_2 + \lambda_3 - 2\lambda_4 + \lambda_7)$, $Y_4(\lambda_2 - \lambda_4 + \lambda_7)$ and $H^2(P_4, Y_4(\tau)) = 0$ for $\tau = 3\lambda_2 - 2\lambda_4 + \lambda_7, \lambda_1 + \lambda_2 + \lambda_3 - 2\lambda_4 + \lambda_7, \lambda_2 - \lambda_4 + \lambda_7$. Hence $H^2(P, Y_4(\lambda_2 - \lambda_4) \otimes Y_4(2\lambda_2 - \lambda_4 + \lambda_7)) = 0$ and so $Ext^1_P(J_4/J_5, Y_{1,4}(-2\lambda_1 + 2\lambda_2 - \lambda_4 + \lambda_7)) = 0$. From (2) we obtain that $Ext^L_P(J_6/J_7, Y_{1,4}(-2\lambda_1 + 2\lambda_2 - \lambda_4 + \lambda_7))$ is isomorphic to $Ext^2_P(A', Y_{1,4}(-2\lambda_1 + 2\lambda_2 - \lambda_4 + \lambda_7)) =$ $Ext^2_P(Y_4(\lambda_1 - \lambda_4 + \lambda_5) \otimes -2\lambda_1, Y_{1,4}(-2\lambda_1 + 2\lambda_2 - \lambda_4 + \lambda_7))$ and so isomorphic to $H^2(P, Y_4(\lambda_3 - 2\lambda_4 + \lambda_6) \otimes Y_{1,4}(2\lambda_2 - \lambda_4 + \lambda_7))$. This last cohomology group is isomorphic to $H^2(P_4, Y_4(\lambda_3 - 2\lambda_4 + \lambda_6) \otimes Y_4(2\lambda_2 - \lambda_4 + \lambda_7))$. However $Y_4(\lambda_3 - 2\lambda_4 + \lambda_6) \otimes Y_4(2\lambda_2 - \lambda_4 + \lambda_7)$ has a filtration with quotients $Y_4(2\lambda_2 + \lambda_3 - 3\lambda_4 + \lambda_6 + \lambda_7)$, $Y_4(2\lambda_2 + \lambda_3 - 3\lambda_4 + \lambda_5)$, $Y_4(\lambda_1 + \lambda_2 - 2\lambda_4 + \lambda_6 + \lambda_7)$, $Y_4(\lambda_1 + \lambda_2 - 2\lambda_4 + \lambda_5)$ and

$H^2(P_4, Y_4(\tau)) = 0$ for $\tau = 2\lambda_2 + \lambda_3 - 3\lambda_4 + \lambda_6 + \lambda_7, 2\lambda_2 + \lambda_3 - 3\lambda_4 + \lambda_5,$ $\lambda_1 + \lambda_2 - 2\lambda_4 + \lambda_6 + \lambda_7, \lambda_1 + \lambda_2 - 2\lambda_4 + \lambda_5$ so that $H^2(P_4, Y_4(\lambda_3 - 2\lambda_4 + \lambda_6) \otimes Y_4(2\lambda_2 - \lambda_4 + \lambda_7))$ is 0. Thus $Ext^1_P(J_6/J_7, Y_{1,4}(-2\lambda_1 + 2\lambda_2 - \lambda_4 + \lambda_7))$ is 0. We have now shown that $Ext^1_P(J_r/J_{r+1}, Y_{1,4}(-2\lambda_1 + 2\lambda_2 - \lambda_4 + \lambda_7)) = 0$ for $0 \le r \le 6$. Hence $Ext^1_P(J/J_7, Y_{1,4}(-2\lambda_1 + 2\lambda_2 - \lambda_4 + \lambda_7)) = 0$ and J_7/J_8 is a non-split extension of $Y_{1,4}(-2\lambda_1 + 2\lambda_2 - \lambda_4 + \lambda_7)$ by $Y_{1,4}(\lambda_2 - \lambda_4 + \lambda_7)$. We now have that, for $0 \le r \le 8$ $R^iG(J_r/J_{r+1})$ has a good filtration for $i = 0$ and is 0 for $i > 0$ by (5.18), (2.1.5) and (1.7.4). Hence $R^iG(J)$ has a good filtration if $i = 0$ and is 0 if $i > 0$. The proof of (10.6.1), in the case $j = 4$, is complete.

We now move on to the most complicated case $j = 5$ and set $P = P_{1,5}$ and $G = Ind_B^{P_2}$. There is a P submodule, say K, of $Y_1(-2\lambda_1 + \lambda_8)$ with a good filtration such that $Y_1(-2\lambda_1 + \lambda_8)/K$ has a filtration with quotients $Y_{1,5}(-2\lambda_1 + \lambda_8)$, $Y_{1,5}(-2\lambda_1 + \lambda_4 - \lambda_5 + \lambda_6)$, $Y_{1,5}(-2\lambda_1 + \lambda_3 - \lambda_5 + \lambda_7 + \lambda_8)$, $Y_{1,5}(-2\lambda_1 + \lambda_2 - \lambda_5 + \lambda_6 + \lambda_7)$, $_{1,5}(-2\lambda_1 + \lambda_3 + \lambda_4 - \lambda_5)$, $Y_{1,5}(-2\lambda_1 + \lambda_2 - \lambda_5 + 2\lambda_8)$, $_{1,5}(-2\lambda_1 + \lambda_2)$, $Y_{1,5}(-2\lambda_1 + \lambda_2 + \lambda_4 - 2\lambda_5 + \lambda_6 + \lambda_8)$, $_{1,5}(-2\lambda_1 + \lambda_2 + \lambda_3 - \lambda_5 + \lambda_7)$, $Y_{1,5}(-2\lambda_1 + 2\lambda_2 - \lambda_5 + \lambda_8)$. Moreover $^iG(K)$ has a good filtration for $i = 0$ and is 0 for $i > 0$. Thus, o prove (10.1.6) in the case $j = 5$, it suffices to show that $R^iG(J)$ as a good filtration for $i = 0$ and is 0 for $i > 0$, where $= \bar{E}/K$. The module J has a descending filtration $= J_0, J_1, \ldots, J_{22} = 0$ with quotients, in order, $Y_{1,5}(-\lambda_1 + \lambda_7)$, $_{1,5}(-\lambda_1 + \lambda_4 - \lambda_5 + \lambda_8)$, $Y_{1,5}(-\lambda_1 + \lambda_3 - \lambda_5 + \lambda_6)$, $_{1,5}(-\lambda_1 + \lambda_2 - \lambda_5 + \lambda_7 + \lambda_8)$, $Y_{1,5}(-\lambda_5 + \lambda_6 + \lambda_7)$, $_{1,5}(-\lambda_1 + \lambda_2 + \lambda_4 - \lambda_5)$, $Y_{1,5}(-\lambda_5 + 2\lambda_8)$, $Y_{1,5}(0)$, $_{1,5}(\lambda_4 - 2\lambda_5 + \lambda_6 + \lambda_8)$, $Y_{1,5}(\lambda_3 - \lambda_5 + \lambda_7)$, $Y_{1,5}(\lambda_2 - \lambda_5 + \lambda_8)$, $_{,5}(\lambda_1 - \lambda_5 + \lambda_6)$, $Y_{1,5}(-2\lambda_1 + \lambda_8)$, $Y_{1,5}(-2\lambda_1 + \lambda_4 - \lambda_5 + \lambda_6)$, $_{,5}(-2\lambda_1 + \lambda_3 - \lambda_5 + \lambda_7 + \lambda_8)$, $Y_{1,5}(-2\lambda_1 + \lambda_2 - \lambda_5 + \lambda_6 + \lambda_7)$, $_{,5}(-2\lambda_1 + \lambda_3 + \lambda_4 - \lambda_5)$, $Y_{1,5}(-2\lambda_1 + \lambda_2 - \lambda_5 + 2\lambda_8)$, $Y_{1,5}(-2\lambda_1 + \lambda_2)$,

$Y_{1,5}(-2\lambda_1 + \lambda_2 + \lambda_4 - 2\lambda_5 + \lambda_6 + \lambda_8)$, $Y_{1,5}(-2\lambda_1 + \lambda_2 + \lambda_3 - \lambda_5 + \lambda_7)$, $Y_{1,5}(-2\lambda_1 + 2\lambda_2 - \lambda_5 + \lambda_8)$ and J/J_{13} is a non-split extension of J_{12}/J_{13} by J/J_{12} . It is easy to check that $Ext^1_P(Y_{1,5}(-\lambda_1 + \lambda_7), Y_{1,5}(-2\lambda_1 + \lambda_8)) = 0$ and this, together with Lemma 3.2.1, shows that the filtration maybe rechosen so that J_o/J_1 is unchanged, J_1/J_2 is a non-split extension of $Y_{1,5}(-2\lambda_1 + \lambda_8)$ by $Y_{1,5}(-\lambda_1 + \lambda_4 - \lambda_5 + \lambda_8)$, J_2/J_3 is an extension of $Y_{1,5}(-2\lambda_1 + \lambda_4 - \lambda_5 + \lambda_6)$ by $Y_{1,5}(-\lambda_1 + \lambda_3 - \lambda_5 + \lambda_6)$, J_3/J_4 is an extension of $Y_{1,5}(-2\lambda_1 + \lambda_3 - \lambda_5 + \lambda_7 + \lambda_8)$ by $Y_{1,5}(-\lambda_1 + \lambda_2 - \lambda_5 + \lambda_7 + \lambda_8)$, J_4/J_5 is an extension of $Y_{1,5}(-2\lambda_1 + \lambda_2 - \lambda_5 + \lambda_6 + \lambda_7)$ by $Y_{1,5}(-\lambda_5 + \lambda_6 + \lambda_7)$, J_5/J_6 is an extension of $Y_{1,5}(-2\lambda_1 + \lambda_3 + \lambda_4 - \lambda_5)$ by $Y_{1,5}(-\lambda_1 + \lambda_2 + \lambda_4 - \lambda_5)$, J_6/J_7 is an extension of $Y_{1,5}(-2\lambda_1 + \lambda_2 - \lambda_5 + 2\lambda_8)$ by $Y_{1,5}(-\lambda_5 + 2\lambda_8)$, J_7/J_8 is an extension of $Y_{1,5}(-2\lambda_1 + \lambda_2)$ by $Y_{1,5}(0)$, J_8/J_9 is an extension of $Y_{1,5}(-2\lambda_1 + \lambda_2 + \lambda_4 - 2\lambda_5 + \lambda_6 + \lambda_8)$ by $Y_{1,5}(\lambda_4 - 2\lambda_5 + \lambda_6 + \lambda_8)$, J_9/J_{10} is an extension of $Y_{1,5}(-2\lambda_1 + \lambda_2 + \lambda_3 - \lambda_5 + \lambda_7)$ by $Y_{1,5}(\lambda_3 - \lambda_5 + \lambda_7)$, J_{10}/J_{11} is an extension of $Y_{1,5}(-2\lambda_1 + 2\lambda_2 - \lambda_5 + \lambda_8)$ by $Y_{1,5}(\lambda_2 - \lambda_5 + \lambda_8)$, J_{11} is isomorphic to $Y_{1,5}(\lambda_1 - \lambda_5 + \lambda_6)$ and $J_{12} = 0$.

There is only one non-split extension of $Y_{1,5}(-2\lambda_1 + \lambda_8)$ by $Y_{1,5}(-\lambda_1 + \lambda_4 - \lambda_5 + \lambda_8)$ so that J_1/J_2 is isomorphic to $Y_5(\lambda_4 - \lambda_5 + \lambda_8) \otimes -\lambda_1$. It is easy to show that $Ext^1_P(J/J_2, Y_{1,5}(-2\lambda_1 + \lambda_4 - \lambda_5 + \lambda_6)) = 0$ and since $Y_{1,5}(-2\lambda_1 + \lambda_4 - \lambda_5 + \lambda_6)$ is not an image of $\bar{E}$ this implies that J_2/J_3 is a non-split extension of $Y_{1,5}(-2\lambda_1 + \lambda_4 - \lambda_5 + \lambda_6)$ by $Y_{1,5}(-\lambda_1 + \lambda_3 - \lambda_5 + \lambda_6)$ and therefore isomorphic to $Y_5(\lambda_3 - \lambda_5 + \lambda_6) \otimes -\lambda_1$. Similar observations reveal that J_3/J_4 is isomorphic to $Y_5(\lambda_2 - \lambda_5 + \lambda_7 + \lambda_8) \otimes -\lambda_1$, J_4/J_5 is isomorphic to $Y_5(\lambda_1 - \lambda_5 + \lambda_6 + \lambda_7) \otimes -\lambda_1$, J_5/J_6 is isomorphic to $Y_5(\lambda_2 + \lambda_4 - \lambda_5) \otimes -\lambda_1$, J_6/J_7 is isomorphic to

$Y_5(\lambda_1 - \lambda_5 + 2\lambda_8) \otimes -\lambda_1$, J_7/J_8 is isomorphic to $Y_5(\lambda_1) \otimes -\lambda_1$ and the extension J_8/J_9 , of $Y_{1,5}(-2\lambda_1 + \lambda_2 + \lambda_4 - 2\lambda_5 + \lambda_6 + \lambda_8)$ by $Y_{1,5}(\lambda_4 - 2\lambda_5 + \lambda_6 + \lambda_8)$, is non-split. It follows that there is a short exact sequence

$$0 \to Y_5(\lambda_1 - \lambda_5 + \lambda_6 + \lambda_8) \otimes -2\lambda_1 \to Y_5(\lambda_1 + \lambda_4 - 2\lambda_5 + \lambda_6 + \lambda_8) \otimes -\lambda_1$$
$$\to J_8/J_9 \to 0 \qquad (3).$$

From (3) we obtain $Ext_P^1(J_8/J_9, Y_{1,5}(-2\lambda_1 + \lambda_2 + \lambda_3 - \lambda_5 + \lambda_7))$ is isomorphic to $Ext_P^2(Y_5(\lambda_1 - \lambda_5 + \lambda_6 + \lambda_8), Y_{1,5}(\lambda_2 + \lambda_3 - \lambda_5 + \lambda_7))$ and hence to $Ext_{P_5}^2(Y_5(\lambda_1 - \lambda_5 + \lambda_6 + \lambda_8), Y_5(\lambda_2 + \lambda_3 - \lambda_5 + \lambda_7))$ which is 0 by Lemma 3.2.1. It is not difficult to show that $Ext_P^1(J/J_8, Y_{1,5}(-2\lambda_1 + \lambda_2 + \lambda_3 - \lambda_5 + \lambda_7)) = 0$. It follows that $Ext_P^1(J/J_9, Y_{1,5}(-2\lambda_1 + \lambda_2 + \lambda_3 - \lambda_5 + \lambda_7)) = 0$ and J_8/J_{10} is a non-split extension of $Y_{1,5}(-2\lambda_1 + \lambda_2 + \lambda_3 - \lambda_5 + \lambda_7)$ by $Y_{1,5}(\lambda_3 - \lambda_5 + \lambda_7)$. We deduce that there is a short exact sequence of ?-modules

$$0 \to Y_5(\lambda_4 - \lambda_5 + \lambda_7) \otimes -2\lambda_1 \to Y_5(\lambda_1 + \lambda_3 - \lambda_5 + \lambda_7) \otimes -\lambda_1$$
$$\to J_9/J_{10} \to 0 \qquad (4).$$

From (3) we obtain that $Ext_P^1(J_8/J_9, Y_{1,5}(-2\lambda_1 + 2\lambda_2 - \lambda_5 + \lambda_8))$ s isomorphic to $Ext_P^2(Y_5(\lambda_1 - \lambda_5 + \lambda_6 + \lambda_8), Y_{1,5}(2\lambda_2 - \lambda_5 + \lambda_8))$ which s isomorphic to $H^2(P_5, Y_5(\lambda_4 - 2\lambda_5 + \lambda_7 + \lambda_8) \otimes Y_5(2\lambda_2 - \lambda_5 + \lambda_8))$. owever $Y_5(\lambda_4 - 2\lambda_5 + \lambda_7 + \lambda_8) \otimes Y_5(2\lambda_2 - \lambda_5 + \lambda_8)$ has a filtration ith quotients $Y_5(2\lambda_2 + \lambda_4 - 3\lambda_5 + \lambda_7 + \lambda_8)$, $Y_5(2\lambda_2 + \lambda_4 - 2\lambda_5 + \lambda_7)$, $_5(\lambda_1 + \lambda_2 - 2\lambda_5 + \lambda_7 + 2\lambda_8)$, $Y_5(\lambda_1 + \lambda_2 - \lambda_5 + \lambda_7)$, and $^2(P_5, Y_5(\tau)) = 0$ for $\tau = 2\lambda_2 + \lambda_4 - 3\lambda_5 + \lambda_7 + \lambda_8$, $\lambda_2 + \lambda_4 - 2\lambda_5 + \lambda_7$, $\lambda_1 + \lambda_2 - 2\lambda_5 + \lambda_7 + 2\lambda_8$, $\lambda_1 + \lambda_2 - \lambda_5 + \lambda_7$. ence $H^2(P_5, Y_5(\lambda_4 - 2\lambda_5 + \lambda_7 + \lambda_8) \otimes Y_5(2\lambda_2 - \lambda_5 + \lambda_8))$ is 0 and herefore $Ext_P^1(J_8/J_9, Y_{1,5}(-2\lambda_1 + 2\lambda_2 - \lambda_5 + \lambda_8)) = 0$. From (4) we btain that $Ext_P^1(J_9/J_{10}, Y_{1,5}(-2\lambda_1 + 2\lambda_2 - \lambda_5 + \lambda_8))$ is isomorphic to

$Ext^2_P(Y_5(\lambda_4 - \lambda_5 + \lambda_7), Y_{1,5}(2\lambda_2 - \lambda_5 + \lambda_8))$ and hence isomorphic to $Ext^2_{P_5}(Y_5(\lambda_4 - \lambda_5 + \lambda_7), Y_5(2\lambda_2 - \lambda_5 + \lambda_8))$. However this is 0 by Lemma 3.2.1 and so $Ext^1_P(J_9/J_{10}, Y_{1,5}(-2\lambda_1 + 2\lambda_2 - \lambda_5 + \lambda_8))$. It is easy to see that $Ext^1_P(J_r/J_{r+1}, Y_{1,5}(-2\lambda_1 + 2\lambda_2 - \lambda_5 + \lambda_8)) = 0$ for $0 \le r \le 7$ and so $Ext^1_P(J/J_{10}, Y_{1,5}(-2\lambda_1 + 2\lambda_2 - \lambda_5 + \lambda_8)) = 0$. Since $Y_{1,5}(-2\lambda_1 + 2\lambda_2 - \lambda_5 + \lambda_8)$ is not an image of $\bar{E}$, J_{10}/J_{11} must be a non-split extension of $Y_{1,5}(-2\lambda_1 + 2\lambda_2 - \lambda_5 + \lambda_8)$ by $Y_{1,5}(\lambda_2 - \lambda_5 + \lambda_8)$.

We now have that $R^iG(J_r/J_{r+1}) = 0$ for all i and $0 \le r \le 7$ by (2.1.5) and the tensor identity, $R^iG(J_r/J_{r+1}) = 0$ for all i and $8 \le r \le 10$ by (5.18) and $R^iG(J_{11})$ is $Y_5(\lambda_1 - \lambda_5 + \lambda_6)$ for $i = 0$ and 0 for $i > 0$ by the vanishing theorem. Hence $R^iG(J)$ has a good filtration if $i = 0$ and is 0 if $i > 0$. This completes the proof of (10.6.1) in the case $j = 5$.

We now consider the case $j = 6$ and set, $P = P_{1,6}$ and $G = Ind_B^{P_6}$. There is a P submodule, say K, of $Y_1(-2\lambda_1 + \lambda_8)$ with a good filtration such that $Y_1(-2\lambda_1 + \lambda_8)/K$ has a filtration with quotients $Y_{1,6}(-2\lambda_1 + \lambda_8)$, $Y_{1,6}(-2\lambda_1 + \lambda_4 - \lambda_6 + \lambda_7)$, $Y_{1,6}(-2\lambda_1 + \lambda_2 - \lambda_6 + 2\lambda_7)$, $Y_{1,6}(-2\lambda_1 + \lambda_2)$, $Y_{1,6}(-2\lambda_1 + \lambda_2 + \lambda_5 - 2\lambda_6 + \lambda_7)$, $Y_{1,6}(-2\lambda_1 + \lambda_2 + \lambda_3 - \lambda_6)$. Moreover $R^iG(K)$ has a good filtration for $i = 0$ and is 0 for $i > 0$. Thus to prove (10.1.6) in the case $j = 6$, it suffices to show that $R^iG(J)$ has a good filtration for $i = 0$ and is 0 for $i > 0$, where $J = \bar{E}/K$. The module J has a descending filtration $J = J_0, J_1, \ldots, J_{15} = 0$ with quotients, in order, $Y_{1,6}(-\lambda_1 + \lambda_7)$, $Y_{1,6}(-\lambda_1 + \lambda_5 - \lambda_6)$, $Y_{1,6}(-\lambda_1 + \lambda_3 - \lambda_6 + \lambda_7)$, $Y_{1,6}(-\lambda_1 + \lambda_2 - \lambda_6 + \lambda_8)$, $Y_{1,6}(-\lambda_6 + 2\lambda_7)$, $Y_{1,6}(0)$, $Y_{1,6}(\lambda_5 - 2\lambda_6 + \lambda_7)$, $Y_{1,6}(\lambda_3 - \lambda_6)$, $Y_{1,6}(\lambda_1 - \lambda_6 + \lambda_7)$, $Y_{1,6}(-2\lambda_1 + \lambda_8)$, $Y_{1,6}(-2\lambda_1 + \lambda_4 - \lambda_6 + \lambda_7)$, $Y_{1,6}(-2\lambda_1 + \lambda_2 - \lambda_6 + 2\lambda_7)$, $Y_{1,6}(-2\lambda_1 + \lambda_2)$, $Y_{1,6}(-2\lambda_1 + \lambda_2 + \lambda_5 - 2\lambda_6 + \lambda_7)$, $Y_{1,6}(-2\lambda_1 + \lambda_2 + \lambda_3 - \lambda_6)$ and J/J_{10}

is a non-split extension of J_9/J_{10} by J/J_9 . It is easy to check that $Ext^1_P(Y_{1,6}(-\lambda_1 + \lambda_7), Y_{1,6}(-2\lambda_1 + \lambda_8)) = 0$ and this, together with Lemma 3.2.1 implies that the filtration may be rechosen so that J_o/J_1 is unchanged, J_1/J_2 is a non-split extension of $Y_{1,6}(-2\lambda_1 + \lambda_8)$ by $Y_{1,6}(-\lambda_1 + \lambda_5 - \lambda_6)$, J_2/J_3 is an extension of $Y_{1,6}(-2\lambda_1 + \lambda_4 - \lambda_6 + \lambda_7)$ by $Y_{1,6}(-\lambda_1 + \lambda_3 - \lambda_6 + \lambda_7)$, J_3/J_4 is isomorphic to $Y_{1,6}(-\lambda_1 + \lambda_2 - \lambda_6 + \lambda_8)$, J_4/J_5 is an extension of $Y_{1,6}(-2\lambda_1 + \lambda_2 - \lambda_6 + 2\lambda_7)$ by $Y_{1,6}(-\lambda_6 + 2\lambda_7)$, J_5/J_6 is an extension of $Y_{1,6}(-2\lambda_1 + \lambda_2)$ by $Y_{1,6}(0)$, J_6/J_7 is an extension of $Y_{1,6}(-2\lambda_1 + \lambda_2 + \lambda_5 - 2\lambda_6 + \lambda_7)$ by $Y_{1,6}(\lambda_5 - 2\lambda_6 + \lambda_7)$, J_7/J_8 is an extension of $Y_{1,6}(-2\lambda_1 + \lambda_2 + \lambda_3 - \lambda_6)$ by $Y_{1,6}(\lambda_3 - \lambda_6)$, J_8 is isomorphic to $Y_{1,6}(\lambda_1 - \lambda_6 + \lambda_7)$ and $J_9 = 0$. There is only one non-split extension of $Y_{1,6}(-2\lambda_1 + \lambda_8)$ by $Y_{1,6}(-\lambda_1 + \lambda_5 - \lambda_6)$ so that J_1/J_2 is isomorphic to $Y_6(\lambda_5 - \lambda_6) \otimes -\lambda_1$. It is easy to show that $Ext^1_P(J/J_2, Y_{1,6}(-2\lambda_1 + \lambda_4 - \lambda_6 + \lambda_7)) = 0$ and since $Y_{1,6}(-2\lambda_1 + \lambda_4 - \lambda_6 + \lambda_7)$ is not an image of $\bar{E}$ this implies that J_2/J_3 is a non-split extension of $Y_{1,6}(-2\lambda_1 + \lambda_4 - \lambda_6 + \lambda_7)$ by $Y_{1,6}(-\lambda_1 + \lambda_3 - \lambda_6 + \lambda_7)$ and therefore isomorphic to $Y_6(\lambda_3 - \lambda_6 + \lambda_7) \otimes -\lambda_1$. Similar observations reveal that J_4/J_5 is isomorphic to $Y_6(\lambda_1 - \lambda_6 + 2\lambda_7) \otimes -\lambda_1$, J_5/J_6 is isomorphic to $Y_6(\lambda_1) \otimes -\lambda_1$ and J_6/J_7 is a non-split extension of $Y_{1,6}(-2\lambda_1 + \lambda_2 + \lambda_5 - 2\lambda_6 + \lambda_7)$ by $Y_{1,6}(\lambda_5 - 2\lambda_6 + \lambda_7)$. It follows that there is a short exact sequence of P-modules

$$0 \to A_1 \to Y_6(\lambda_1 + \lambda_5 - 2\lambda_6 + \lambda_7) \otimes -\lambda_1 \to J_6/J_7 \to 0 \qquad (5)$$

where A_1 is an extension of $Y_{1,6}(-3\lambda_1 + \lambda_2 - \lambda_6 + \lambda_7 + \lambda_8)$ by $Y_{1,6}(-\lambda_1 - \lambda_6 + \lambda_7 + \lambda_8)$.

It is not difficult to show that $Ext^1_P(J_r/J_{r+1}, Y_{1,6}(-2\lambda_1 + \lambda_2 + \lambda_3 - \lambda_6)) = 0$ for $0 \leq r \leq 5$. Moreover, by (5), $Ext^1_P(J_6/J_7, Y_{1,6}(-2\lambda_1 + \lambda_2 + \lambda_3 - \lambda_6)$ is isomorphic to $Ext^2_P(A_1, Y_{1,6}(-2\lambda_1 + \lambda_2 + \lambda_3 - \lambda_6))$ which is 0 by Lemma 3.2.1. Thus

$Ext^1_P(J/J_7, Y_{1,6}(-2\lambda_1 + \lambda_2 + \lambda_3 - \lambda_6)) = 0$ and since $Y_{1,6}(-2\lambda_1 + \lambda_2 + \lambda_3 - \lambda_6)$ is not an image of $\bar{E}$ we obtain that J_7/J_8 is a non-split extension of $Y_{1,6}(-2\lambda_1 + \lambda_2 + \lambda_3 - \lambda_6)$ by $Y_{1,6}(\lambda_3 - \lambda_6)$. We now have $R^iG(J_r/J_{r+1}) = 0$ for all i and $0 \le r \le 7$ by (2.1.5), the tensor identity and (5.18). Also $R^iG(J_8)$ is isomorphic to $Y_6(\lambda_1 - \lambda_6 + \lambda_7)$ for $i = 0$ and is 0 for $i > 0$. Thus $R^iG(J)$ has a good filtration for $i = 0$ and is 0 for $i > 0$. This completes the proof of (10.6.1) in the case $j = 6$.

We now consider the case $j = 7$ and set $P = P_{1,7}$ and $G = Ind_B^{P_7}$. There is a P submodule, say K, of $Y_1(-2\lambda_1 + \lambda_8)$ with a good filtration such that $Y_1(-2\lambda_1 + \lambda_8)/K$ has a filtration with quotients $Y_{1,7}(-2\lambda_1 + \lambda_8)$, $Y_{1,7}(-2\lambda_1 + \lambda_4 - \lambda_7)$, $Y_{1,7}(-2\lambda_1 + \lambda_2)$, $Y_{1,7}(-2\lambda_1 + \lambda_2 + \lambda_6 - 2\lambda_7)$. Moreover $R^iG(K)$ has a good filtration for $i = 0$ and is 0 for $i > 0$. Thus to prove (10.6.1) in the case $j = 7$, it suffices to show that $R^iG(J)$ has a good filtration for $i = 0$ and is 0 for $i > 0$, where $J = \bar{E}/K$. The module J has a descending filtration $J = J_o, J_1, \ldots, J_{10} = 0$ with quotients, in order, $Y_{1,7}(-\lambda_1 + \lambda_7)$, $Y_{1,7}(-\lambda_1 + \lambda_6 - \lambda_7)$, $J_{1,7}(-\lambda_1 + \lambda_3 - \lambda_7)$, $Y_{1,7}(0)$, $Y_{1,7}(\lambda_6 - 2\lambda_7)$, $Y_{1,7}(\lambda_1 - \lambda_7)$, $Y_{1,7}(-2\lambda_1 + \lambda_8)$, $Y_{1,7}(-2\lambda_1 + \lambda_4 - \lambda_7)$, $Y_{1,7}(-2\lambda_1 + \lambda_2)$, $Y_{1,7}(-2\lambda_1 + \lambda_2 + \lambda_6 - 2\lambda_7)$ and J/J_7 is a non-split extension of J_6/J_7 by J/J_6. It is easy to check that $Ext^1_P(Y_{1,7}(-\lambda_1 + \lambda_7), Y_{1,7}(-2\lambda_1 + \lambda_8)) = 0$ and this, together with Lemma 3.2.1 implies that the filtration may be rechosen so that J_o/J_1 is unchanged, J_1/J_2 is a non-split extension of $Y_{1,7}(-2\lambda_1 + \lambda_8)$ by $J_{1,7}(-\lambda_1 + \lambda_6 - \lambda_7)$, J_2/J_3 is an extension of $Y_{1,7}(-2\lambda_1 + \lambda_4 - \lambda_7)$ by $Y_{1,7}(-\lambda_1 + \lambda_3 - \lambda_7)$, J_3/J_4 is an extension of $Y_{1,7}(-2\lambda_1 + \lambda_2)$ by $Y_{1,7}(0)$, J_4/J_5 is an extension of $Y_{1,7}(-2\lambda_1 + \lambda_2 + \lambda_6 - 2\lambda_7)$ by $Y_{1,7}(\lambda_6 - 2\lambda_7)$, J_5 is $Y_{1,7}(\lambda_1 - \lambda_7)$ and $J_6 = 0$. There is only one non-split extension of $Y_{1,7}(-2\lambda_1 + \lambda_8)$ by $Y_{1,7}(-\lambda_1 + \lambda_6 - \lambda_7)$ so that J_1/J_2 is isomorphic to $Y_7(\lambda_6 - \lambda_7) \otimes -\lambda_1$. It is easy to show that $Ext^1_P(J/J_2, Y_{1,7}(-2\lambda_1 + \lambda_4 - \lambda_7)) = 0$ and since

$Y_{1,7}(-2\lambda_1 + \lambda_4 - \lambda_7)$ is not an image of $\bar{E}$ this implies that J_2/J_3 is a non-split extension of $Y_{1,7}(-2\lambda_1 + \lambda_4 - \lambda_7)$ by $Y_{1,7}(-\lambda_1 + \lambda_3 - \lambda_7)$. It follows that there is a short exact sequence

$$0 \to A_2 \to Y_7(\lambda_3 - \lambda_7) \otimes -\lambda_1 \to J_2/J_3 \to 0 \qquad (6)$$

where A_2 is an extension of $Y_{1,7}(-3\lambda_1 + \lambda_3)$ by $Y_{1,7}(-2\lambda_1 + \lambda_2)$. By (6) we have that $Ext^1_P(J_2/J_3, Y_{1,7}(-2\lambda_1 + \lambda_2))$ is isomorphic to $Ext^2_P(A_2, Y_{1,7}(-2\lambda_1 + \lambda_2))$ which is 0 since $Ext^2_P(Y_{1,7}(\tau), Y_{1,7}(-2\lambda_1 + \lambda_2)) = 0$ for $\tau = -3\lambda_1 + \lambda_3, -2\lambda_1 + \lambda_2$ by Lemma 3.2.1. It is easy to check that $Ext^1_P(J/J_2, Y_{1,7}(-2\lambda_1 + \lambda_2)) = 0$ so that $Ext^1_P(J/J_3, Y_{1,7}(-2\lambda_1 + \lambda_2)) = 0$ and, since $Y_{1,7}(-2\lambda_1 + \lambda_2)$ is not an image of $\bar{E}$, J_3/J_4 is a non-split extension of $Y_{1,7}(-2\lambda_1 + \lambda_2)$ by $Y_{1,7}(0)$. It follows that there is a short exact sequence

$$0 \to -2\lambda_1 + \lambda_7 \to Y_7(\lambda_1) \otimes -\lambda_1 \to J_3/J_4 \to 0 \qquad (7).$$

y (7) we have that $Ext^1_P(J_3/J_4, Y_{1,7}(-2\lambda_1 + \lambda_2 + \lambda_6 - 2\lambda_7))$ is isomor-hic to $Ext^2_P(-2\lambda_1 + \lambda_7, Y_{1,7}(-2\lambda_1 + \lambda_2 + \lambda_6 - 2\lambda_7))$ and so isomorphic o $H^2(P, Y_{1,7}(\lambda_2 + \lambda_6 - \lambda_7))$ which is 0 . We claim that $xt^1_P(J_2/J_3, Y_{1,7}(-2\lambda_1 + \lambda_2 + \lambda_6 - 2\lambda_7))$ is 0 . By (6) it suffices to how that $Ext^2_P(A_2, Y_{1,7}(-2\lambda_1 + \lambda_2 + \lambda_6 - 2\lambda_7))$ is 0 and so it is nough to show that $Ext^2_P(Y_{1,7}(\tau), Y_{1,7}(-2\lambda_1 + \lambda_2 + \lambda_6 - 2\lambda_7)$ is 0 or $\tau = -3\lambda_1 + \lambda_3, -2\lambda_1 + \lambda_2$. For $\tau = -3\lambda_1 + \lambda_3$ this is true by emma 3.2.1. Also, $Ext^2_P(Y_{1,y}(-2\lambda_1 + \lambda_2), Y_{1,7}(-2\lambda_1 + \lambda_2 + \lambda_6 - 2\lambda_7))$ s isomorphic to $H^2(B, \lambda_6 - 2\lambda_7)$, by (5.10), which is isomorphic to $^1(B,k)$ which is 0 . It is easy to check that $t^1_P(J/J_2, Y_{1,7}(-2\lambda_1 + \lambda_2 + \lambda_6 - 2\lambda_7)) = 0$ so we obtain $t^1_P(J/J_4, Y_{1,7}(-2\lambda_1 + \lambda_2 + \lambda_6 - 2\lambda_7)) = 0$ and it follows that J_4/J_5 a non-split extension of $Y_{1,7}(-2\lambda_1 + \lambda_2 + \lambda_6 - 2\lambda_7)$ by $_{,7}(\lambda_6 - 2\lambda_7)$. We now have that $R^i G(J_r/J_{r+1}) = 0$ for all i and $\le r \le 4$ by (2.1.5), the tensor identity and (5.18). Also

$R^iG(J_5)$ is isomorphic to $Y_7(\lambda_1 - \lambda_7)$ for $i = 0$ and 0 for $i > 0$. Thus $R^iG(J)$ has a good filtration for $i = 0$ and is 0 for $i > 0$. This completes the proof of (10.6.1) in the case $j = 7$.

We now turn our attention to the remaining case $j = 8$ and set $P = P_{1,8}$ and $G = Ind_B^{P_8}$. There is a P submodule, say K, of $Y_1(-2\lambda_1 + \lambda_8)$ with a good filtration such that $Y_1(-2\lambda_1 + \lambda_8)/K$ has a filtration with quotients $Y_{1,8}(-2\lambda_1 + \lambda_8)$, $Y_{1,8}(-2\lambda_1 + \lambda_5 - \lambda_8)$, $Y_{1,8}(-2\lambda_1 + \lambda_3 + \lambda_7 - \lambda_8)$, $Y_{1,8}(-2\lambda_1 + \lambda_2)$, $Y_{1,8}(-2\lambda_1 + \lambda_2 + \lambda_5 - 2\lambda_8)$, $Y_{1,8}(-2\lambda_1 + 2\lambda_2 - \lambda_8)$. Moreover $R^iG(K)$ has a good filtration for $i = 0$ and is 0 for $i > 0$. Thus to prove (10.6.1) in the case $j = 8$, it suffices to show that $R^iG(J)$ has a good filtration for $i = 0$ and is 0 for $i > 0$, where $J = \overline{E}/K$. The module J has a descending filtration $J = J_0, J_1, \ldots, J_{12} = 0$ with quotients, in order, $Y_{1,8}(-\lambda_1 + \lambda_7)$, $Y_{1,8}(-\lambda_1 + \lambda_4 - \lambda_8)$, $Y_{1,8}(-\lambda_1 + \lambda_2 + \lambda_7 - \lambda_8)$, $Y_{1,8}(0)$, $Y_{1,8}(\lambda_5 - 2\lambda_8)$, $Y_{1,8}(\lambda_2 - \lambda_8)$, $Y_{1,8}(-2\lambda_1 + \lambda_8)$, $Y_{1,8}(-2\lambda_1 + \lambda_5 - \lambda_8)$, $Y_{1,8}(-2\lambda_1 + \lambda_3 + \lambda_7 - \lambda_8)$, $Y_{1,8}(-2\lambda_1 + \lambda_2)$, $Y_{1,8}(-2\lambda_1 + \lambda_2 + \lambda_5 - 2\lambda_8)$, $Y_{1,8}(-2\lambda_1 + 2\lambda_2 - \lambda_8)$ and J/J_7 is a non-split extension of J_6/J_7 by J/J_6. Lemma 3.2.1 implies that the filtration maybe rechosen so that J/J_1 is a non-split extension of $Y_{1,8}(-2\lambda_1 + \lambda_8)$ by $Y_{1,8}(-\lambda_1 + \lambda_7)$, J_1/J_2 is an extension of $Y_{1,8}(-2\lambda_1 + \lambda_5 - \lambda_8)$ by $Y_{1,8}(-\lambda_1 + \lambda_4 - \lambda_8)$, J_2/J_3 is an extension of $Y_{1,8}(-2\lambda_1 + \lambda_3 + \lambda_7 - \lambda_8)$ by $Y_{1,8}(-\lambda_1 + \lambda_2 + \lambda_7 - \lambda_8)$, J_3/J_4 is an extension of $Y_{1,8}(-2\lambda_1 + \lambda_2)$ by $Y_{1,8}(0)$, J_4/J_5 is an extension of $Y_{1,8}(-2\lambda_1 + \lambda_2 + \lambda_5 - 2\lambda_8)$ by $Y_{1,8}(\lambda_5 - 2\lambda_8)$ and J_5/J_6 is an extension of $Y_{1,8}(-2\lambda_1 + 2\lambda_2 - \lambda_8)$ by $Y_{1,8}(\lambda_2 - \lambda_8)$. There is only one non-split extension of $Y_{1,8}(-2\lambda_1 + \lambda_8)$ by $Y_{1,8}(-\lambda_1 + \lambda_7)$ so that J/J_1 is isomorphic to $Y_8(\lambda_7) \otimes -\lambda_1$. Since $Ext_P^1(J/J_1, Y_{1,8}(-2\lambda_1 + \lambda_5 - \lambda_8)) = 0$ and $Y_{1,8}(-2\lambda_1 + \lambda_5 - \lambda_8)$ is not an image of $\overline{E}$, J_1/J_2 is a non-split extension of $Y_{1,8}(-2\lambda_1 + \lambda_5 - \lambda_8)$ by $Y_{1,8}(-\lambda_1 + \lambda_4 - \lambda_8)$ and therefore isomorphic to $Y_8(\lambda_4 - \lambda_8) \otimes -\lambda_1$. Similar considerations reveal that J_2/J_3 is a non-split extension of $Y_{1,8}(-2\lambda_1 + \lambda_3 + \lambda_7 - \lambda_8)$ by $Y_{1,8}(-\lambda_1 + \lambda_2 + \lambda_7 - \lambda_8)$ and it follows that there is a short exact sequence

$$0 \to Y_8(\lambda_2) \otimes -2\lambda_1 \to Y_8(\lambda_2 + \lambda_7 - \lambda_8) \otimes -\lambda_1 \to J_2/J_3 \to 0 \quad (8).$$

By (8) we have that $Ext^1_P(J_2/J_3, Y_{1,8}(-2\lambda_1 + \lambda_2))$ is isomorphic to $Ext^2_P(Y_8(\lambda_2), Y_{1,8}(\lambda_2))$ and hence to $Ext^2_P(Y_8(\lambda_2), Y_8(\lambda_2))$, which is 0. Also $Ext^1_P(J/J_2, Y_{1,8}(-2\lambda_1 + \lambda_2)) = 0$ so that $Ext^1_P(J/J_3, Y_{1,8}(-2\lambda_1 + \lambda_2)) = 0$. It follows that J_3/J_4 is a non-split extension of $Y_{1,8}(-2\lambda_1 + \lambda_2)$ by $Y_{1,8}(0)$ and therefore isomorphic to $Y_8(\lambda_1) \otimes -\lambda_1$.

By (8) we also have that $Ext^1_P(J_2/J_3, Y_{1,8}(-2\lambda_1 + \lambda_2 + \lambda_5 - 2\lambda_8))$ is isomorphic to $Ext^2_P(Y_8(\lambda_2), Y_{1,8}(\lambda_2 + \lambda_5 - 2\lambda_8))$. This is isomorphic to $Ext^1_P(Y_8(\lambda_2), Y_{1,8}(\lambda_2))$, which is 0. It is easy to check that $Ext^1_P(J_r/J_{r+1}, Y_{1,8}(-2\lambda_1 + \lambda_2 + \lambda_5 - 2\lambda_8)) = 0$ for $r = 0,1,3$ so that $Ext^1_P(J/J_4, Y_{1,8}(-2\lambda_1 + \lambda_2 + \lambda_5 - 2\lambda_8)) = 0$ and so J_4/J_5 is a non-split extension of $Y_{1,8}(-2\lambda_1 + \lambda_2 + \lambda_5 - 2\lambda_8)$ by $Y_{1,8}(\lambda_5 - 2\lambda_8)$. It follows that there is a short exact sequence

$$0 \to A_3 \to Y_8(\lambda_1 + \lambda_5 - 2\lambda_8) \otimes -\lambda_1 \to J_4/J_5 \to 0 \quad (9)$$

where A_3 is a non-split extension of $Y_{1,8}(-2\lambda_1 + \lambda_2 + \lambda_6 - \lambda_8)$ by $Y_{1,8}(\lambda_6 - \lambda_8)$.

By (8) we have that $Ext^1_P(J_2/J_3, Y_{1,8}(-2\lambda_1 + 2\lambda_2 - \lambda_8))$ is isomorphic to $Ext^2_P(Y_8(\lambda_2), Y_{1,8}(2\lambda_2 - \lambda_8))$ which is isomorphic to $Ext^2_{P_8}(Y_8(\lambda_2), Y_8(2\lambda_2 - \lambda_8))$. This last Ext group is isomorphic to $Ext^2_{P_\Sigma}(Y_\Sigma(\lambda_2), Y_\Sigma(2\lambda_2 - \lambda_8))$ for $\Sigma = \Lambda \backslash \{\alpha_1, \alpha_2\}$, and so isomorphic to $H^2(P_\Sigma, Y_\Sigma(\lambda_2 - \lambda_8))$, which is 0. By (9), $Ext^1_P(J_4/J_5, Y_{1,8}(-2\lambda_1 + 2\lambda_2 - \lambda_8))$ is isomorphic to $Ext^2_P(A_3, Y_{1,8}(-2\lambda_1 + 2\lambda_2 - \lambda_8))$. There is a unique non-split extension of $Y_{1,8}(-2\lambda_1 + \lambda_2 + \lambda_6 - \lambda_8)$ by $Y_{1,8}(\lambda_6 - \lambda_8)$ so there is a short exact sequence

$$0 \to A_4 \to Y_8(\lambda_1 + \lambda_6 - \lambda_8) \otimes -\lambda_1 \to A_3 \to 0 \qquad (10)$$

where A_4 is a non-split extension of $Y_{1,8}(-3\lambda_1 + \lambda_2 + \lambda_7)$ by $Y_{1,8}(-\lambda_1 + \lambda_7)$. From (10) we obtain that $Ext^2_P(A_3, Y_{1,8}(-2\lambda_1 + 2\lambda_2 - \lambda_8))$ is isomorphic to $Ext^3_P(A_4, Y_{1,8}(-2\lambda_1 + 2\lambda_2 - \lambda_8))$. There is a unique non-split extension of $Y_{1,8}(-3\lambda_1 + \lambda_2 + \lambda_7)$ by $Y_{1,8}(-\lambda_1 + \lambda_7)$ and so there is a short exact sequence

$$0 \to Y_8(\lambda_1 + \lambda_8) \otimes -3\lambda_1 \to Y_8(\lambda_1 + \lambda_7) \otimes -2\lambda_1 \to A_4 \to 0 \qquad (11).$$

It follows that $Ext^3_P(A_4, Y_{1,8}(-2\lambda_1 + 2\lambda_2 - \lambda_8))$ is isomorphic to $Ext^4_P(Y_8(\lambda_1 + \lambda_8) \otimes -3\lambda_1, Y_{1,8}(-2\lambda_1 + 2\lambda_2 - \lambda_8))$, which is isomorphic to $Ext^4_P(Y_8(\lambda_1 + \lambda_8), Y_{1,8}(\lambda_1 + 2\lambda_2 - \lambda_8))$ and therefore 0 . Thus $Ext^1_P(J_r/J_{r+1}, Y_{1,8}(-2\lambda_1 + 2\lambda_2 - \lambda_8)) = 0$ for $r = 2,4$. It is easy to check that $Ext^1_P(J_r/J_{r+1}, Y_{1,8}(-2\lambda_1 + 2\lambda_2 - \lambda_8)) = 0$ for $r = 0,1,3$ so that $Ext^1_P(J/J_5, Y_{1,8}(-2\lambda_1 + 2\lambda_2 - \lambda_8)) = 0$. It follows that J_5/J_6 is a non-split extension of $Y_{1,8}(-2\lambda_1 + 2\lambda_2 - \lambda_8)$ by $Y_{1,8}(\lambda_2 - \lambda_8)$. We now have $R^iG(J_r/J_{r+1})$ for all i and $r = 0,1$ by (2.1.5) and the tensor identity. Also, $R^iG(J_2/J_3) = 0$ for all i by (8) and $R^iG(J_r/J_{r+1}) = 0$ for $r > 2$ by (5.18). Hence $R^iG(J) = 0$ for all i and the proof of (10.6.1) is complete.

10.7 $R^i Ind_B^{P_\Sigma}(\overline{F})$

In this section we deal with $\overline{F} = F/E$ (in the notation of section 10.3). In order to analyse $R^i Ind_B^{P_\Sigma}(\overline{F})$ we shall need to know:

(10.7.1) *$\overline{F}$ is a non-split extension of* $Y_1(-2\lambda_1 + \lambda_8)$ *by* $Y_1(-\lambda_1 + \lambda_2)$.

By definition $\overline{F}$ is an extension of $Y_1(-2\lambda_1 + \lambda_8)$ by

$Y_1(-\lambda_1+\lambda_2)$. The lowest weight of $Y_1(-\lambda_1+\lambda_2)$ is $2\lambda_1-\lambda_2$ so that if $\vec{F}$ is the split extension $((Y/E)\otimes(-2\lambda_1+\lambda_2))^B \neq 0$, where $Y = Y(\lambda_7)$. Thus, by the long exact sequence of cohomology, it suffices to show that $H^1(B,E\otimes(-2\lambda_1+\lambda_2)) = 0$. We shall do this by showing that both $H^1(B,D\otimes(-2\lambda_1+\lambda_2))$ and $H^1(B,\bar{E}\otimes(-2\lambda_1+\lambda_2))$ are 0 .

It follows from (5.5), (2.1.5) and the tensor identity $R^i Ind_B^G(Y(\lambda_1)\otimes(-3\lambda_1+\lambda_2)) = 0$ for all i so that $H^i(B,Y(\lambda_1)\otimes(-3\lambda_1+\lambda_2)) = 0$ for all i by (1.1.8). Thus, by (10.3.3), to show that $H^1(B,D\otimes(-2\lambda_1+\lambda_2))$ is 0 it suffices to show that $H^2(B,M\otimes(-3\lambda_1+\lambda_2)) = 0$. Now M has a filtration with quotients $Y_1(0)$, $Y_1(-2\lambda_1+\lambda_2)$ and $Y_1(-\lambda_1)$. We have $H^2(B,-3\lambda_1+\lambda_2) = H^2(B,-4\lambda_1+\lambda_2) = 0$ by (5.4) and (2.1.6) so it suffices to show that $H^2(B,Y_1(-2\lambda_1+\lambda_2)\otimes(-3\lambda_1+\lambda_2)) = 0$. However $H^2(B,Y_1(-2\lambda_1+\lambda_2)\otimes(-3\lambda_1+\lambda_2))$ is isomorphic, by (2.1.3), to $H^2(B,Y_1(-2\lambda_1+\lambda_2)\otimes Y_1(-3\lambda_1+\lambda_2))$. Moreover $Y_1(-2\lambda_1+\lambda_2)\otimes Y_1(-3\lambda_1+\lambda_2)$ has a good filtration with quotients $Y_1(\tau)$ for $\tau\in V = \{-5\lambda_1+2\lambda_2, -4\lambda_1+\lambda_3, -3\lambda_1+\lambda_7, -2\lambda_1\}$ (see section 10.3). It follows from (2.1.3) and (5.4) that $H^2(B,Y_1(\tau)) = 0$ for $\tau\in V$ and so $H^2(B,Y_1(-2\lambda_1+\lambda_2)\otimes Y_1(-3\lambda_1+\lambda_2)) = 0$. Hence $H^1(B,D\otimes(-2\lambda_1+\lambda_2)) = 0$.

We now show that $H^1(B,\bar{E}\otimes(-2\lambda_1+\lambda_2)) = 0$. We have, by (5.10), that $Ext^1_{P_1}(Y_1(-\lambda_1+\lambda_7),Y_1(-2\lambda_1+\lambda_8))$ is isomorphic to $Ext^1_{P_{1,8}}(Y_{1,8}(-\lambda_1+\lambda_7),Y_{1,8}(-2\lambda_1+\lambda_8))$, which is isomorphic to $H^1(P_{1,8},Y_{1,8}(\lambda_2-\lambda_8)\otimes(-2\lambda_1+\lambda_8)) = H^1(P_{1,8},Y_{1,8}(-2\lambda_1+\lambda_2))$. This is one dimensional by (2.1.3) and (1.4.7). Thus, by (5.11), $\bar{E}$ is the unique non-split extension of $Y_1(-2\lambda_1+\lambda_8)$ by $Y_1(-\lambda_1+\lambda_7)$ and it follows from (5.13) that $\bar{E}$ is isomorphic to $G(Y_1(\lambda_7)\otimes -\lambda_1)$ where $G = Ind_{P_{1,8}}^{P_1}$. Thus $H^1(B,\bar{E}\otimes(-2\lambda_1+\lambda_2))$ is isomorphic to $H^1(B,G(Y_8(\lambda_7)\otimes -\lambda_1)\otimes(-2\lambda_1+\lambda_2))$ and, by (2.1.3), this is isomorphic to $H^1(P_1,G(Y_8(\lambda_2)\otimes -\lambda_1)\otimes Y_1(-2\lambda_1+\lambda_2))$. Now $Y_8(\lambda_7)\otimes -\lambda_1$ has a

$P_{1,8}$ filtration with quotients $Y_{1,8}(\lambda_7 - \lambda_1)$, $Y_{1,8}(-2\lambda_1 + \lambda_8)$ so that $R^iG(Y_8(\lambda_7) \otimes -\lambda_1) = 0$ for $i > 0$ by (1.7.4) and (2.1.2). It now follows from (1.1.8) that $H^1(P_1, G(Y_8(\lambda_7) \otimes -\lambda_1) \otimes Y_1(-2\lambda_1 + \lambda_2))$ and hence $H^1(B, \bar{E} \otimes (-2\lambda_1 + \lambda_2))$ is isomorphic to $H^1(P_{1,8}, Y_8(\lambda_7) \otimes Y_1(-3\lambda_1 + \lambda_2))$. It follows from (10.5.8) that $Y_1(-3\lambda_1 + \lambda_2)$ has a $P_{1,8}$ filtration with quotients $Y_{1,8}(-3\lambda_1 + \lambda_2)$, $Y_{1,8}(-2\lambda_1 + \lambda_6 - \lambda_8)$, $Y_{1,8}(-2\lambda_1 + \lambda_3 - \lambda_8)$, $Y_{1,8}(-\lambda_1 + \lambda_7 - \lambda_8)$. Thus to show that $H^1(B, \bar{E} \otimes (-2\lambda_1 + \lambda_2)) = 0$ it suffices to show that $H^1(P_{1,8}, Y_8(\lambda_7) \otimes Y_{1,8}(\tau)) = 0$ for $\tau \in V = \{-3\lambda_1 + \lambda_2, -2\lambda_1 + \lambda_6 - \lambda_8, -2\lambda_1 + \lambda_3 - \lambda_8, -\lambda_1 + \lambda_7 - \lambda_8\}$. Let $F = Ind_{P_{1,8}}^{P_8}$. It follows from (2.1.2), (5.5) and (2.1.5) that $R^iF(Y_{1,8}(\tau)) = 0$ for all i and all $\tau \in V$. Hence by (1.1.8) and the tensor identity $H^1(P_{1,8}, Y_8(\lambda_7) \otimes Y_{1,8}(\tau)) = 0$ for all $\tau \in V$ and $H^1(B, \bar{E} \otimes (-2\lambda_1 + \lambda_2))$. This completes the proof of (10.7.1).

The main purpose of this section is to prove:

(10.7.2) $R^iInd_B^{P_\Sigma}(\bar{F})$ *has a good filtration for* $i = 0$ *and is* 0 *for* $i > 0$ *, where* $\Sigma = \Delta\backslash\{\alpha_j\}$ *for some* $2 \le j \le 8$.

The method of proof is the same as that used in proving (9.7.1) and (10.6.1). As in section 10.6 we shall, for the most part, omit references and minor cohomological calculations. We shall, in particular, use (10.5.8) and (10.5.10) without reference.

We shall identify $Y_\Omega(-2\lambda_1 + \lambda_3)$ with a submodule of $\bar{F}$. Let $j = 2$, $P = P_{1,2}$, $G = Ind_B^{P_2}$ and $\Gamma = \Delta\backslash\{\alpha_1, \alpha_2\}$. Let $K = M_\Gamma(-2\lambda_1 + \lambda_3)$. Then $R^iG(K)$ has a good filtration if $i = 0$ and is 0 if $i > 0$. Thus, to prove (10.7.2) for $j = 2$, it suffices to show that the non-split extension $J = \bar{F}/K$ of $Y_{1,2}(-2\lambda_1 + \lambda_3)$ by $Y_1(-\lambda_1 + \lambda_2)$ has the property that $R^iG(J)$ has a good filtration for $i = 0$ and is 0 for $i > 0$. Now $Y_1(-\lambda_1 + \lambda_2)|_P$ has a filtration with quotients $Y_{1,2}(-\lambda_1 + \lambda_2)$, $Y_{1,2}(-\lambda_2 + \lambda_3)$, $Y_{1,2}(2\lambda_1 - \lambda_2)$. Hence J has a descending filtration $J = J_0, J_1, J_2, J_3, J_4 = 0$ with quotients, in order, $Y_{1,2}(-\lambda_1 + \lambda_2)$, $Y_{1,2}(-\lambda_2 + \lambda_3)$,

$Y_{1,2}(2\lambda_1 - \lambda_2)$, $Y_{1,2}(-2\lambda_1 + \lambda_3)$ such that J is a non-split extension of J_3 by J/J_3 . We have $Ext^1_P(Y_{1,2}(-\lambda_1 + \lambda_2), Y_{1,2}(-2\lambda_1 + \lambda_3)) \cong H^1(P_{1,2}, Y_{1,2}(-\lambda_1 - \lambda_2 + \lambda_3))$, which is 0 . Moreover $Ext^1_P(Y_{1,2}(2\lambda_1 - \lambda_2), Y_{1,2}(-2\lambda_1 + \lambda_3)) = 0$ by Lemma 3.2.1 so the filtration may be rechosen with J/J_1 unchanged, J_1/J_2 a non-split extension of $Y_{1,2}(-2\lambda_1 + \lambda_3)$ by $Y_{1,2}(-\lambda_2 + \lambda_3)$ and hence isomorphic to $Y_2(\lambda_1 - \lambda_2 + \lambda_3) \otimes -\lambda_1$, J_2 isomorphic to $Y_{1,2}(2\lambda_1 - \lambda_2)$ and $J_3 = 0$. Thus $R^i G(J_r/J_{r+1}) = 0$ for all i and $r = 0,1$ and $R^i G(J_2)$ is $Y_2(2\lambda_1 - \lambda_2)$ for $i = 0$ and 0 for $i > 0$. Hence $R^i G(J)$ has a good filtration for $i = 0$ and is 0 for $i > 0$.

Now let $j = 3$, $P = P_{1,3}$ and $G = Ind_B^{P_3}$. There is a P sub-module, say K , of $Y_1(-2\lambda_1 + \lambda_3)$ with a good filtration such that $Y_1(-2\lambda_1 + \lambda_3)/K$ has a filtration with quotients $Y_{1,3}(-2\lambda_1 + \lambda_3)$, $Y_{1,3}(-2\lambda_1 + \lambda_2 - \lambda_3 + \lambda_4)$, $Y_{1,3}(-2\lambda_1 + 2\lambda_2 - \lambda_3 + \lambda_7)$. Moreover $R^i G(K)$ has a good filtration for $i = 0$ and is 0 for $i > 0$. Thus to prove (10.7.2) in the case $j = 3$ it suffices to show that $R^i G(J)$ has a good filtration for $i = 0$ and is 0 for $i > 0$, where $J =$ $J = \bar{F}/K$. The module J has a descending filtration $J = J_0, J_1, \ldots, J_7 = 0$ with quotients, in order, $Y_{1,3}(-\lambda_1 + \lambda_2)$, $Y_{1,3}(-\lambda_3 + \lambda_4)$, $Y_{1,3}(\lambda_2 - \lambda_3 + \lambda_7)$, $Y_{1,3}(\lambda_1 - \lambda_3 + \lambda_8)$, $Y_{1,3}(-2\lambda_1 + \lambda_3)$, $Y_{1,3}(-2\lambda_1 + \lambda_2 - \lambda_3 + \lambda_4)$, $Y_{1,3}(-2\lambda_1 + 2\lambda_2 - \lambda_3 + \lambda_7)$ and J/J_5 is a non-split extension of J_4/J_5 by J/J_4 . Lemma 3.2.1 mplies that we may rechoose the filtration so that J/J_1 is a non-plit extension of $Y_{1,3}(-2\lambda_1 + \lambda_3)$ by $Y_{1,3}(-\lambda_1 + \lambda_2)$, J_1/J_2 is an xtension of $Y_{1,3}(-2\lambda_1 + \lambda_2 - \lambda_3 + \lambda_4)$ by $Y_{1,3}(-\lambda_3 + \lambda_4)$, J_2/J_3 is n extension of $Y_{1,3}(-2\lambda_1 + 2\lambda_2 - \lambda_3 + \lambda_7)$ by $Y_{1,3}(\lambda_2 - \lambda_3 + \lambda_7)$, $_3$ is isomorphic to $Y_{1,3}(\lambda_1 - \lambda_3 + \lambda_8)$ and $J_4 = 0$. There is a nique non-split extension of $Y_{1,3}(-2\lambda_1 + \lambda_3)$ by $Y_{1,3}(-\lambda_1 + \lambda_2)$ so hat J/J_1 is isomorphic to $Y_3(\lambda_2) \otimes -\lambda_1$. Since $xt^1_P(J/J_1, Y_{1,3}(-2\lambda_1 + \lambda_2 - \lambda_3 + \lambda_4)) = 0$ and $Y_{1,3}(-2\lambda_1 + \lambda_2 - \lambda_3 + \lambda_4)$ s not an image of $\bar{F}$, J_1/J_2 is a non-split extension of

$Y_{1,3}(-2\lambda_1 + \lambda_2 - \lambda_3 + \lambda_4)$ by $Y_{1,3}(-\lambda_3 + \lambda_4)$ and therefore isomorphic to $Y_3(\lambda_1 - \lambda_3 + \lambda_4) \otimes -\lambda_1$. Similar considerations imply that J_2/J_3 is a non-split extension of $Y_{1,3}(-2\lambda_1 + 2\lambda_2 - \lambda_3 + \lambda_7)$ by $Y_{1,3}(\lambda_2 - \lambda_3 + \lambda_7)$. We now obtain $R^iG(J_r/J_{r+1}) = 0$ for all i and $r = 0,1,2$ by (2.1.5), the tensor identity and (5.18). Also $R^iG(J_3)$ is $Y_3(\lambda_1 - \lambda_3 + \lambda_8)$ for $i = 0$ and 0 for $i > 0$. Thus $R^iG(J)$ has a good filtration for $i = 0$ and is 0 for $i > 0$. This completes the proof of (10.7.2) in the case $j = 3$.

Let $j = 4$, $P = P_{1,4}$ and $G = Ind_B^{P_4}$. There is a P submodule, say K , of $Y_1(-2\lambda_1 + \lambda_3)$ with a good filtration such that $Y_1(-2\lambda_1 + \lambda_3)/K$ has a filtration with quotients $Y_{1,4}(-2\lambda_1 + \lambda_3)$, $Y_{1,4}(-2\lambda_1 + \lambda_2 - \lambda_4 + \lambda_5)$, $Y_{1,4}(-2\lambda_1 + \lambda_2 + \lambda_3 - \lambda_4 + \lambda_7)$, $Y_{1,4}(-2\lambda_1 + 2\lambda_2 - \lambda_4 + \lambda_8)$. Moreover $R^iG(K)$ has a good filtration for $i = 0$ and is 0 for $i > 0$. Thus to prove (10.7.2) in the case $j = 4$ it suffices to show that $R^iG(J)$ has a good filtration for $i = 0$ and is 0 for $i > 0$, where $J = \bar{F}/K$. The module J has a descending filtration $J = J_0, J_1, \ldots, J_{10} = 0$ with quotients, in order, $Y_{1,4}(-\lambda_1 + \lambda_2)$, $Y_{1,4}(-\lambda_4 + \lambda_5)$, $Y_{1,4}(\lambda_3 - \lambda_4 + \lambda_7)$, $Y_{1,4}(\lambda_2 - \lambda_4 + \lambda_8)$, $Y_{1,4}(\lambda_1 - \lambda_4 + \lambda_6)$, $Y_{1,4}(\lambda_1 + \lambda_3 - \lambda_4)$, $Y_{1,4}(-2\lambda_1 + \lambda_3)$, $Y_{1,4}(-2\lambda_1 + \lambda_2 - \lambda_4 + \lambda_5)$, $Y_{1,4}(-2\lambda_1 + \lambda_2 + \lambda_3 - \lambda_4 + \lambda_7)$, $Y_{1,4}(-2\lambda_1 + 2\lambda_2 - \lambda_4 + \lambda_8)$ and J/J_7 is a non-split extension of J_6/J_7 by J/J_6 . Using Lemma 3.2.1 we may rechoose the filtration such that J/J_1 is a non-split extension of $Y_{1,4}(-2\lambda_1 + \lambda_3)$ by $Y_{1,4}(-\lambda_1 + \lambda_2)$, J_1/J_2 is an extension of $Y_{1,4}(-2\lambda_1 + \lambda_2 - \lambda_4 + \lambda_5)$ by $Y_{1,4}(-\lambda_4 + \lambda_5)$, J_2/J_3 is an extension of $Y_{1,4}(-2\lambda_1 + \lambda_2 + \lambda_3 - \lambda_4 + \lambda_7)$ by $Y_{1,4}(\lambda_3 - \lambda_4 + \lambda_7)$, J_3/J_4 is an extension of $Y_{1,4}(-2\lambda_1 + 2\lambda_2 - \lambda_4 + \lambda_8)$ by $Y_{1,4}(\lambda_2 - \lambda_4 + \lambda_8)$, J_4/J_5 is isomorphic to $Y_{1,4}(\lambda_1 - \lambda_4 + \lambda_6)$, J_5 is isomorphic to $Y_{1,4}(\lambda_1 + \lambda_3 - \lambda_4)$ and $J_6 = 0$. There is only one non-split extension of $Y_{1,4}(-2\lambda_1 + \lambda_3)$ by $Y_{1,4}(-\lambda_1 + \lambda_2)$ so that J/J_1 is isomorphic to $Y_4(\lambda_2) \otimes -\lambda_1$. It follows that
$Ext_P^1(J/J_1, Y_{1,4}(-2\lambda_1 + \lambda_2 - \lambda_4 + \lambda_5) = 0$ and, since

$Y_{1,4}(-2\lambda_1 + \lambda_2 - \lambda_4 + \lambda_5)$ is not an image of $\overline{F}$, J_1/J_2 is a non-split extension of $Y_{1,4}(-2\lambda_1 + \lambda_2 - \lambda_4 + \lambda_5)$ by $Y_{1,4}(-\lambda_4 + \lambda_5)$, and hence isomorphic to $Y_4(\lambda_1 - \lambda_4 + \lambda_5) \otimes -\lambda_1$. Similar considerations show that J_2/J_3 is a non-split extension of $Y_{1,4}(-2\lambda_1 + \lambda_2 + \lambda_3 - \lambda_4 + \lambda_7)$ by $Y_{1,4}(\lambda_3 - \lambda_4 + \lambda_7)$. It follows that there is a short exact sequence of P-modules

$$0 \to Y_4(\lambda_1 + \lambda_7) \otimes -2\lambda_1 \to Y_4(\lambda_1 + \lambda_3 - \lambda_4 + \lambda_7) \otimes -\lambda_1$$
$$\to J_2/J_3 \to 0 \qquad (1).$$

By (1), $Ext^1_P(J_2/J_3, Y_{1,4}(-2\lambda_1 + 2\lambda_2 - \lambda_4 + \lambda_8))$ is isomorphic to $Ext^2_P(Y_4(\lambda_1 + \lambda_7), Y_{1,4}(2\lambda_2 - \lambda_4 + \lambda_8))$ which is isomorphic to $Ext^2_{P_4}(Y_4(\lambda_1 + \lambda_7), Y_4(2\lambda_2 - \lambda_4 + \lambda_8))$. However, this is 0 by Lemma 3.2.1. It is not difficult to show that $Ext^1_P(J/J_2, Y_{1,4}(-2\lambda_1 + 2\lambda_2 - \lambda_4 + \lambda_8)) = 0$ and so $Ext^1_P(J/J_3, Y_{1,4}(-2\lambda_1 + 2\lambda_2 - \lambda_4 + \lambda_8)) = 0$. Since $Y_{1,4}(-2\lambda_1 + 2\lambda_2 - \lambda_4 + \lambda_8)$ is not an image of $\overline{F}$, J_3/J_4 is a non-split extension of $Y_{1,4}(-2\lambda_1 + 2\lambda_2 - \lambda_4 + \lambda_8)$ by $Y_{1,4}(\lambda_2 - \lambda_4 + \lambda_8)$. We now obtain $R^iG(J_r/J_{r+1}) = 0$ for $0 \le r \le 3$ and all i by (2.1.5), the tensor identity and (5.18). Also $R^iG(J_4/J_5)$ (resp. $R^iG(J_5)$) is $Y_4(\lambda_1 - \lambda_4 + \lambda_6)$ (resp. $Y_4(\lambda_1 + \lambda_3 - \lambda_4)$) for $i = 0$ and 0 for $i > 0$. Hence $R^iG(J)$ has a good filtration for $i = 0$ and is 0 for $i > 0$. This completes the proof of (10.7.2) in the case $j = 4$.

We now come to the most complicated case, $j = 5$, $P = P_{1,5}$ and $= Ind_B^{P_5}$. There is a P-submodule, say K of $Y_1(-2\lambda_1 + \lambda_3)$ with good filtration such that $Y_1(-2\lambda_1 + \lambda_3)$ has a filtration with quotients $Y_{1,5}(-2\lambda_1 + \lambda_3)$, $Y_{1,5}(-2\lambda_1 + \lambda_2 - \lambda_5 + \lambda_6 + \lambda_8)$, $Y_{1,5}(-2\lambda_1 + \lambda_2 + \lambda_4 - \lambda_5 + \lambda_7)$, $Y_{1,5}(-2\lambda_1 + \lambda_2 + \lambda_3 - \lambda_5 + \lambda_8)$, $Y_{1,5}(-2\lambda_1 + 2\lambda_2 - \lambda_5 + \lambda_6)$. Moreover $R^iG(K)$ has a good filtration for $i = 0$ and is 0 for $i > 0$. Thus, to prove (10.7.2) in the case $j = 5$, it suffices to show that $R^iG(J)$ has a good filtration for $i = 0$ and is 0 for $i > 0$, where $J = \overline{F}/K$. The module J has a descending filtration $J = J_o, J_1, \ldots, J_{12} = 0$ with quotients, in order $Y_{1,5}(-\lambda_1 + \lambda_2)$, $Y_{1,5}(-\lambda_5 + \lambda_6 + \lambda_8)$, $Y_{1,5}(\lambda_4 - \lambda_5 + \lambda_7)$, $Y_{1,5}(\lambda_3 - \lambda_5 + \lambda_8)$, $Y_{1,5}(\lambda_2 - \lambda_5 + \lambda_6)$, $Y_{1,5}(\lambda_1 - \lambda_5 + \lambda_7 + \lambda_8)$, $Y_{1,5}(\lambda_1 + \lambda_4 - \lambda_5)$, $Y_{1,5}(-2\lambda_1 + \lambda_3)$, $Y_{1,5}(-2\lambda_1 + \lambda_2 - \lambda_5 + \lambda_6 + \lambda_8)$, $Y_{1,5}(-2\lambda_1 + \lambda_2 + \lambda_4 - \lambda_5 + \lambda_7)$, $Y_{1,5}(-2\lambda_1 + \lambda_2 + \lambda_3 - \lambda_5 + \lambda_8)$, $Y_{1,5}(-2\lambda_1 + 2\lambda_2 - \lambda_5 + \lambda_7)$ and J/J_8 is a non-split extension of J_7/J_8 by J/J_7 . Using Lemma 3.2.1 we may rechoose the filtration such that J/J_1 is a non-split extension of $Y_{1,5}(-2\lambda_1 + \lambda_3)$ by $Y_{1,5}(-\lambda_1 + \lambda_2)$, J_1/J_2 is an extension of $Y_{1,5}(-2\lambda_1 + \lambda_2 - \lambda_5 + \lambda_6 + \lambda_8)$ of $Y_{1,5}(-\lambda_5 + \lambda_6 + \lambda_8)$, J_2/J_3 is an extension of $Y_{1,5}(-2\lambda_1 + \lambda_2 + \lambda_4 - \lambda_5 + \lambda_7)$ by $Y_{1,5}(\lambda_4 - \lambda_5 + \lambda_7)$, J_3/J_4 is an extension of $Y_{1,5}(-2\lambda_1 + \lambda_2 + \lambda_3 - \lambda_5 + \lambda_8)$ by $Y_{1,5}(\lambda_3 - \lambda_5 + \lambda_8)$, J_4/J_5 is an extension of $Y_{1,5}(-2\lambda_1 + 2\lambda_2 - \lambda_5 + \lambda_6)$ by $Y_{1,5}(\lambda_2 - \lambda_5 + \lambda_6)$, J_5/J_6 is isomorphic to $Y_{1,5}(\lambda_1 - \lambda_5 + \lambda_7 + \lambda_8)$, J_6 is isomorphic to $Y_{1,5}(\lambda_1 + \lambda_4 - \lambda_5)$ and $J_7 = 0$. There is only one non-split extension of $Y_{1,5}(-2\lambda_1 + \lambda_3)$ by $Y_{1,5}(-\lambda_1 + \lambda_2)$ so that J/J_1 is isomorphic to $Y_5(\lambda_2) \otimes -\lambda_1$.

We have $\mathit{Ext}^1_P(J/J_1, Y_{1,5}(-2\lambda_1 + \lambda_2 - \lambda_5 + \lambda_6 + \lambda_8)) = 0$ and $Y_{1,5}(-2\lambda_1 + \lambda_2 - \lambda_5 + \lambda_6 + \lambda_8)$ is not an image of $\overline{F}$ so that J_1/J_2 is a non-split extension of $Y_{1,5}(-2\lambda_1 + \lambda_2 - \lambda_5 + \lambda_6 + \lambda_8)$ by $Y_{1,5}(-\lambda_5 + \lambda_6 + \lambda_8)$ and hence isomorphic to $Y_5(\lambda_1 - \lambda_5 + \lambda_6 + \lambda_8) \otimes -\lambda_1$. Similar considerations imply that J_2/J_3 is a non-split extension of $Y_{1,5}(-2\lambda_1 + \lambda_2 + \lambda_4 - \lambda_5 + \lambda_7)$ by $Y_{1,5}(\lambda_4 - \lambda_5 + \lambda_7)$. It follows that there is a short exact sequence

$$0 \to Y_5(\lambda_1 + \lambda_7) \otimes -2\lambda_1 \to Y_5(\lambda_1 + \lambda_4 - \lambda_5 + \lambda_7) \otimes -\lambda_1 \to J_2/J_3 \to 0 \qquad (2).$$

We obtain, from (2), that $\mathit{Ext}^1_P(J_2/J_3, Y_{1,5}(-2\lambda_1 + \lambda_2 + \lambda_3 - \lambda_5 + \lambda_8))$ is isomorphic to $\mathit{Ext}^2_P(Y_5(\lambda_1 + \lambda_7), Y_{1,5}(\lambda_2 + \lambda_3 - \lambda_5 + \lambda_8))$ and hence isomorphic to

$Ext^2_{P_5}(Y_5(\lambda_1 + \lambda_7), Y_5(\lambda_2 + \lambda_3 - \lambda_5 + \lambda_8))$ and therefore 0 by Lemma 3.2.1. Also $Ext^1_P(J_r/J_{r+1}, Y_{1,5}(-2\lambda_1\ 1\ \lambda_2 + \lambda_3 - \lambda_5 + \lambda_8)) = 0$ for $r = 0,1$ so that $Ext^1_P(J/J_3, Y_{1,5}(-2\lambda_1 + \lambda_2 + \lambda_3 - \lambda_5 + \lambda_8)) = 0$. Since $Y_{1,5}(-2\lambda_1 + \lambda_2 + \lambda_3 - \lambda_5 + \lambda_8)$ is not an image of $\overline{F}$ the extension J_3/J_4 of $Y_{1,5}(-2\lambda_1 + \lambda_2 + \lambda_3 - \lambda_5 + \lambda_8)$ by $Y_{1,5}(\lambda_3 - \lambda_5 + \lambda_8)$ is non-split. It follows that there is a short exact sequence

$$0 \to A_1 \to Y_5(\lambda_1 + \lambda_3 - \lambda_5 + \lambda_8) \otimes -\lambda_1 \to J_3/J_4 \to 0 \qquad (3)$$

where A_1 is an extension of $Y_{1,5}(-3\lambda_1 + \lambda_2 + \lambda_4 - \lambda_5 + \lambda_8)$ by $Y_{1,5}(-\lambda_1 + \lambda_4 - \lambda_5 + \lambda_8)$.

We obtain, from (2), that $Ext^1_P(J_2/J_3, Y_{1,5}(-2\lambda_1 + 2\lambda_2 - \lambda_5 + \lambda_6))$ is isomorphic to $Ext^2_P(Y_5(\lambda_1 + \lambda_7), Y_{1,5}(2\lambda_2 - \lambda_5 + \lambda_6))$ and so to $Ext^2_{P_5}(Y_5(\lambda_1 + \lambda_7), Y_5(2\lambda_2 - \lambda_5 + \lambda_6))$, which is 0 by Lemma 3.2.1. Also, from (3), $Ext^1_P(J_3/J_4, Y_{1,5}(-2\lambda_1 + 2\lambda_2 - \lambda_5 + \lambda_6))$ is isomorphic to $Ext^2_P(A_1, Y_{1,5}(-2\lambda_1 + 2\lambda_2 - \lambda_5 + \lambda_6))$. However this is 0 since, by Lemma 3.2.1, $Ext^2_P(Y_{1,5}(\tau), Y_{1,5}(-2\lambda_1 + 2\lambda_2 - \lambda_5 + \lambda_6))$ is 0 for $\tau = -3\lambda_1 + \lambda_2 + \lambda_4 - \lambda_5 + \lambda_8, -\lambda_1 + \lambda_4 - \lambda_5 + \lambda_8$. However $Ext^1_P(J_r/J_{r+1}, Y_{1,5}(-2\lambda_1 + 2\lambda_2 - \lambda_5 + \lambda_6))$ is 0 for $r = 0,1$ so that $Ext^1_P(J/J_4, Y_{1,5}(-2\lambda_1 + 2\lambda_2 - \lambda_5 + \lambda_6)) = 0$. Since $Y_{1,5}(-2\lambda_1 + 2\lambda_2 - \lambda_5 + \lambda_6)$ is not an image of $\overline{F}$ the extension J_4/J_5 of $Y_{1,5}(-2\lambda_1 + 2\lambda_2 - \lambda_5 + \lambda_6)$ by $Y_{1,5}(\lambda_2 - \lambda_5 + \lambda_6)$ is non-split.

We have $R^iG(J_r/J_{r+1}) = 0$ for $0 \le r \le 4$ and all i by (2.1.5), the tensor identity and (5.18). Also $R^iG(J_5/J_6)$ (resp. $R^iG(J_6)$) is $Y_5(\lambda_1 - \lambda_5 + \lambda_7 + \lambda_8)$ (resp. $Y_5(\lambda_1 + \lambda_4 - \lambda_5)$) for $i = 0$ and 0 for $i > 0$. Hence $R^iG(J)$ has a good filtration for $i = 0$ and is 0 for $i > 0$. This completes the proof of (10.7.2) in the case $j = 5$.

Let $j = 6$, $P = P_{1,6}$ and $G = Ind_B^{P_6}$. There is a P submodule, say K, of $Y_1(-2\lambda_1 + \lambda_3)$ with a good filtration such that $Y_1(-2\lambda_1 + \lambda_3)/K$ has a filtration with quotients $Y_{1,6}(-2\lambda_1 + \lambda_3)$,

$Y_{1,6}(-2\lambda_1 + \lambda_2 - \lambda_6 + \lambda_7 + \lambda_8)$, $Y_{1,6}(-2\lambda_1 + \lambda_2 + \lambda_4 - \lambda_6)$, $Y_{1,4}(-2\lambda_1 + 2\lambda_2 - \lambda_6 + \lambda_7)$. Moreover $R^iG(K)$ has a good filtration for $i = 0$ and is 0 for $i > 0$. Thus to prove (10.7.2) in the case $j = 6$ it suffices to show that $R^iG(J)$ has a good filtration for $i = 0$ and is 0 for $i > 0$, where $J = \overline{F}/K$. The module J has a descending filtration $J = J_0, J_1, \ldots, J_9 = 0$ with quotients, in order, $Y_{1,6}(-\lambda_1 + \lambda_2)$, $Y_{1,6}(-\lambda_6 + \lambda_7 + \lambda_8)$, $Y_{1,6}(\lambda_4 - \lambda_6)$, $Y_{1,6}(\lambda_2 - \lambda_6 + \lambda_7)$, $Y_{1,6}(\lambda_1 - \lambda_6 + \lambda_8)$, $Y_{1,6}(-2\lambda_1 + \lambda_3)$, $Y_{1,6}(-2\lambda_1 + \lambda_2 - \lambda_6 + \lambda_7 + \lambda_8)$, $Y_{1,6}(-2\lambda_1 + \lambda_2 + \lambda_4 - \lambda_6)$, $Y_{1,6}(-2\lambda_1 + 2\lambda_2 - \lambda_6 + \lambda_7)$ and J/J_6 is a non-split extension of J_5/J_6 by J/J_5 . Using Lemma 3.2.1 we may rechoose the filtration such that J/J_1 is a non-split extension of $Y_{1,6}(-2\lambda_1 + \lambda_3)$ by $Y_{1,6}(-\lambda_1 + \lambda_2)$, J_1/J_2 is an extension of $Y_{1,6}(-2\lambda_1 + \lambda_2 - \lambda_6 + \lambda_7 + \lambda_8)$ by $Y_{1,6}(-\lambda_6 + \lambda_7 + \lambda_8)$ J_2/J_3 is an extension of $Y_{1,6}(-2\lambda_1 + \lambda_2 + \lambda_4 - \lambda_6)$ by $Y_{1,6}(\lambda_4 - \lambda_6)$, J_3/J_4 is an extension of $Y_{1,6}(-2\lambda_1 + 2\lambda_2 - \lambda_6 + \lambda_7)$ by $Y_{1,6}(\lambda_2 - \lambda_6 + \lambda_7)$, J_4 is isomorphic to $Y_{1,6}(\lambda_1 - \lambda_6 + \lambda_8)$ and $J_5 = 0$. There is only one non-split extension of $Y_{1,6}(-2\lambda_1 + \lambda_3)$ by $Y_{1,6}(-\lambda_1 + \lambda_2)$ so that J/J_1 is isomorphic to $Y_6(\lambda_2) \otimes -\lambda_1$. It follows that $Ext^1_P(J/J_1, Y_{1,6}(-2\lambda_1 + \lambda_2 - \lambda_6 + \lambda_7 + \lambda_8)) = 0$ and, since $Y_{1,6}(-2\lambda_1 + \lambda_2 - \lambda_6 + \lambda_7 + \lambda_8)$ is not an image of $\overline{F}$, J_1/J_2 is a non-split extension of $Y_{1,6}(-2\lambda_1 + \lambda_2 - \lambda_6 + \lambda_7 + \lambda_8)$ by $Y_{1,6}(-\lambda_6 + \lambda_7 + \lambda_8)$. It follows that there is a short exact sequence

$$0 \to Y_6(\lambda_1 + \lambda_7) \otimes -2\lambda_1 \to Y_6(\lambda_1 - \lambda_6 + \lambda_7 + \lambda_8) \otimes -\lambda_1 \to J_1/J_2 \to 0 \tag{4}$$

of P-modules. From (4) we obtain that $Ext^1_P(J_1/J_2, Y_{1,6}(-2\lambda_1 + \lambda_2 + \lambda_4 - \lambda_6))$ is isomorphic to $Ext^2_P(Y_6(\lambda_1 + \lambda_7), Y_{1,6}(\lambda_2 + \lambda_4 - \lambda_6))$ and so isomorphic to $Ext^2_{P_6}(Y_6(\lambda_1 + \lambda_7), Y_6(\lambda_2 + \lambda_4 - \lambda_6))$ which is 0 by Lemma 3.2.1. Also $Ext^1_P(J/J_1, Y_{1,6}(-2\lambda_1 + \lambda_2 + \lambda_4 - \lambda_6))$ is 0 so that

$Ext^1_P(J/J_2, Y_{1,6}(-2\lambda_1 + \lambda_2 + \lambda_4 - \lambda_6)) = 0$ and, since $Y_{1,6}(-2\lambda_1 + \lambda_2 + \lambda_4 - \lambda_6)$ is not an image of $\overline{F}$, J_2/J_3 is a non-split extension of $Y_{1,6}(-2\lambda_1 + \lambda_2 + \lambda_4 - \lambda_6)$ by $Y_{1,6}(\lambda_4 - \lambda_6)$. It follows that there is a short exact sequence

$$0 \to A_2 \to Y_6(\lambda_1 + \lambda_4 - \lambda_6) \otimes -\lambda_1 \to J_2/J_3 \to 0 \tag{5}$$

where A_2 is an extension of $Y_{1,6}(-3\lambda_1 + \lambda_2 + \lambda_5 - \lambda_6)$ by $Y_{1,6}(-\lambda_1 + \lambda_5 - \lambda_6)$.

We obtain, from (4), that $Ext^1_P(J_1/J_2, Y_{1,6}(-2\lambda_1 + 2\lambda_2 - \lambda_6 + \lambda_7))$ is isomorphic to $Ext^2_P(Y_6(\lambda_1 + \lambda_7), Y_{1,6}(2\lambda_2 - \lambda_6 + \lambda_7))$ and so isomorphic to $Ext^2_{P_6}(Y_6(\lambda_1 + \lambda_7), Y_6(2\lambda_2 - \lambda_6 + \lambda_7))$ which is 0 by Lemma 3.2.1. From (5) we obtain that $Ext^1_P(J_2/J_3, Y_{1,6}(-2\lambda_1 + 2\lambda_2 - \lambda_6 + \lambda_7))$ is isomorphic to $Ext^2_P(A_2, Y_{1,6}(-2\lambda_1 + 2\lambda_2 - \lambda_6 + \lambda_7))$ which is 0 since $Ext^2_P(Y_{1,6}(\tau), Y_{1,6}(-2\lambda_1 + 2\lambda_2 - \lambda_6 + \lambda_7)) = 0$ for $\tau = -3\lambda_1 + \lambda_2 + \lambda_5 - \lambda_6, -\lambda_1 + \lambda_5 - \lambda_6$ by Lemma 3.2.1. Also $Ext^1_P(J/J_1, Y_{1,6}(-2\lambda_1 + 2\lambda_2 - \lambda_6 + \lambda_7)) = 0$ so that $Ext^1_P(J/J_3, Y_{1,6}(-2\lambda_1 + 2\lambda_2 - \lambda_6 + \lambda_7)) = 0$. Since $_{1,6}(-2\lambda_1 + 2\lambda_2 - \lambda_6 + \lambda_7)$ is not an image of $\overline{F}$ the extension J_3/J_4 f $Y_{1,6}(-2\lambda_1 + 2\lambda_2 - \lambda_6 + \lambda_7)$ by $Y_{1,6}(\lambda_2 - \lambda_6 + \lambda_7)$ is non-split.

We have $R^iG(J_r/J_{r+1}) = 0$ for $0 \le r \le 3$ and all i by (2.1.5), he tensor identity and (5.18). Also $R^iG(J_4)$ is $Y_6(\lambda_1 - \lambda_6 + \lambda_8)$ or $i = 0$ and 0 for $i > 0$. Hence $R^iG(J)$ has a good filtration or $i = 0$ and is 0 for $i > 0$. This completes the proof of 10.7.2) in the case $j = 6$.

Let $j = 7$, $P = P_{1,7}$ and $G = Ind_B^{P_7}$. There is a P submodule, ay, K, of $Y_1(-2\lambda_1 + \lambda_3)$ with a good filtration such that $(-2\lambda_1 + \lambda_3)/K$ has a filtration with quotients $Y_{1,7}(-2\lambda_1 + \lambda_3)$, $_{,7}(-2\lambda_1 + \lambda_2 - \lambda_7 + \lambda_8)$, $Y_{1,7}(-2\lambda_1 + 2\lambda_2 - \lambda_7)$. Moreover $R^iG(K)$ s a good filtration for $i = 0$ and is 0 for $i > 0$. Thus to prove 0.7.2) in the case $j = 7$ it suffices to show that $R^iG(J)$ has a good filtration for $i = 0$ and is 0 for $i > 0$, where $J = \overline{F}/K$. The module J has a descending filtration $J = J_o, J_1, \ldots, J_6 = 0$ with quotients, in order, $Y_{1,7}(-\lambda_1 + \lambda_2)$, $Y_{1,7}(-\lambda_7 + \lambda_8)$, $Y_{1,7}(\lambda_2 - \lambda_7)$, $Y_{1,7}(-2\lambda_1 + \lambda_3)$, $Y_{1,7}(-2\lambda_1 + \lambda_2 - \lambda_7 + \lambda_8)$, $Y_{1,7}(-2\lambda_1 + 2\lambda_2 - \lambda_7)$ and J/J_4 is a non-split extension of J_3/J_4 by J/J_3. Using Lemma 3.2.1 we may rechoose the filtration such that J/J_1 is a non-split extension of $Y_{1,7}(-2\lambda_1 + \lambda_3)$ by $Y_{1,7}(-\lambda_1 + \lambda_2)$, J_1/J_2 is an extension of $Y_{1,7}(-2\lambda_1 + \lambda_2 - \lambda_7 + \lambda_8)$ by $Y_{1,7}(-\lambda_7 + \lambda_8)$, J_2/J_2 is an extension of $Y_{1,7}(-2\lambda_1 + 2\lambda_2 - \lambda_7)$ by $Y_{1,7}(\lambda_2 - \lambda_7)$. There is only one non-split extension of $Y_{1,7}(-2\lambda_1 + \lambda_3)$ by $Y_{1,7}(-\lambda_1 + \lambda_2)$ so there is a short exact sequence

$$0 \to A_3 \to Y_7(\lambda_2) \otimes -\lambda_1 \to J/J_1 \to 0 \tag{6}$$

where A_3 is an extension of $Y_{1,7}(-3\lambda_1 + \lambda_2 + \lambda_7)$ by $Y_{1,7}(\lambda_1 + \lambda_7)$. This extension is non-split since the B-socle of $Y_7(\lambda_2)$ is simple and so there is a short exact sequence

$$0 \to 3\lambda_1 + 2\lambda_7 \to Y_7(\lambda_1 + \lambda_7) \otimes -2\lambda_1 \to A_3 \to 0 \tag{7}.$$

From (6) it follows that $Ext^1_P(J/J_1, Y_{1,7}(-2\lambda_1 + \lambda_2 - \lambda_7 + \lambda_8))$ is isomorphic to $Ext^2_P(A_3, Y_{1,7}(-2\lambda_1 + \lambda_2 - \lambda_7 + \lambda_8))$ which is 0 since by Lemma 3.2.1, $Ext^2_P(Y_{1,7}(\tau), Y_{1,7}(-2\lambda_1 + \lambda_2 - \lambda_7 + \lambda_8)) = 0$ for $\tau = -3\lambda_1 + \lambda_2 + \lambda_7, -\lambda_1 + \lambda_7$. Since $Y_{1,7}(-2\lambda_1 + \lambda_2 - \lambda_7 + \lambda_8)$ is not an image of $\overline{F}$, the extension J_1/J_2 of $Y_{1,7}(-2\lambda_1 + \lambda_2 - \lambda_7 + \lambda_8)$ by $Y_{1,7}(-\lambda_7 + \lambda_8)$ is non-split. It follows that there is a short exact sequence

$$0 \to A_4 \to Y_7(\lambda_1 - \lambda_7 + \lambda_8) \otimes -\lambda_1 \to J_1/J_2 \to 0 \tag{8}$$

where A_4 has a filtration with quotients $Y_{1,7}(-\lambda_1 + \lambda_6 - \lambda_7)$, $Y_{1,7}(-3\lambda_1 + \lambda_2 + \lambda_6 - \lambda_7)$, $Y_{1,7}(-2\lambda_1 + \lambda_3)$, $Y_{1,7}(-3\lambda_1 + \lambda_6)$.

From (6) we obtain that $Ext^1_P(J/J_1, Y_{1,7}(-2\lambda_1 + 2\lambda_2 - \lambda_7))$ is isomorphic to $Ext^2_P(A_3, Y_{1,7}(-2\lambda_1 + 2\lambda_2 - \lambda_7))$ which is 0 since, by Lemma 3.2.1, $Ext^2_P(Y_{1,y}(\tau), Y_{1,7}(-2\lambda_1 + 2\lambda_2 - \lambda_7)) = 0$ for $\tau = -3\lambda_1 + \lambda_2 + \lambda_7, -\lambda_1 + \lambda_7$. From (8) we obtain that

$Ext^1_P(J_1/J_2, Y_{1,7}(-2\lambda_1 + 2\lambda_2 - \lambda_7))$ is isomorphic to $Ext^2_P(A_4, Y_{1,7}(-2\lambda_1 + 2\lambda_2 - \lambda_7))$ which is 0 since, by Lemma 3.2.1, $Ext^2_P(Y_{1,7}(\tau), Y_{1,7}(-2\lambda_1 + 2\lambda_2 - \lambda_7)) = 0$ for $\tau = -\lambda_1 + \lambda_6 - \lambda_7$, $-3\lambda_1 + \lambda_2 + \lambda_6 - \lambda_7$, $-2\lambda_1 + \lambda_8$, $-3\lambda_1 + \lambda_6$. Hence $Ext^1_P(J/J_2, Y_{1,7}(-2\lambda_1 + 2\lambda_2 - \lambda_7))$ is 0 and, since $Y_{1,7}(-2\lambda_1 + 2\lambda_2 - \lambda_7)$ is not an image of $\bar{F}$, the extension J_2/J_3 of $Y_{1,7}(-2\lambda_1 + 2\lambda_2 - \lambda_7)$ by $Y_{1,7}(\lambda_2 - \lambda_7)$ is non-split.

We have, by (6), that $R^iG(J/J_1)$ is isomorphic to $R^{i+1}G(A_3)$. However $R^mG(Y_7(\lambda_1 + \lambda_7) \otimes -2\lambda_1) = 0$ for all m by the tensor identity, (2.3.1) and (2.1.5). Thus (7) gives that $R^{i+1}G(A_3)$ is isomorphic to $R^{i+2}G(-3\lambda_1 + 2\lambda_7)$ which is 0 for all i by (5.5), (2.3.1) and (2.1.5). Hence $R^iG(J/J_1) = 0$ for all i. Also $R^iG(J_r/J_{r+1}) = 0$ for all i and $r = 1,2$ by (5.18). Hence $R^iG(J) = 0$ for all i and the proof of (10.7.2), in the case $j = 7$, is complete.

We now deal with the remaining case $j = 8$. Let $P = P_{1,8}$ and $G = Ind_B^{P_8}$. There is a P submodule, say K, of $Y_1(-2\lambda_1 + \lambda_3)$ with good filtration such that $Y_1(-2\lambda_1 + \lambda_3)/K$ has a filtration with quotients $Y_{1,8}(-2\lambda_1 + \lambda_3)$, $Y_{1,8}(-2\lambda_1 + \lambda_2 + \lambda_6 - \lambda_8)$, $Y_{1,8}(-2\lambda_1 + \lambda_2 + \lambda_3 - \lambda_8)$. Moreover $R^iG(K)$ has a good filtration for $i = 0$ and is 0 for $i > 0$. Thus to prove (10.7.2) in the case $j = 8$ it suffices to show that $R^iG(J)$ has a good filtration for $i = 0$ and is 0 for $i > 0$, where $J = \bar{F}/K$. The module J has a descending filtration $J = J_0, J_1, \ldots, J_7 = 0$ with quotients, in order, $Y_{1,8}(-\lambda_1 + \lambda_2)$, $Y_{1,8}(\lambda_6 - \lambda_8)$, $Y_{1,8}(\lambda_3 - \lambda_8)$, $Y_{1,8}(\lambda_1 + \lambda_7 - \lambda_8)$, $Y_{1,8}(-2\lambda_1 + \lambda_3)$, $Y_{1,8}(-2\lambda_1 + \lambda_2 + \lambda_6 - \lambda_8)$, $Y_{1,8}(-2\lambda_1 + \lambda_2 + \lambda_3 - \lambda_8)$ and J/J_5 is a non-split extension of J_4/J_5 by J/J_4. Using Lemma 3.2.1 we may rechoose the filtration so that J/J_1 is a non-split extension of $Y_{1,8}(-2\lambda_1 + \lambda_3)$ by $Y_{1,8}(-\lambda_1 + \lambda_2)$, J_1/J_2 is an extension of $Y_{1,8}(-2\lambda_1 + \lambda_2 + \lambda_6 - \lambda_8)$ by $Y_{1,8}(\lambda_6 - \lambda_8)$, J_2/J_3 is an extension of $Y_{1,8}(-2\lambda_1 + \lambda_2 + \lambda_3 - \lambda_8)$ by $Y_{1,8}(\lambda_3 - \lambda_8)$, J_3 is isomorphic to $Y_{1,8}(\lambda_1 + \lambda_7 - \lambda_8)$. There is only one non-split exten-

sion of $Y_{1,8}(-2\lambda_1 + \lambda_3)$ by $Y_{1,8}(-\lambda_1 + \lambda_2)$ so that J/J_1 is isomorphic to $Y_8(\lambda_2) \otimes -\lambda_1$. We deduce that
$Ext^1_P(J/J_1, Y_{1,8}(-2\lambda_1 + \lambda_2 + \lambda_6 - \lambda_8)) = 0$ and, since
$Y_{1,8}(-2\lambda_1 + \lambda_2 + \lambda_6 - \lambda_8)$ is not an image of $\bar{F}$, the extension J_1/J_2 of $Y_{1,8}(-2\lambda_1 + \lambda_2 + \lambda_6 - \lambda_8)$ by $Y_{1,8}(\lambda_6 - \lambda_8)$ is non-split. It follows that there is a short exact sequence

$$0 \to A_5 \to Y_8(\lambda_1 + \lambda_6 - \lambda_8) \otimes -\lambda_1 \to J_1/J_2 \to 0 \tag{9}$$

where A_5 is an extension of $Y_{1,8}(-3\lambda_1 + \lambda_2 + \lambda_7)$ by $Y_{1,8}(-\lambda_1 + \lambda_7)$.

From (9) we obtain that $Ext^1_P(J_1/J_2, Y_{1,8}(-2\lambda_1 + \lambda_2 + \lambda_3 - \lambda_8))$ is isomorphic to $Ext^2_P(A_5, Y_{1,8}(-2\lambda_1 + \lambda_2 + \lambda_3 - \lambda_8))$ which is 0 since, by Lemma 3.2.1, $Ext^2_P(Y_{1,8}(\tau), Y_{1,8}(-2\lambda_1 + \lambda_2 + \lambda_3 - \lambda_8)) = 0$ for $\tau = -3\lambda_1 + \lambda_2 + \lambda_7, -\lambda_1 + \lambda_7$. Also,
$Ext^1_P(J/J_1, Y_{1,8}(-2\lambda_1 + \lambda_2 + \lambda_3 - \lambda_8))$ is 0 so that
$Ext^1_P(J/J_2, Y_{1,8}(-2\lambda_1 + \lambda_2 + \lambda_3 - \lambda_8)) = 0$ and, since
$Y_{1,8}(-2\lambda_1 + \lambda_2 + \lambda_3 - \lambda_8)$ is not an image of $\bar{F}$, the extension J_2/J_3 of $Y_{1,8}(-2\lambda_1 + \lambda_2 + \lambda_3 - \lambda_8)$ by $Y_{1,8}(\lambda_3 - \lambda_8)$ is non-split.

We now have $R^iG(J_r/J_{r+1}) = 0$ for $0 \le r \le 2$ and all i by (2.1.5), the tensor identity and (5.18). Also $R^iG(J_3)$ is $Y_8(\lambda_1 + \lambda_7 - \lambda_8)$ for $i = 0$ and 0 for $i > 0$. Hence $R^iG(J)$ has a good filtration for $i = 0$ and is 0 for $i > 0$. This completes the proof of (10.7.1) in the final case $j = 8$.

10.8 The case $p = 7$

Let $2 \le j \le 8$ and $\Sigma = \Delta\backslash\{\alpha_j\}$. We have that $R^i Ind_B^{P_\Sigma}(M)$ has a good filtration for $i = 0$ and is 0 for $i > 0$ for $M = D$, E/D and F/E by (10.4.1), (10.6.1) and (10.7.1). Also, the same holds for $M = Y(\lambda_7)/F$ by (5.2) so that $R^i Ind_B^{P_\Sigma}(Y(\ _7))$ has a good filtration, as a P_Σ-module, for $i = 0$ and is 0 for $i > 0$. However $Ind_B^{P_\Sigma}(Y(\lambda_7))$ is isomorphic to $Y(\lambda_7)|_{P_\Sigma}$ which therefore has a good filtration. Combining this with (10.3.13) we obtain:

(10.8.1) $Y(\lambda_7)|_{P_\Sigma}$ *has a good filtration for all maximal proper subsets* Σ *of* Δ .

However, if Γ is a proper non-maximal subset of Δ then $\Gamma \subset \Lambda \subset \Delta$ for a proper maximal subset Λ such that Φ_Λ does not have type E_7 . Thus $Y(\lambda_7)|_{P_\Lambda}$ has a good filtration by (10.8.1) and so $Y(\lambda_7)|_{P_\Gamma}$, which is isomorphic to $(Y(\lambda_7)|_{P_\Lambda})|_{P_\Gamma}$ has a good filtration by $\underline{H}(\Phi_\Lambda)$. Thus we obtain:

(10.8.2) $Y(\lambda_7)|_{P_\Sigma}$ *has a good filtration for every* $\Sigma \subseteq \Delta$.

From Proposition 3.5.3 we obtain:

(10.8.3) *For every* G-*module* V *with a good filtration,* $V \otimes Y(\lambda_7)$ *has a good filtration.*

We now assume that the characteristic p of k is 7 . Let S denote the set of rational G-modules V such that $V \otimes V'$ has a good filtration for every rational G-module V' with a good filtration and $V|_{P_\Sigma}$ has a good filtration for every $\Sigma \subseteq \Delta$. The only reason that we were prevented, in the argument leading up to (10.2.9), from deducing $Y(\lambda_8) \in S$ in the case $p = 7$ is that we did not know that $Y(\lambda_7)$ belongs to S . Thus we now obtain, thanks to (10.8.2) and (10.8.3), $Y(\lambda_8) \in S$. By considering exterior powers $(\Lambda^i Y(\lambda_1)$ for $1 \le i \le 4$, $\Lambda^2 Y(\lambda_7)$ and $\Lambda^2 Y(\lambda_8))$ as in section 9.3 we now obtain that $Y(\tau) \in S$ for all $\tau \in X^+$. Thus we now have that $\underline{H}(G)$ is true when the characteristic p of k is 7 . Combining this with (10.2.15) we obtain:

(10.8.4) *If* $p \neq 2$ *then* $\underline{H}(G)$ *is true.*

Combining now (8.18), (9.8.3) and (10.8.4) we obtain, via Corollary 3.4.5 and Corollary 3.4.7, our strongest result to date.

(10.8.5) <u>Theorem</u> *Let* G *be a connected algebraic group over an algebraically closed field k and assume that either the characteristic of k is not* 2 *or that the root system of* G *involves no component of type* E_7 *or* E_8 . *Then* $\underline{H}(G)$ *is true, that is, the tensor product, of modules induced from one dimensional representations of a Borel subgroup, has a good filtration and the restriction to a parabolic subgroup, of any module induced from a one dimensional representation of a Borel subgroup, has a good filtration.*

11. An example and applications

11.1 An example

Let G be a connected, affine algebraic group over an algebraically closed field k. So far our analysis of $V|_H$ for a closed subgroup H of G and a G-module V with a good filtration has been confined to cases in which H is parabolic or, when G is reductive, a Levi factor of a parabolic subgroup. The reader will no doubt be wondering whether this is unnecessarily restrictive and may perhaps be thinking that $V|_H$ will have a good filtration for every module V with a good filtration for a reductive group G and reductive subgroup H. We now give an example to show that this is not the case.

Let k be an algebraically closed field of characteristic $p > 0$ and let k_0 be the prime subfield of k. Suppose that H is an affine, reductive, non-abelian group over k, defined over k_0 and containing a split maximal torus T, for example $H = GL_n(k)$, for $n > 1$ with the obvious k_0-form and T the subgroup consisting of the diagonal matrices. In what follows weights and characters are computed with respect to the split torus T. Let Y be a finite dimensional rational H-module such that the representation $H \to GL(Y)$ induces an isomorphism between H and the image. Now let Z be any finite dimensional rational representation of H such that the derived subgroup of H does not act trivially on Z. We choose an integer n such that Z^{F^n} does not have a good filtration, where F denotes the Frobenias operator on rational H-modules. (Let λ be a maximal weight of Z not fixed by the Weyl group, if Z^{F^n} has a good filtration then $\dim Z \geq \dim Y(p^n\lambda)$, but by Weyl's Character Formula $\dim Y(p^n\lambda)$ tends to infinity with n so that Z^{F^n} does not have a good filtration for n large - actually $n = 1$ will do). We let $V = Y \oplus Z^{F^n}$ and put $G = GL(V)$. By the choice of Y the representation $H \to GL(V)$ induces an isomorphism of H with the image, we thus identify H with a closed subgroup of GL(V). The natural module V for GL(V) has a good filtration (by Proposition 3.2.7(iii) it is enough to prove that V as an SL(V)-module has a good filtration; however $V|_{SL(V)}$ is the module induced from the fundamental weight corresponding to a terminal vertex of the Dynkin diagram). By Corollary 3.2.5 and the construction $V|_H$

does not have a good filtration.

11.2 Cohomology

We obtain, for most reductive groups, the following generalisation of a result of Cline, Parshall, Scott and van der Kallen (13.3) Corollary of [20]).

(11.2.1) Theorem *Let* G *be a reductive group over an algebraically closed field* k *and suppose that the characteristic of* k *is not* 2 *or that* G *involves no component of type* E_7 *or* E_8. *Let* Ω *be any subset of the base* Δ *let* $\Gamma(r)$ $(r = 2,2,\ldots,n)$ *be subsets of* Δ *containing* Ω *and let, for* $1 \le r \le n$, V_r *be a* $G_{\Gamma(r)}$*-module with a good filtration. Then for all* $i > 0$,

$$H^i(G_\Omega, V_1|_{G_\Omega} \otimes V_2|_{G_\Omega} \otimes \ldots \otimes V_n|_{G_\Omega}) = 0$$

in particular, for $\lambda_1,\lambda_2,\ldots,\lambda_n \in X^+$ *and* $i > 0$,

$$H^i(G,Y(\lambda_1) \otimes Y(\lambda_2) \otimes \ldots \otimes Y(\lambda_n)) = 0 .$$

Proof Since cohomology commutes with direct limits it is enough to prove the result for V_r finite dimensional $(1 \le r \le n)$. By the long exact sequence of cohomology it is enough to prove the result with

$V_r = Ind_{B_{\Gamma(r)}}^{P_{\Gamma(r)}}(\lambda_r)$ for some $\lambda_r \in X_{\Gamma(r)}$. But G_Ω is a Levi factor of a parabolic subgroup of $G_{\Gamma(r)}$ so that, by Proposition 3.3.2 and (10.8.5) Theorem each $V_r|_{G_\Omega}$ has a good filtration. Hence by (10.8.5) Theorem $V_1|_{G_\Omega} \otimes V_2|_{G_\Omega} \otimes \ldots \otimes V_n|_{G_\Omega}$ has a good filtration. Thus by the long exact sequence of cohomology it suffices to note that $H^i(G_\Omega,Y_\Omega(\lambda)) = 0$ for all $\lambda \in X_\Omega$. This is true by (3.3) Corollary of [20] (listed here as (2.1.4)). This completes the proof of the Theorem.

It is interesting to note that, for a reductive group G, the in particular part of the Theorem with $i = 1$ and $n = 3$ is equivalent to $\underline{H}_1(G)$. If $\underline{H}_1(G)$ holds then $Y(\lambda_1) \otimes Y(\lambda_2) \otimes Y(\lambda_3)$ has a good filtration so that, by the long exact sequence of cohomology and (3.3) Corollary of [20], $H^1(G_1Y(\lambda_1) \otimes Y(\lambda_2) \otimes Y(\lambda_3)) = 0$. If on the other hand $H^1(G, Y(\lambda_1) \otimes Y(\lambda_2) \otimes Y(\lambda_3)) = 0$ for all $\lambda_1, \lambda_2, \lambda_3 \in X^+$ then from the cohomological criterion, Corollary 1.3 of [25], $Y(\lambda_1) \otimes Y(\lambda_2)$ has a good filtration. Thus $\underline{H}_1(G)$ is true. It is difficult to see why the transition from Cline, Parshall, Scott and van der Kallen's elegant result ((3.3) Corollary of [20]) on the acyclicity of the tensor products $Y(\lambda) \otimes Y(\mu)$ to the acyclicity of the three fold tensor products $Y(\lambda) \otimes Y(\mu) \otimes Y(\tau)$ should be so complicated.

11.3 Homomorphisms between Weyl modules

Homomorphisms between Weyl modules for a semisimple group have been studied in [15], [5] and [16]. Since the dual of a Weyl module is a module induced from a one dimensional module for a Borel subgroup the study of homomorphisms between Weyl modules is equivalent to the study of $Hom_G(Y(\lambda), Y(\mu))$ for $\lambda, \mu \in X^+$. In this section we offer the following.

(11.3.1) Proposition *Let* G *be a reductive group over an algebraically closed field* k *and suppose that either the characteristic of* k *is not* 2 *or* G *involves no component of type* E_7 *or* E_8. *If* $\lambda - \mu \in \mathbb{Z}\Omega$ *for* $\lambda, \mu \in X^+$ *and some* $\Omega \subseteq \Delta$, *then*

$$Hom_G(Y(\lambda), Y(\mu)) \cong Hom_{G_\Omega}(Y_\Omega(\lambda), Y_\Omega(\mu)) .$$

Proof By (5.10), $Hom_G(Y(\lambda), Y(\mu))$ is isomorphic to $Hom_{P_\Omega}(Y_\Omega(\lambda), Y_\Omega(\mu))$. But U^Ω acts trivially on $Y_\Omega(\lambda)$ and $Y_\Omega(\mu)$ so, by the Levi decomposition, $Hom_{P_\Omega}(Y_\Omega(\lambda), Y_\Omega(\mu)) = Hom_{G_\Omega}(Y_\Omega(\lambda), Y_\Omega(\mu))$.

For the analogous result (without assumptions on the reductive group G) on composition multiplicities, see [27].

11.4 Canonical Products

Let G be an affine algebraic group over an algebraically closed field k, H a closed subgroup of G and L a closed subgroup containing H. Let V, V' be rational H-modules. Then we have canonical maps $\phi: Ind_H^G(V) \otimes Ind_H^G(V') \to Ind_H^G(V \otimes V')$ and $\psi: Ind_H^G(V) \to Ind_H^L(V)$

satisfying $\phi(f \otimes f')(g) = f(g) \otimes f'(g)$ for $f \in Ind_H^G(V)$, $f' \in Ind_H^G(V')$, $g \in G$ and $\psi(f)(x) = f(x)$ for $x \in L$.

The following result was proved for the classical groups by Lakshmibai, Musili and Seshadri, [44], in the context of sheaf cohomology on P/B. The argument of proof was also noticed by Wang Jian-pan, Lemma of section 6 of [52].

(11.4.1) Theorem *Let G be a reductive group over an algebraically closed field k and assume that either the characteristic of k is not 2 or that G involves no component of type E_7 or E_8. Let $\lambda, \mu \in X^+$ and let Ω be a subset of the base Δ. Then the canonical maps $\phi: Y(\lambda) \otimes Y(\mu) \to Y(\lambda+\mu)$ and $\psi: Y(\lambda) \to Y_\Omega(\lambda)$ are surjective.*

Proof The unique maximal weight of $Y(\lambda) \otimes Y(\mu)$ is $\lambda + \mu$. By (10.8.5) Theorem $Y(\lambda) \otimes Y(\mu)$ has a good filtration and so by Lemma 3.2.3 there is a submodule, say M, such that $(Y(\lambda) \otimes Y(\lambda))/M$ is isomorphic to $Y(\lambda+\mu)$. Now $Y(\lambda+\mu)$ has a simple B-socle $w_o(\lambda+\mu)$ and $w_o(\lambda+\mu)$ is not a weight of M so that $\phi(M) = 0$. Since ϕ is non zero it induces a non-zero map $\bar{\phi}: (Y(\lambda) \otimes Y(\mu))/M \to Y(\lambda+\mu)$ which must be an isomorphism by (1.5.3). Hence ϕ is surjective.

The proof that ψ is surjective is similar.

11.5 Filtrations over $\mathbb{Z}$

I am grateful to J.C. Jantznn for pointing out the application of our results, to filtrations over $\mathbb{Z}$, with which this section is concerned. I am also grateful to M. Koppinen for shortening the proof of (11.5.2) Lemma.

Let $\underline{g}$ be a finite dimensional, complex, semisimple Lie algebra with Cartan subalgebra $\underline{h}$ and root system Φ. Let $\underline{b}$ be a Borel subalgebra and $\Phi^+ = \{-\alpha: \alpha \in \Phi^-\}$, where Φ^- is the set of non-zero weights of the adjoint action of $\underline{h}$ on $\underline{b}$. Let $\{X_\alpha, H_i: \alpha \in \Phi, 1 \le i \le \ell\}$ be a Chevalley basis of $\underline{g}$ and let $U_{\mathbb{Z}}$ denote the Kostant $\mathbb{Z}$-form of $U(\underline{g})$ - the enveloping algebra of $\underline{g}$ - defined by the Chevalley basis. The Hopf algebra structure on $U(\underline{g})$ gives naturally the structure of a $\mathbb{Z}$-Hopf algebra on $U_{\mathbb{Z}}$ and hence the structure of a k-Hopf algebra on the hyperalgebra $U_k = U_{\mathbb{Z}} \otimes_{\mathbb{Z}} k$, for any field k. Thus it makes sense to regard the tensor product of $U_{\mathbb{Z}}$ (respectively U_k) modules as a $U_{\mathbb{Z}}$ (respectively U_k) module.

Let X denote the set of integral weights of $\underline{h}$ and X^+ the set of dominant integral weights. For $\lambda \in X^+$ we have a simple, finite dimensional $U(\underline{g})$-module of highest weight λ. Let $0 \neq v$ be an element of $V(\lambda)$ of weight λ and put $V_{\mathbb{Z}}(\lambda) = U_{\mathbb{Z}}v$ - a $U_{\mathbb{Z}}$ submodule of $V(\lambda)$ which is free of finite rank as a $\mathbb{Z}$-module. We define, for a field k, $V_k(\lambda)$ to be the U_k-module $V_{\mathbb{Z}}(\lambda) \otimes_{\mathbb{Z}} k$.

We denote by U_k^0 the k-subalgebra of U_k generated by the elements $H_{i,b} = \binom{H_i}{b} \otimes 1$ $(1 \leq i \leq \ell,\ b > 0)$. Each $\lambda \in X$ defines a k-algebra map $\tilde{\lambda}: U_k^0 \to k$ satisfying $\tilde{\lambda}(H_{i,b}) = \binom{\lambda(H_i)}{b} 1_k$. The map $\lambda \to \tilde{\lambda}$ from X to the set of k-algebra homomorphisms from U_k^0 to k is a monomorphism so we identify X with a subset of $Hom_{k-alg}(U_k^0, k)$. For a U_k-module V and $\mu \in X$ we write

$$V^{\mu} = \{v \in V\colon hv = \mu(h)v \quad \text{for all} \quad h \in U^0\}$$

- the μ weight space of V. Any finite dimensional U_k-module is the direct sum of its weight spaces.

Now suppose that k is algebraically closed and let G_k be the universal Chevalley group defined by a $U_{\mathbb{Z}}$ lattice in some finite dimensional $U(\underline{g})$-module. Let T_k and B_k denote the torus and negative Borel subgroup arising via the Chevalley construction. Then X may be naturally identified with the character group of T_k. There is an equivalence of categories between finite dimensional rational G_k-modules and finite dimensional U_k-modules (by Verma's conjecture, proved by Cline, Parshall and Scott in [19] and Sullivan in [50]). Moreover the weight spaces correspond under this equivalence of categories. Thus we may restate Jantzen's result, 5.2(2) of [41] as:

(11.5.1) *If* V *is a finite dimensional* U_k*-module with a Weyl filtration and* v *is a non-zero weight vector of maximal weight* λ *then* $U_k v$ *is isomorphic to* $V_k(\lambda)$ *and* $V/U_k v$ *has a Weyl filtration.*

By a Weyl filtration for a U_k-module we mean a descending filtration in which successive quotients are zero or isomorphic to $V_k(\lambda)$ for some $\lambda \in X^+$. The dual of $V_k(\lambda)$ $(\lambda \in X^+)$ as a G_k-module is, by §1 of [40], isomorphic to $Y(\mu)$ for some dominant weight μ. Thus we deduce from (10.8.5) Theorem by dualizing:

(11.5.2) *If* $\underline{g}$ *has no component of type* E_7 *or* E_8 *or the characteristic of* k *is not* 2 *then, for every* $\lambda, \mu \in X^+$, $V_k(\lambda) \otimes V_k(\mu)$ *has a Weyl filtration.*

By a Weyl filtration for a $U_{\mathbb{Z}}$-module we mean a descending filtration in which the successive quotients are zero or $V_{\mathbb{Z}}(\lambda)$ for some $\lambda \in X^+$. We wish to show that the above result remains true when k is replaced by $\mathbb{Z}$. To show this we need the following.

(11.5.3) Lemma *Let* M *be a* $U_{\mathbb{Z}}$*-module which is free of finite rank as a* $\mathbb{Z}$*-module. If the* U_k*-module* $M \otimes_{\mathbb{Z}} k$ *has a Weyl filtration for each algebraically closed field* k *then* M *has a Weyl filtration.*

Proof For a $U_{\mathbb{Z}}$-module V we denote by V_k the U_k-module $V \otimes_{\mathbb{Z}} k$, where k is a field. For $\lambda \in X$ we denote by V^λ the subgroup

$$\{v \in V: \binom{H_i}{b} v = \binom{\lambda(H_i)}{b} v \quad \text{for all} \quad 1 \le i \le \ell,\ b > 0\}\ .$$

If V is free of finite rank as a $\mathbb{Z}$-module then it follows from Corollary 1, p.17 of [49] that V is the direct sum of the subgroups V^λ.

We prove the Lemma by induction on the rank of M. Suppose the rank of M is greater than 0 and the Lemma holds for all modules of smaller rank. Let λ be a maximal weight of M. Then

$$M^\lambda = \mathbb{Z}v \oplus D \tag{1}$$

for some $v \in M^\lambda \setminus \{0\}$ and $\mathbb{Z}$-submodule D. Let $R = U_{\mathbb{Z}}v$. Since tensoring with C over $\mathbb{Z}$ is exact, $R_{\mathbb{C}}$ is naturally identified with the submodule $U_{\mathbb{C}}(v \otimes 1)$ of $M_{\mathbb{C}}$. Hence $R_{\mathbb{C}} \cong V_{\mathbb{C}}(\lambda)$ by (11.5.1). Furthermore, since R is free over $\mathbb{Z}$, it can be identified with the submodule $Uv \otimes 1 = U(v \otimes 1)$ of $R_{\mathbb{C}}$, where $U = U(\underline{g})$. Thus $R \cong V_{\mathbb{Z}}(\lambda)$.

Let k be an algebraically closed field. Consider the map $R_k \to M_k$ induced by the inclusion $R \to M$. Its image $U_k(v \otimes 1)$ is non-zero by (1), so it is isomorphic to $V_k(\lambda)$ and $M_k/U_k(v \otimes 1)$ has a Weyl filtration by (11.5.1). Since R_k is also isomorphic to $V_k(\lambda)$ the sequence

$$0 \to R_k \to M_k \to (M/R)_k \to 0$$

is exact. This shows that $Tor_1(M/R,k) = 0$ for any algebraically closed field k, and consequently M/R is torsion free. According to the induction hypothesis M/R has a Weyl filtration since $(M/R)_k \cong M_k/U_k(v \otimes 1)$ has one. The Lemma follows.

Combining (11.5.2) and (11.5.3) we obtain:

(11.5.4) Proposition *Suppose that* $\underline{g}$ *has no component of type* E_7 *or type* E_8. *Then for any* $\lambda, \mu \in X^+$, $V_{\mathbb{Z}}(\lambda) \otimes V_{\mathbb{Z}}(\mu)$ *has a Weyl filtration.*

For an arbitrary complex, semisimple Lie algebra one can obtain the corresponding result for Weyl modules over $U_{\mathbb{Z}[\frac{1}{2}]} = U_{\mathbb{Z}} \otimes \mathbb{Z}[\frac{1}{2}]$, where $\mathbb{Z}[\frac{1}{2}]$ is the localisation of $\mathbb{Z}$ at 2.

One may also obtain under the hypothesis of (11.5.4) that the restriction of $V_{\mathbb{Z}}(\lambda)$ to each sub Kostant $\mathbb{Z}$-form, defined by a subset of Δ, has a Weyl filtration.

12. Miscellany

12.1 Injective modules for parabolic subgroups

In this chapter we allow ourselves leisure to reflect on some of the key ideas which occur in the monograph. We start here by generalising to parabolic subgroups of reductive groups the main results of [25] on injective modules for semisimple groups. It was shown in [25] that the injective indecomposable modules for a semisimple, simply connected group have a good filtration (this was also announced by L.L. Scott in [46]). The construction of a filtration, given in [25], rests on acyclicity results, proved by Cline, Parshall, Scott and van der Kallen in [20], for certain modules. These acyclicity results are valid, by the same arguments, for reductive groups (this is spelled out in (2.1.4)). We shall therefore appeal directly to [25] for filtration results for reductive, rather than semisimple, groups and leave it to the reader to make the appropriate modifications.

Let G be a connected, reductive, affine algebraic group over an algebraically closed field k. We adopt our usual conventions and notations. Thus, for example, T denotes a maximal torus in G, $X = X(T)$ the weight lattice, B a Borel subgroup containing T, Φ the root lattice and Δ a base (determined by B). For a subset Ω of Δ, P_Ω denotes the corresponding parabolic subgroup. For $\lambda \in X_\Omega$ we denote by $E_\Omega(\lambda)$ the injective envelope of $Y_\Omega(\lambda)$ as a P_Ω-module and by $J_\Omega(\lambda)$ the injective envelope of $Y_\Omega(\lambda)$ as a G_Ω-module. Since $L_\Omega(\lambda)$ is an essential submodule of $Y_\Omega(\lambda)$, $E_\Omega(\lambda)$ is also the injective envelope of $L_\Omega(\lambda)$ as a P_Ω-module and $J_\Omega(\lambda)$ is the injective envelope of $L_\Omega(\lambda)$ as a G_Ω-module.

Let V be a P_Ω-module with a good filtration $\mathcal{V} = \{0 = V_0, V_1, \ldots\}$ and $\lambda \in X_\Omega$. By the multiplicity of $Y_\Omega(\lambda)$ in V with respect to $\mathcal{V}$ we understand the cardinality of $\{i > 0 : V_i/V_{i-1} \cong Y_\Omega(\lambda)\}$ and denote this by $(V : Y_\Omega(\lambda))_{\mathcal{V}}$. Note that:

(12.1.1) *The multiplicity* $(V : Y_\Omega(\lambda))_{\mathcal{V}}$ *is equal to* *dim* $(V \otimes Y_\Omega(-s_\Omega\lambda))^{G_\Omega}$. *In particular it is independent of the choices of good filtration and will therefore be denoted simply by* $(V : Y_\Omega(\lambda))$

This follows from Proposition 2.2 of [25] and the fact that the G_Ω fixed point functor commutes with direct limits. We deduce from

Theorem 1.1' of [25] (a restatement of a result of Cline, Parshall, Scott and van der Kallen), the fact that G_Ω cohomology commutes with direct limits and (12.1.1):

(12.1.2) *For* $\lambda \in X_\Omega$, $(-, Y_\Omega(\lambda))$ *is additive on short exact sequences of* P_Ω*-modules with a good filtration.*

We now show:

(12.1.3) <u>Theorem</u> *Any rationally injective* P_Ω*-module of countable dimension has a good filtration.*

Let E be any injective P_Ω-module of countable dimension. By (1.5.h) of [29], E is a direct sum of countably many injective indecomposable modules. By Proposition 3.1.1 we may assume that E is indecomposable so that, by (1.5g)(i) of [29] we may assume that $E = E_\Omega(\lambda)$ for some $\lambda \in X_\Omega$.

Recall the function $\delta_\Omega : \mathbb{R} \otimes_{\mathbb{Z}} X \to \mathbb{R}$ (see section 3.3). This is defined to be the $\mathbb{R}$-linear map such that $\delta_\Omega(\alpha)$ is 0 if $\alpha \in \Omega$ and 1 if $\alpha \in \Delta\backslash\Omega$ and δ_Ω is 0 on the orthogonal complement to the $\mathbb{R}$ span of Δ . For a rational T-module V and $r \in \mathbb{Q}$, $F_\Omega^{(r)}(V)$ is the sum of the weight spaces V^τ $(\tau \in X)$ such that $\delta_\Omega(\tau) = r$ and $F_\Omega^{r}(V)$ is the sum of the $F_\Omega^{(s)}(V)$ for $s \leq r$.

Let $r_0 = \delta_\Omega(\lambda)$. We claim:

(12.1.4) *If* $F_\Omega^{(s)}(E) \neq 0$ *then* $s \geq r_0$.

If $F_\Omega^{(s)}(E) \neq 0$ then $F_\Omega^{s}(E) \neq 0$ and since E is an essential extension of $L_\Omega(\lambda)$, $L_\Omega(\lambda) \leq F_\Omega^{s}(E)$. But λ is a weight of $L_\Omega(\lambda)$ so we have, from the definition, $\delta_\Omega(\lambda) = r_0 \leq s$.

Let E' be the sum of all the weight spaces E^τ such that $\tau \in \lambda + \mathbb{Z}\Phi$. It is easy to see that E' is a direct summand of E containing $L_\Omega(\lambda)$ and thus $E' = E$. It follows that if $F_\Omega^{(r)}(E) \neq 0$ then $r - r_0 \in \mathbb{Z}$. We obtain from (12.1.4) and Proposition 3.3.1:

(12.1.5) $E = \bigoplus_{n=0}^{\infty} F_\Omega^{(r_0+n)}(E)$ *as* G_Ω*-modules.*

Thus, by Proposition 3.1.1, in order to establish (12.1.3) it suffices to prove that $F_\Omega^{r_o+n+1}(E)/F_\Omega^{r_o+n}(E)$ has a good filtration, for $n \geq 0$. However the unipotent radical U^Ω of P_Ω acts trivially on $F_\Omega^{r_o+n+1}(E)/F_\Omega^{r_o+n}(E)$ so that a subspace is a P_Ω submodule if and only if it is a G_Ω submodule. However, by (12.1.5), as a G_Ω module, $F_\Omega^{r_o+n+1}(E)/F_\Omega^{r_o+n}(E)$ is isomorphic to $F_\Omega^{(r_o+n+1)}(E)$. Now P_Ω is the semidirect product of G_Ω and U^Ω so the quotient space P_Ω/G_Ω is isomorphic to U^Ω - in particular it is affine. Hence by 4.3 Theorem and 2.1 Proposition of [18], E is an injective G_Ω-module. By (12.1.5), $F_\Omega^{(r_o+n+1)}(E)$ is a direct summand of E as a G_Ω-module and

3.1.1, $O_\pi(V) = \bigcup_{n=0}^{\infty} U_n$ has a good filtration. Also $Q_n \cong V_n/(O_\pi(V) \cap V_n) = V_n/U_n$ and so has a good filtration by Proposition 3.2.4. Thus $V/O_\pi(V) = \bigcup_{n=0}^{\infty} Q_n$ has a good filtration by Proposition 3.1.1. Finally, if $(V/O_\pi(V) : Y_\Omega(\tau)) \neq 0$ then $(Q_n : Y_\Omega(\tau)) \neq 0$ for some n and so $(V_n/O_\pi(V_n) : Y_\Omega(\tau)) \neq 0$ which implies that $\tau \notin \pi$ by the finite dimensional case.

(12.1.7) *If, for* $\lambda, \tau \in X_\Omega$, $(E_\Omega(\lambda) : Y_\Omega(\tau)) \neq 0$ *then* $\tau \geq \lambda$.

Let $\pi = \{\xi \in X_\Omega : \xi \leq \tau\}$. Then, by (12.1.2) and (12.1.6), $(O_\pi(E) : Y_\Omega(\tau)) \neq 0$, where $E = E_\Omega(\lambda)$, in particular $O_\pi(E) \neq 0$. Hence $L_\Omega(\lambda)$ is a composition factor of $Y_\Omega(\xi)$ for some $\xi \in \pi$. But ξ is the unique highest weight of $Y_\Omega(\xi)$ so that $\lambda \leq \xi \leq \tau$, as required.

(12.1.8) *For any* $\lambda, \tau \in X_\Omega$, $(E_\Omega(\lambda) : Y_\Omega(\tau))$ *is finite.*

We fix λ and define, for $\mu \in X_\Omega$, $S(\mu) = \{\xi \in X_\Omega : \lambda \leq \xi \leq \tau\}$. We shall prove (12.1.8) by induction on $|S(\tau)|$. If $|S(\tau)| = 0$ then $(E: Y_\Omega(\tau)) = 0$ (where $E = E_\Omega(\lambda)$) by (12.1.7). Now suppose that $|S(\tau)| > 0$ and the assertion is true for $\tau' \in X_\Omega$ with $|S(\tau')| < |S(\tau)|$. Let $\pi = \{\xi \in X_\Omega : \xi \leq \tau\}$ and $\pi_o = \pi \backslash \{\tau\}$. For $\xi \in \pi_o$, $|S(\xi)| < |S(\tau)|$ so that $M = O_{\pi_o}(E)$ is finite dimensional by (12.1.6), (12.1.2) and the inductive hypothesis. We have $Ext^1_{P_\Omega}(L_\Omega(\tau), M) \cong H^1(P_\Omega, L_\Omega(\tau)^* \otimes M)$, which is isomorphic to $H^1(B, L_\Omega(\tau)^* \otimes M)$ by (2.1.3), and so is finite dimensional by (3.8) Corollary of [20] (or rather, since G is reductive but not necessarily semisimple, the argument of (3.8) Corollary of [20]). Let N be the dimension of $Ext^1_{P_\Omega}(L_\Omega(\tau), M)$ and suppose, for a contradiction, that $(E: Y_\Omega(\tau)) \geq N + 1$. By (12.1.6) and (12.1.2), $(E: Y_\Omega(\tau)) = (O_\pi(E) : Y_\Omega(\tau))$ which, again by (12.1.2), is equal to $(O_\pi(E)/M: Y_\Omega(\tau))$. Moreover, by (12.1.6), $(O_\pi(E)/M: Y_\Omega(\xi)) = 0$ for all $\xi \in \pi_o$ so that $O_\pi(E)/M$ has a good filtration in which each non-zero successive quotient is isomorphic to $Y_\Omega(\tau)$ and there are at least $N+1$ of these. Thus, by Lemma 3.2.1, there is a submodule K of E containing M such that K/M is isomorphic to a direct sum of $N+1$ copies of $Y_\Omega(\tau)$. Thus, by Schur's Lemma, $\dim \mathrm{Hom}_{P_\Omega}(L_\Omega(\tau),E/M) \geq N+1$. But applying $\mathrm{Hom}_{P_\Omega}(L_\Omega(\tau),-)$ to the short exact sequence $0 \to M \to E \to E/M \to 0$ gives $\mathrm{Hom}_{P_\Omega}(L_\Omega(\tau),E/M) \cong \mathrm{Ext}^1_{P_\Omega}(L_\Omega(\tau),M)$. The latter has dimension N and the former at least $N+1$. Hence $(E: Y_\Omega(\tau)) \leq N$ (actually the argument shows that $(E: Y_\Omega(\tau)) = N$) in particular the multiplicity is finite.

From (12.1.8) together with (12.1.7) and (12.1.6) we now obtain:

(12.1.9) *For any bounded saturated subset* π *of* X_Ω *and any* $\lambda \in X_\Omega$, $O_\pi(E_\Omega(\lambda))$ *is a finite dimensional* P_Ω*-module with a good filtration.*

One may recognize that certain modules have the form $O_\pi(E_\Omega(\lambda))$ by means of the following.

(12.1.10) *Let* $\lambda \in X_\Omega$ *and let* π *be a saturated subset of* X_Ω *containing* λ. *For a non-zero* (P_Ω,π) *submodule* V *of* $E = E_\Omega(\lambda)$ *the following are equivalent*

(i) $V = O_\pi(E)$;

(ii) $\mathrm{Ext}^i_{P_\Omega}(L_\Omega(\tau),V) = 0$ *for all* $\tau \in \pi$ *and* $i > 0$;

(iii) $\mathrm{Ext}^1_{P_\Omega}(L_\Omega(\tau),V) = 0$ *for all* $\tau \in \pi$.

For (i) implies (ii) we note that $\mathrm{Ext}^i_{P_\Omega}(L_\Omega(\tau),O_\pi(E)) \cong \mathrm{Ext}^{i-1}_{P_\Omega}(L_\Omega(\tau),E/O_\pi(E))$ by the long exact sequence of cohomology. If the latter is non-zero then by (12.1.6), the long exact sequence of cohomology and the fact that rational cohomology commutes with direct limits we obtain $\mathrm{Ext}^{i-1}_{P_\Omega}(L_\Omega(\tau),Y_\Omega(\xi)) \neq 0$ for some $\xi \in X_\Omega \backslash \pi$. This implies, by the argument of (2.1.4) that $\tau \geq \xi$ and so $\tau \notin \pi$.

Clearly (ii) implies (iii).

We now deduce (i) from (iii). Since V is a (P_Ω,π)-module, $V \leq O_\pi(E)$. Suppose $O_\pi(E)/V \neq 0$ and let M/V be a simple submodule of $O_\pi(E)/V$. Then $M/V \cong L_\Omega(\tau)$ for some $\tau \in \pi$ and, by the long exact sequence of cohomology, $Ext^1_{P_\Omega}(L_\Omega(\tau),V) \cong Hom_{P_\Omega}(L_\Omega(\tau),E/V)$ is non-zero. Thus $O_\pi(E)/V = 0$ and $V = O_\pi(E)$.

In an earlier draught of this work (12.1.9) and (12.1.10) were used to show that various modules, for parabolic subgroups of the exceptional groups, have good filtrations. We were subsequently able to do without these results, and indeed to shorten the calculations by proceeding by other means. However it is not inconceivable that these results will be useful in connexion with similar problems at some later date.

12.2 Kempf's Vanishing Theorem for rank one groups

The Vanishing Theorem of Kempf is of fundamental importance to our work - almost every cohomological calculation in the monograph depends in some way upon it. The existing proofs of the theorem rely, to a greater or lesser extent, on properties of the cohomology of sheaves on projective varieties, indeed it is usually stated in these terms. The statement of the theorem given in section 1.6 is a result on the derived functors of induction. In view of this and the extreme importance of the theorem to the representation theorist it is highly desirable to have a purely representation theoretic proof. Best of all would be to have a purely representation theoretic proof of moderate length which is equicharacteristic, i.e. does not distinguish between the cases $char\ k = 0$ and $char\ k > 0$. Unfortunately, to the author's knowledge, no such proof exists. Kempf's proof, [43], is equicharacteristic and entirely geometric. The proofs of Haboush, [31], and Andersen, [6] are much shorter than Kempf's and depend on properties of the Steinberg modules in characteristic p . They also use much less algebraic geometry than Kempf's proof (and so are a great improvement *to the representation theorist* over the original proof), virtually the only geometric fact they use is the ampleness of a certain line bundle on G/B .

We give here, for amusement only, representation theoretic proofs of two special cases of the Vanishing Theorem. In this section we treat rank one groups in arbitrary characteristic and in the next we use the category $\mathcal{O}$ of Bernstein, Gelfand and Gelfand to treat semisimple, simply connected groups in characteristic 0 .

Let G be a connected, reductive group of rank one over an algebraically closed field k . We assume the usual notations for subgroups of G and their representations.

(12.2.1) *For* $\mu \in X^+$, $Y(\mu)$ *has dimension* $(\mu,\alpha^\vee) + 1$ *and weights* $\mu, \mu - \alpha, \dots, \mu - (\mu,\alpha^\vee)\alpha$ *each with multiplicity one.*

Since $Y(\mu)$ has a simple socle $w_0\mu$, $Y(\mu)$ may be identified with a B submodule of the injective envelope $E(w_0\mu)$ of $w_0\mu$. Now (1.2.4) and (1.5.2) imply that the weights of $Y(\mu)$ are a subset of those listed above and each weight occurs with multiplicity one. Thus $\dim Y(\mu) \le (\mu,\alpha^\vee) + 1$ and, to prove (12.2.1), it suffices to show that $\dim Y(\mu) \ge (\mu,\alpha^\vee) + 1$. If $(\mu,\alpha^\vee) = 0$ then $\dim Y(\mu) = 1$ by (1.5.2). We next consider the case $\mu = \alpha$. Let G_1 denote the derived subgroup of G . We claim that $Y(\alpha)$ is isomorphic to the dual $Lie(G_1)^*$ of $Lie(G_1)$ on which G acts by the adjoint action. Now G_1 is a rank one semisimple group and so isomorphic to $SL_2(k)$ or $PSL_2(k)$. From this it may be easily checked that $Lie(G_1)^*$ has a simple $B_1 = B \cap G_1$ socle and so a simple B socle. Moreover the weights of $Lie(G_1)^*$ are α , 0 and $-\alpha$ so that $Lie(G_1)^*$ embeds in $Y(\alpha)$ by (1.5.4). Thus $\dim Y(\alpha) \ge 3$ and (12.2.1) holds for $\mu = \alpha$. We now argue, for general $\mu \in X^+$, by induction on $(\mu,\alpha^\vee)$. We have already done the case $(\mu,\alpha^\vee) = 0$. If $(\mu,\alpha^\vee) = 1$ we have $\dim Y(\mu) = 2$ by (1.5.4) so (12.2.1) is true for μ . If $(\mu,\alpha^\vee) = 2$, it follows from the universal property of induction that $Y(\mu) \cong Y(\alpha) \otimes Y(\mu-\alpha)$ and so $\dim Y(\mu) = \dim Y(\alpha) \,.\, \dim Y(\mu-\alpha) = 3$. Now suppose $(\mu,\alpha^\vee) > 2$ and the result holds for $\tau \in X^+$ with $(\tau,\alpha^\vee) < (\mu,\alpha^\vee)$. Consider first the case in which $(\mu,\alpha^\vee)$ is even. By reciprocity $Y(\mu-\alpha) \otimes Y(\alpha)$ is isomorphic to $Ind_B^G((\mu-\alpha) \otimes Y(\alpha))$. The weights of $(\mu-\alpha) \otimes Y(\alpha)$ are μ , $\mu - \alpha$ and $\mu - 2\alpha$. Since induction is left exact we obtain that $\dim Y(\mu-\alpha) \otimes Y(\alpha)$ is less than or equal to $\dim Y(\alpha) + \dim Y(\mu-\alpha) + \dim Y(\mu - 2\alpha)$. This gives $\dim Y(\mu) \ge 3((\mu-\alpha,\alpha^\vee) + 1) - ((\mu-\alpha,\alpha^\vee) + 1) - ((\mu-2\alpha,\alpha^\vee) + 1)$ i.e. $\dim Y(\mu) \ge (\mu,\alpha^\vee) + 1$ as required. Now suppose $(\mu,\alpha^\vee) = 2m+1$ is odd and let $\lambda = \mu - m\alpha$. Applying the above argument to $Y(\mu-\lambda) \otimes Y(\lambda)$ gives $\dim Y(\mu) \ge (\mu,\alpha^\vee) + 1$ and thus completes the proof of (12.2.1).

(12.2.2) *For* $\mu \in X^+$, $(R^i Ind_B^G)\mu = 0$ *for all* $i > 0$.

For $n \ge 0$ the B socle of $Y(n\alpha)$ is $-n\alpha$ so we may, and do, identify $E_n = Y(n\alpha) \otimes (\mu + n\alpha)$ with a B submodule of $E = E(\mu)$. We let $Q = E/\mu$ and $Q_n = E_n/\mu$. By considering the long exact sequence of cohomology arising from the short exact sequence $0 \to \mu \to E \to Q \to 0$ we see that, to prove $(RInd_B^G)\mu = 0$, it suffices to show that this natural map $Ind_B^G(E) \to Ind_B^G(Q)$ is surjective. Hence it is enough to show that $Ind_B^G(E_n) \to Ind_B^G(Q_n)$ is surjective for each $n > 0$. The short exact sequence $0 \to \mu \to E_n \to Q_m \to 0$ gives rise to an exact sequence

$$0 \to Y(\mu) \to Ind_B^G(E_n) \to Ind_B^G(Q_n) .$$

Since $E_n = Y(n\alpha) \otimes (\mu + n\alpha)$ by the tensor identity $Ind_B^G(E_n)$ is isomorphic to $Y(n\alpha) \otimes Y(\mu + n\alpha)$ which has dimension $(2n + 1)(2n + m + 1)$, where $m = (\mu,\alpha^\vee)$. It follows from (12.2.1) that the weights to Q_n are $\mu + \alpha, \mu + 2\alpha, \ldots, \mu + 2n\alpha$. Thus by (12.2.1) and the left exactness of induction the dimension of $Ind_B^G(Q_n)$ is at most $(m + 3) + (m + 5) + \ldots + (m + 4n + 1)$ which is $2n(m + 2n + 2)$. However $(2n + 1)(2n + m + 1) = (m + 1) + 2n(m + 2n + 2)$ so that $dim\ Ind_B^G(E_n) \geq dim\ Y(\mu) + dim\ Ind_B^G(Q_n)$ and it follows that the map $nd_B^G(E_n) \to Ind_B^G(Q_n)$ is surjective. Hence $(RInd_B^G)\mu = 0$ for all $\in X^+$. We can now easily complete the proof of (12.2.2) by induction n i. Suppose $i > 1$ and $(R^{i-1}Ind_B^G)\tau = 0$ for all $\tau \in X^+$. it ollows by the long exact sequence of cohomology that $(R^{i-1}Ind_B^G)M = 0$ or any finite dimensional B-module such that every weight is dominant nd then one obtains that $(R^{i-1}Ind_B^G)M = 0$ for a B-module of arbitrary imension with every weight dominant, by (1.1.1). But, for $\mu \in X^+$, very weight of $Q = E(\mu)/\mu$ is dominant, by (1.2.4) so that $R^iInd_B^G) \cong (R^{i-1}Ind_B^G)Q = 0$, as required.

To complete the picture we deduce (1.6.1) and (1.6.2) for G from e Vanishing Theorem.

2.2.3) *For every finite dimensional* B-*module* M *and* $i \geq 0$, $^iInd_B^G)M$ *is finite dimensional. For* $i > 1$, $R^iInd_B^G = 0$.

By the long exact sequence of cohomology and (1.1.1) it suffices to show that, for $\mu \in X$, $(R^i Ind_B^G)\mu$ is 0 if $i > 1$ and finite dimensional for $i = 0,1$. For $i = 0$, $(R^i Ind_B^G)\mu = Ind_B^G\mu$ is finite dimensional by (1.5.1) and (12.2.1). By the Vanishing Theorem it suffices to show that, for $\mu \in X$ with $(\mu,\alpha^\vee) > 0$, $(R^i Ind_B^G)(-\mu)$ is finite dimensional for $i = 1$ and is 0 for $i > 1$. If $(\mu,\alpha^\vee) = 1$ the short exact sequence

$$0 \to -\mu \to Y(-\mu + \alpha) \to -\mu + \alpha \to 0$$

Kempf's Vanishing Theorem and the tensor identity show that $(R^i Ind_B^G)(-\mu) = 0$ for all i. For $(\mu,\alpha^\vee) > 0$, embeddings in $E(s_\alpha\mu - \mu + \alpha)$ give a short exact sequence of B-modules

$$0 \to Y(\mu - \alpha) \otimes -\mu \to Y(\mu) \otimes (-\mu + \alpha) \to Q \to 0 .$$

The weights of Q are 0 and α so that $(R^i Ind_B^G)Q = 0$ for all $i > 0$. The long exact sequence of cohomology and the tensor identity gives an exact sequence

$$Ind_B^G Q \to Y(\mu - \alpha) \otimes (RInd_B^G)(-\mu) \to Y(\mu) \otimes (RInd_B^G)(-\mu + \alpha)$$

and isomorphisms $Y(\mu - \alpha) \otimes (R^i Ind_B^G)(-\mu) \to Y(\mu) \otimes (R^i Ind_B^G)(-\mu + \alpha)$ for $i > 1$. Thus $(RInd_B^G)(-\mu)$ is finite dimensional if $(RInd_B^G)(-\mu + \alpha)$ is and $(R^i Ind_B^G)(-\mu)$ is 0, for $i > 1$, provided that $(R^i Ind_B^G)(-\mu + \alpha)$ is 0. This gives a proof by induction on $(\mu,\alpha^\vee)$.

12.3 The Vanishing Theorem in characteristic 0

In characteristic 0 the Vanishing Theorem is a weak version of Bott's Theorem which describes all the $(R^i Ind_B^G)\lambda$, $\lambda \in X$, $i \geq 0$. A very short proof of Bott's Theorem was given by Demazure, [22]. Here we present a rather whimsical proof of the Vanishing Theorem in which the result is deduced from properties of the category $\mathcal{O}$ of Bernstein, Gelfand and Gelfand, [11]. The idea is to describe induction in terms of a functor on $\mathcal{O}$ taking a module to its largest finite dimensional quotient and use properties of the projectives in $\mathcal{O}$ to say something about the derived functors.

We first recall the definition of $\mathcal{O}$ and some of its main feature (see [39] for example). Let G denote a semisimple, simple connected algebraic group over a field k of characteristic 0 and let $\underline{g}$ be the Lie algebra of G. We use the usual notational conventions, for subgroups of G, Lie subalgebras of $\underline{g}$ (see section 11.5) and their representations, with one important exception. We depart from our usual convention in that B (resp. $\underline{b}$) now denotes a positive Borel subgroup (resp. subalgebra). This is done to bring us into line with the established notation pertaining to $\mathcal{O}$. The objects in $\mathcal{O}$ are the $U(\underline{g})$-modules M which have the following properties:

(i) M is the direct sum of its $\underline{h}$ weight-spaces;
(ii) For every $m \in M$ we have $\dim U(\underline{b})m < \infty$;
(iii) M is a finitely generated $U(\underline{g})$-module.

The morphisms in $\mathcal{O}$ are the $\underline{g}$-module homomorphisms. For any finite dimensional $\underline{b}$-module V we obtain $I(V) = U(\underline{g}) \otimes_{U(\underline{b})} V$ in $\mathcal{O}$. The $\underline{g}$-module obtained from the one dimensional $\underline{b}$-module of weight $\lambda \in X$ is denoted by $M(\lambda)$. The category $\mathcal{O}$ has enough projectives, indeed every object has a projective cover. The projective cover of $M(\lambda)$ is denoted by $Q(\lambda)$. Each $Q(\lambda)$ has a filtration of finite length in which each quotient has the form $M(\mu)$ for some $\mu \in X$ and the multiplicity $(Q(\lambda): M(\mu))$ of $M(\mu)$ in any such filtration is equal to the composition multiplicity $[M(\mu): L(\lambda)]$, where $L(\lambda)$ denotes the simple $U(\underline{g})$-module of highest weight λ. Moreover whenever $[M(\mu): L(\lambda)] \neq 0$ then $\mu \geq \lambda$ and $\lambda = w.\mu$ for some w in the Weyl group W. From these facts one easily obtains:

(12.3.1) $M(\lambda)$ *is projective for* $\lambda \in X^+$.

An important subcategory of $\mathcal{O}$ is the category of finite dimensional $U(\underline{g})$-modules. A finite dimensional G-module V may be regarded naturally as a $U(\underline{g})$-module which we also denote by V. Moreover this process of regarding finite dimensional G-modules as $U(g)$-modules determines an equivalence of categories between M'_G, the category of finite dimensional rational G-modules, and the category of finite dimensional $U(g)$-modules.

Suppose that $V \in \mathcal{O}$. The conditions (i)-(iii) above ensure that the set of weights of V is bounded above and that each weight space of V is finite dimensional. It follows that the sum, say V^+, of the weight spaces V^μ for which μ is conjugate under the Weyl group to a dominant weight, is finite dimensional. However every weight of a finite dimensional $U(g)$-module M is conjugate to a dominant weight of

M. It follows that, for any finite dimensional quotient M of V, $\dim M \le \dim V^+$ and so there is a unique, largest finite dimensional quotient of V, which we denote by $Q(V)$. In this way we obtain a right exact functor $Q\colon \mathcal{O} \to M_G$. We denote by L^iQ $(i \ge 0)$ the left derived functors. These functors satisfy the following tensor identity.

(12.3.2) *For* $V, M \in \mathcal{O}$ *with* M *finite dimensional,* $L^iQ(V \otimes M) \cong L^iQ(V) \otimes M$ *for all* $i \ge 0$.

This is easily demonstrated in dimension 0 and follows in arbitrary dimension by tensoring a projective resolution of V by M to calculate the derived functors on $V \otimes M$.

For a finite dimensional module V we denote by V^* the dual module $Hom_k(V,k)$.

(13.3.3) *For every* $i \ge 0$ *the functors* $F^i, G^i\colon M'_B \to M_G$, *defined by* $F^iV = (R^i Ind_B^G)V$ *and* $G^iV = (L^iQ(IV^*))^*$, *are naturally isomorphic.*

Here M'_B denotes the category of finite dimensional B-modules and, in the definition of G^iV, V^* is regarded in the natural way as a $U(\underline{b})$-module. The first thing to do is establish the isomorphism in dimension 0. We define a map $\theta\colon V^* \to Q(IV^*)$ by the rule that, for $\alpha \in V^*$, $\theta(\alpha)$ is the image of $1 \otimes \alpha$ under the natural map $IV^* \to Q(IV^*)$. Let $\eta_V\colon (Q(IV^*))^* \to V$ be the map obtained by composing the dual of θ with the natural isomorphism $V^{**} \to V$.

Suppose $M \in M'_G$ and $\phi\colon M \to V$ is a B-map. Then the dual map $\phi^*\colon V^* \to M^*$ gives rise to a $U(\underline{g})$-map $\check{\phi}^*\colon IV^* \to M^*$ satisfying $\check{\phi}^*(a \otimes \alpha) = a\phi^*(\alpha)$, for $a \in U(\underline{g})$, $\alpha \in V^*$ and so to $Q(\check{\phi}^*)\colon Q(IV^*) \to M^*$. Composing the natural isomorphism $M \to M^{**}$ with the dual of $Q(\check{\phi}^*)$ we obtain a map, say $\tilde{\phi}$, from M to $Q(IV^*)^*$ such that $\eta_V \circ \tilde{\phi} = \phi$. Moreover it is not difficult to show that $\tilde{\phi}$ is the unique G-map with this property. Comparing this universal property with that for induction, (1.1.2), gives an isomorphism $\psi_V\colon (Ind_B^G)V \to (Q(IV^*))^*$ which is easily checked to be natural.

It would be pleasant if at this point we could remark that F^i and G^i are isomorphic because they are the ith derived functors of the isomorphic functors F^0 and G^0. However the problem is that the category M'_B of finite dimensional B-modules does not have enough injectives and so we cannot form derived functors of left exact functors. However the next best thing is to show that $(F^i)_{i\geq 0}$ and $(G^i)_{i\geq 0}$ are universal δ-functors and this is what we shall do. To show that $(F^i)_{i\geq 0}$ is universal we use an auxilary category $\tilde{M}_B$. The objects of $\tilde{M}_B$ are B-modules M such that each weight space of M is finite dimensional and the weights of M are bounded above (i.e. there exist elements $\mu_1,\ldots,\mu_n$ of X such that each weight of λ of M satisfies $\lambda \leq \mu_i$ for some $1 \leq i \leq n$). It follows from (1.2.4) (when referring to results in Chapter 1 the reader must take into account that the Borel subgroup is negative there and positive here) that $\tilde{M}_B$ contains enough injectives and it is not difficult to show that a module in $\tilde{M}_B$ is injective if and only if it is injective in M'_B. From this we conclude that the restriction to $\tilde{M}_B$ of the nth derived functor of a left exact functor $U:M_B \to M_G$ coincides with the nth derived functor of the restriction of U to $\tilde{M}_B$. We shall make no notational distinct-ion therefore between the derived functors of U and their restriction to $\tilde{M}_B$. For a module V in $\tilde{M}_B$ and $n \geq 0$ we denote by V_n the span of the weight spaces V^λ for λ satisfying $(\lambda,\beta_0^\vee) \geq -n$, where β_0 is the highest short root. We have that $V_0 \subseteq V_1 \subseteq V_2 \ldots$, V is he union of the V_n, by (1.4.3) V_n is a B submodule of V and from he definition of $\tilde{M}_B$ we obtain $V_n \in M_B$. For a functor $U:M_B \to M_G$ e define $\tilde{U}:\tilde{M}_B \to M_G$ by $\tilde{U}(V) = \varinjlim U(V_n)$. The restriction of $\tilde{U}$ to $'_B$ is again U. A morphism $\phi:U \to V$ gives rise in the natural way to morphism $\tilde{\phi}:\tilde{U} \to \tilde{V}$ and $\tilde{\phi}$ restricts to ϕ on M'_B, in particular $= \tilde{\psi}$ implies that $\phi = \psi$. Now suppose $(U^i)_{i\geq 0}$ is a δ-functor rom M'_B to M_G and $f^0:F^0 \to U^0$ a morphism. We obtain a morphism $:\tilde{F}^0 \to \tilde{U}^0$. However $\tilde{F}^i = R^i Ind_B^G$ since the derived functors of nduction commute with direct limits. Thus $(\tilde{F}^i)_{i\geq 0}$ is universal Corollary 1.4, p.206 of [32]) and so there exists a unique sequence of rphisms $g^i:\tilde{F}^i \to \tilde{U}^i$ which commute with the $\tilde{\delta}^i$, with $g^0 = \tilde{f}^0$. By striction we obtain a sequence of morphisms $f^i:F^i \to U^i$ starting with e given f^0. If f'^i is another such sequence we obtain $\tilde{f}'^i = g^i$

by the uniqueness of the g^i and so $\gamma'^i = \gamma^i$ and $\delta'^i = \delta^i$. Hence $(F^i)_{i\geq 0}$ is a universal δ-functor.

We shall now show that the functor G^i is effaceable for $i > 0$ and therefore $(G^i)_{i\geq 0}$ is universal (by II, 2.2.1 of [30]). What we must show is that for each $V \in M_B$ there is a monomorphism $V \to M$, for some $M \in M_B$, such that the induced map $G^iV \to G^iM$ is 0. Since V is finite dimensional V embeds in a finite direct sum of injective B-modules and so into a finite direct sum $V_1 \oplus V_2 \oplus \ldots \oplus V_m$ in M_B where each V_j has a simple B socle. Thus it suffices to prove that each V_j embeds in a finite dimensional B-module in such a way that G^i of the embedding is the zero map. This will certainly be the case if each injective indecomposable B-module is a union of finite dimensional submodules on which G^i is 0 and we shall now show, by using an argument of Cline, Parshall and Scott (see p.107 of [19]), that this is the case. For a dominant weight λ we regard the simple $\underline{g}$-module $V(\lambda)$ of highest weight λ as a G-module and so as a B-module. For $\mu \in X$ and $n \geq 0$ the B-module $V(2n\rho) \otimes (-2n\rho+\mu)$ has simple socle μ and so may be identified with a submodule of $E(\mu)$. Here ρ denotes, as usual, half the sum of the positive roots. Moreover it follows from Kostant's multiplicity formula (see p.106 of [19]) that for any finite subset S of X and n sufficiently large

$dim(V(2n\rho) \otimes (-2n\rho+\mu))^{\lambda} = dim\, E(\mu)^{\lambda}$ for all $\lambda \in S$ (p.107 of [19]). Hence any finite dimensional submodule of $E(\mu)$ is contained in $V(2n\rho) \otimes (-2n\rho+\mu)$ for n large and it suffices to show that

$G^i(V(2n\rho) \otimes (-2n\rho+\mu)) = 0$ for n large and $i > 0$. However $I((V(2n\rho) \otimes (-2n\rho+\mu))^*) \cong I(V(2n\rho)^* \otimes (2n\rho-\mu))$ which is isomorphic to $V(2n\rho)^* \otimes M(2n\rho-\mu)$ by the tensor identity for I. Thus, by (12.3.2) and the definition of G^i, $G^i(V(2n\rho) \otimes (-2n\rho+\mu))$ is isomorphic to $(V(2n\rho)^* \otimes L^iQ(M(2n\rho-\mu)))^*$. But for n large $2n\rho - \mu \in X^+$ so that $M(2n\rho-\mu)$ is projective, by (12.3.1), and so $L^iQ(M(2n\rho-\mu)) = 0$ for all $i > 0$. Hence $G^i(V(2n\rho) \otimes (-2n\rho+\mu)) = 0$ for all n large, as required. This completes the proof that $(G^i)_{i\geq 0}$ is a universal δ-functor. Now by universality the isomorphism $F^0 \to G^0$ extends to an isomorphism $F^i \to G^i$ in every degree and the proof of (12.3.3) is complete.

Now if $\lambda \in X^+$, $I(\lambda) = M(\lambda)$ is projective, by (12.3.1), and so acyclic and so $G^i(-\lambda) = 0$ for all $i > 0$. Thus we obtain from (12.3.3) the following formulation of Kempf's Vanishing Theorem for the positively chosen Borel subgroup B.

(12.3.4) $R^i Ind_B^G(-\lambda) = 0$ *for every* $\lambda \in X^+$ *and* $i > 0$.

It seems to us that the main obstacle to obtaining a similar proof in characteristic p is the lack of a suitable category, with enough projectives, analogous to $\mathcal{O}$.

12.4 Induced modules and affine quotients

We have used from time to time the theorem of Cline, Parshall and Scott, [18], that induction from a closed subgroup H, of an affine algebraic group G, to G is exact if and only if the quotient variety G/H is affine. To the proof given in [18] that G/H affine implies that induction is exact (which is actually the part of the theorem we have used) based on the faithful flatness of the quotient map $\pi : G \to G/H$, we have nothing to add. We present here an alternative to the proof given in [18] of the result:

12.4.1) *If* Ind_H^G *is exact then* G/H *is affine.*

The proof is in two parts. First it is shown that G/H is quasi-ffine (an open subvariety of an affine variety) and then G/H is shown o be affine by using an exercise in algebraic geometry (Chapter II, x. 2.17 of [32]) that a variety X is affine if there exist functions $_1, f_2, \ldots, f_n$ in the algebra $k[X]$ of regular functions on X such hat each principal open set X_{f_i} is affine and the ideal generated by $_1, f_2, \ldots, f_n$ is $k[X]$. The idea of proving (12.4.1) in these steps s due to W. Ferrer Santos (see [28], which also contains a proof of 2.4.1) along these lines).

There is a great similarity between the theorem of Cline, Parshall d Scott and Serre's criterion that a variety X is affine if and ly if $H^i(X, F) = 0$ for all quasi-coherent sheaves F and $i > 0$ - uivalently the global section functor Γ on quasi-coherent sheaves exact if and only if X is affine. It is interesting to note that e same exercise (Chapter II, 2.17 of [32]) is used in proving that Γ act implies that X is affine (see Chapter III, Theorem 3.7 of [32]).

It is not difficult to reduce to the case in which G is irreducible and we leave this task to the reader. We say a rational H-module M is co-extendible if there is a short exact sequence of H-modules

$$0 \to U \to V|_H \to M \to 0$$

for some rational G-module V and H-module U. By [12], Theorems 1 and 4, to establish quasi-affineness it is enough to prove that, for a one dimensional co-extendible H-module L, the dual module L^* is also co-extendible. Suppose that L is co-extendible and $0 \to U \to V|_H \to L \to 0$ is a short exact sequence of H-modules with V a G-module. Tensoring with L^* we obtain a short exact sequence

$$0 \to U \otimes L^* \to V|_H \otimes L^* \to k \to 0$$

where k denotes the one dimensional trivial module. Let $Ind = Ind_H^G$. Now $Ind(k) \neq 0$ $((Ind(k))^G \cong k^H = k)$ so by exactness $Ind(V|_H \otimes L^*) \neq 0$ and so by the tensor identity $Ind(L^*) \neq 0$. Let θ_1 and θ_2 denote respectively the zero and identity endomorphisms of $Ind(L^*)$ and let $e: Ind(L^*) \to L^*$ be the natural map. By the universal property of induction (1.1.2), $e \circ \theta_1 \neq e \circ \theta_2$. Thus $e \neq 0$ and we have a short exact sequence

$$0 \to K \to Ind(L^*) \to L^* \to 0$$

where K is the kernel of e, and L^* is co-extendible. Hence G/H is quasi-affine.

Let X be the space G/H, of cosets of the form Hg $(g \in G)$. We identify the algebra $k[X]$ of regular functions on X with $\{f \in k[G]: f(hg) = f(g)$ for all $h \in H$, $g \in G\}$. Since X is quasi-affine we can choose $0 \neq f \in k[X]$ such that X_f is affine. Let $V = kGf$ be the G submodule of $k[X]$ generated by f. Let $\varepsilon: k[X] \to k$ be the map given by $\varepsilon(a) = a(1)$. Then ε is an H-map and $\varepsilon(V) \neq 0$ so that ε is surjective. By exactness $Ind(\varepsilon): Ind(V) \to Ind(k)$ is surjective also and so there is some $A \in Ind(V)$ such that $(\varepsilon \circ A)(g) = 1$ for all $g \in G$. However there is an isomorphism $\phi: k[X] \otimes V \to Ind(V)$ taking $a \otimes v$ to $\phi_{a,v}$ where $\phi_{a,v}(g) = a(g)gv$ for $g \in G$ (see the proof of Proposition 1.5 of [18]). Hence $A = \sum_{i=1}^{n} \phi_{a_i, f_i}$ for $a_i \in k[X]$ and translates f_i of f. Hence $\sum_{i=1}^{n} \varepsilon \circ \phi_{a_i, f_i}(g) = 1$ for all $g \in G$, i.e. $\sum_{i=1}^{n} a_i(g) f_i(g) = 1$

and so $\sum_{i=1}^{n} a_i f_i = 1$. Thus X is covered by affine open sets $X_{f_1}, X_{f_2}, \ldots, X_{f_n}$ such that $f_1, f_2, \ldots, f_n$ generate the unit ideal $k[X]$. Hence $X = G/H$ is affine.

References

1. H.H. ANDERSEN, Cohomology of line bundles on G/B, *Ann. Sci. Ecole Norm. Sup.* 12 (1979), 85-100.

2. H.H. ANDERSEN, The first cohomology group of a line bundle on G/B, *Invent. Math.* 51 (1979), 287-296.

3. H.H. ANDERSEN, Vanishing theorems and induced representations, *J. Alg.* 62 (1980), 86-100.

4. H.H. ANDERSEN, The strong linkage principle, *J. Reine Ang. Math.* 315 (1980), 53-59.

5. H.H. ANDERSEN, On the structure of Weyl modules, *Math. Z.* 170 (1980), 1-14.

6. H.H. ANDERSEN, The Frobenius morphism on the cohomology of homogeneous vector bundles on G/B, *Ann. of Math.* 112 (1980), 113-121.

7. H.H. ANDERSEN, On the structure of the cohomology of line bundles on G/B, *J. Alg.* 71 (1981), 245-258.

8. H.H. ANDERSEN, Line bundles on flag manifolds, (to appear).

9. H.H. ANDERSEN, Extensions of modules for algebraic groups, *Am. J. Math.* (to appear).

10. J.W. BALLARD, Injective modules for restricted enveloping algebras, *Math. Z.* 163 (1978), 57-63.

11. I.N. BERNSTEIN, I.M. GEL'FAND and S.I. GEL'FAND, Category of $\underline{g}$-modules, *Funct. Anal. Appl.* 10 (1976), 87-92.

12. A. BIALYNICKI-BIRULA, G. HOCHSCHILD and G. MOSTOW, Extensions of representations of algebraic linear groups, *Am. J. Math.* 85 (1963), 131-144.

13. N. BOURBAKI, Groupes et algèbres de Lie, Chapitres 7 et 8, Hermann, Paris 1975.

14. R. BRAUER, Sur la multiplication des charactéristiques des groupes continus et semi-simple, *C.R. Acad. Sci. Paris*, Sér. A 204 (1937), 1784-1786.

15. R.W. CARTER and G. LUSZTIG, On the modular representations of the general linear and symmetric groups, *Math. Z.* 136 (1978), 193-242.

16. R.W. CARTER and M.T.J. PAYNE, On homomorphisms between Weyl modules and Specht modules, *Math. Proc. Camb. Philos. Soc.* 87 (1980), 419-425.

17. C. CHEVALLEY, Seminaire sur la classification des groupes de Lie algébriques, Paris: *Ecole Norm. Sup.* 1956-1958.

18. E. CLINE, B. PARSHALL and L. SCOTT, Induced modules and affine quotients, *Math. Ann.* 230 (1977), 1-14.

19. E. CLINE, B. PARSHALL and L. SCOTT, Cohomology, hyperalgebras and representations, *J. Alg.* 63 (1980), 98-123.

20. E. CLINE, B. PARSHALL, L. SCOTT and W. van der KALLEN, Rational and generic cohomology, *Invent. Math.* 39 (1977) 143-163.

21. C.W. CURTIS and I. REINER, *Representation theory of finite groups and associative algebras*, Pure and Appl. Math. Vol. II, Interscience, New York, 1962.

22. M. DEMAZURE, A very simple proof of Bott's theorem, *Invent. Math.* 33 (1974), 271-272.

23. S. DONKIN, Hopf complements and injective comodules for algebraic groups, *Proc. London Math. Soc.* (3) 40 (1980), 298-319.

24. S. DONKIN, Rationally injective modules for algebraic groups as direct limits, *Bull. London Math. Soc.* 35 (1980), 99-102.

25. S. DONKIN, A filtration for rational modules, *Math. Z.*, 177 (1981), 1-8.

26. S. DONKIN, On Ext^1 for semisimple groups and infinitesimal subgroups, *Math. Proc. Camb. Philos. Soc.* 92 (1982), 231-238.

27. S. DONKIN, A note on decomposition numbers for reductive algebraic groups, *J. Alg.* 80 (1983), 226-234.

28. W.R. FERRER SANTOS, A note on affine quotients, (to appear).

29. J.A. GREEN, Locally finite representations, *J. Alg.* 41 (1976) 137-171.

30. A. GROTHENDIECK, Sur quelques points d'algèbre homologique, *Tohoku Math. J.* 9 (1957), 119-221.

31. W. HABOUSH, A short characteristic p proof of Kempf's vanishing theorem, *Invent. Math.* 56 (1980), 109-112.

32. R. HARTSHORNE, *Algebraic Geometry*, Graduate Texts in Mathematics , Berlin-Heidelberg-New York, Springer 1977.

33. G. HOCHSCHILD, Cohomology of algebraic linear groups, *Ill. J. Math.* (1961), 492-519.

34. J.E. HUMPHREYS, *Introduction to Lie algebras and representation theory*, Graduate Texts in Mathematics 9, Berlin-Heidelberg-New York,

Springer 1972.

35. J.E. HUMPHREYS, *Linear algebraic groups*, Graduate Texts in Mathematics 21, Berlin-Heidelberg-New York, Springer 1975.

36. J.E. HUMPHREYS, On the hyperalgebra of a semisimple algebraic groups, in: *Contributions to algebra*, A collection of papers dedicated to Ellis Kolchin, New York-San Fransisco-London, Academic Press 1977.

37. J.E. HUMPHREYS and J.C. JANTZEN, Blocks and indecomposable modules for semisimple algebraic groups, *J. Alg.* 54 (1978), 494-503.

38. N. JACOBSON, *Lie algebras*, New York: Interscience 1962.

39. J.C. JANTZEN, *Moduln mit einem höchsten Gewicht*, Lectures Notes in Mathematics 750, Berlin-Heidelberg-New York, Springer 1979.

40. J.C. JANTZEN, Darstellungen halbeinfacher Gruppen und kontravariante Formen, *J. Reine Ang. Math.* 290 (1977), 117-141.

41. J.C. JANTZEN, Darstellungen halbeinfacher Gruppen und ihrer Frobenius-Kerne, *J. Reine Ang. Math.* 317 (1980), 157-199.

42. J.C. JANTZEN, Zur Reduction modulo p unipotenter Charactere endlicher Chevalley-Gruppen, *Math. Z.* 181 (1982), 97-128.

43. G. KEMPF, Linear systems on homogeneous spaces, *Ann. of Math.* 103 (1976), 557-591.

44. V. LAKSHMIBAI, C. MUSILI and C.S. SESHADRI, Geometry of G/B, *Bull. A.M.S.*, 1 New Series (1979), 432-435.

45. J. O'HALLORAN, A vanishing theorem for the cohomology of Borel subgroups, *Comm. in Alg.* 11 (14), (1983), 1603-1606.

46. L.L. SCOTT, Representations in characteristic p . In: *Proceedings of the A.M.S. Summer Institute on Finite Groups*, Santa Cruz, 1979.

47. I.R. SHAFAREVICH, *Basic algebraic geometry*, Grundlehren 213, Springer-Verlag, Heidelberg (1974).

48. T.A. SPRINGER, Weyl's character formula for algebraic groups, *Invent. Math.* 5 (1968), 85-105.

49. R. STEINBERG, *Lectures on Chevalley groups*, mimeographed lecture notes, Yale Univ. Math. Dept. 1968.

50. J. SULLIVAN, Simply connected groups, the hyperalgebra and Verma's conjecture, *Am. J. Math.* 100 (1978), 1015-1019.

51. J. TITS, *Tabellen zu den einfachen Lie Gruppen und ihren Darstellungen*, Lecture Notes in Mathematics 40, Springer-Verlag, Berlin 1967.

52. WANG JIAN-PAN, Sheaf cohomology on G/B and tensor products of Weyl modules, *J. Alg.* 77 (1982), 162–185.

INDEX